A

M

P

C

The Exploitation of Mammal Populations

The Exploitation of Mammal Populations

Edited by

Victoria J. Taylor

Development Officer,
Universities Federation for Animal Welfare, UK

and

Nigel Dunstone

Department of Biological Sciences,
University of Durham, UK

 CHAPMAN & HALL
London · Weinheim · New York · Tokyo · Melbourne · Madras

Published by Chapman & Hall, 2–6 Boundary Row, London SE1 8HN, UK

Chapman & Hall, 2–6 Boundary Row, London SE1 8HN, UK

Chapman & Hall, GmbH, Pappelallee 3, 69469 Weinheim, Germany

Chapman & Hall USA, 115 Fifth Avenue, New York, NY 10003, USA

Chapman & Hall Japan, ITP-Japan, Kyowa Building, 3F, 2-2-1 Hirakawacho, Chiyoda-ku, Tokyo 102, Japan

Chapman & Hall Australia, 102 Dodds Street, South Melbourne, Victoria 3205, Australia

Chapman & Hall India, R. Seshadri, 32 Second Main Road, CIT East, Madras 600 035, India

First edition 1996

© 1996 Chapman & Hall

Typeset in 10/12pt Palatino by Columns Design Ltd, Reading, Berks, UK

Printed in Great Britain by St Edmundsbury Press, Bury St Edmunds, Suffolk, UK

ISBN 0 412 64420 7

A catalogue record for this book is available from the British Library

Library of Congress Catalog Card Number 96–83440

Contents

Part Three
Hunting and Its Impact on Wildlife

Contributors

Dr Anatolij I. Blizniuk
Chernye Zemli Biosphere Reserve
Elista, Kalmykia, CIS

Mr David Bowles
Royal Society for the Prevention of
Cruelty to Animals
Causeway
Horsham
West Sussex RH12 1HG
UK

Ms Carolyn J. Callaghan
Department of Zoology
University of Guelph
Guelph
Ontario
Canada N1G 2W1

Dr Kenneth L.I. Campbell
Natural Resources Institute
Central Avenue
Chatham Maritime ME4 4TB
UK

Dr Jacob V. Cheeran
College of Veterinary and Animal
Sciences
Kerala Agricultural University
Mannuthy
Trichur
India

Dr Nigel Dunstone
Department of Biological Sciences
University of Durham
South Road
Durham City DH1 3LE
UK

Dr Marion L. East
Max-Planck-Institut für
Verhaltensphysiologie
D-82319 Seewiesen Post Starnberg
Germany

Dr Peter G.H. Evans
Department of Zoology
University of Oxford
South Parks Road
Oxford OX1 3PS
UK

Dr John H. Fanshawe
Wildlife and Conservation
Research Unit
Department of Zoology
University of Oxford
South Parks Road
Oxford OX1 3PS
UK

Dr Clare D. FitzGibbon*
Large Animal Research Group
Department of Zoology
University of Cambridge
Cambridge CB2 3EJ
UK

and
Estación Biológica de Doñana
(CSIC)
Apdo. 1056
41080 Sevilla
Spain

Ms Nilofer Ghaffar
Department of Anthropology
University College London
Gower Street
London WC1E 6BT
UK

Dr Heribert Hofer
Max-Planck-Institut für
Verhaltensphysiologie
D-82319 Seewiesen Post Starnberg
Germany

Ms Sally A. Huish

Dr Paul J. Johnson
Wildlife Conservation Research
Unit
Department of Zoology
University of Oxford
South Parks Road
Oxford OX1 3PS
UK

Dr Michael D. Kock
Veterinary Unit
Department of National Parks and
Wild Life Management
PO Box CY 140
Causeway
Harare
Zimbabwe
Africa

Dr Jochen Langbein*
Deer Management Research Group
Department of Biology
Southampton University
Bassett Crescent East
Southampton SO16 7PX
UK

Professor David M. Lavigne*
Department of Zoology
University of Guelph
Guelph
Ontario
Canada N1G 2W1

Mrs Ashley Leiman
The Orangutan Foundation
7 Kent Terrace
London NW1 4RP
UK

Dr Alastair McNeilage
School of Biological Sciences
University of Bristol
Woodland Road
Bristol BS8 1UG
UK

Dr David W. Macdonald
Wildlife Conservation Research
Unit
Department of Zoology
University of Oxford
South Parks Road
Oxford OX1 3PS
UK

Dr Hezron Mogaka
Kenya Indigenous Forest
Conservation Project
Kenya Forestry Research Institute
PO Box 20412
Nairobi
Kenya
Africa

Dr José Roberto Moreira
EMBRAPA/CENARGEN
SAIN – Parque Rural
Caixa Postal 02372
70770–900 Brasília – DF
Brazil
South America

Dr Valeri M. Neronov
Institute of Evolutionary
Morphology and Animal Ecology
Russian Academy of Sciences
13 Fersman Street
Moscow 117312
Russia

Dr Jane N. O'Sullivan
Department of Agriculture
University of Queensland
Brisbane QLD 4072
Australia

Mr Vassili Papastavrou
School of Biological Sciences
University of Bristol
Woodland Road
Bristol BS8 1UG
UK

Dr Trevor B. Poole
UFAW
8 Hamilton Close
South Mimms
Potters Bar
Herts EN6 3QD
UK

Mr Robert and Mrs Christine
Prescott-Allen
PADATA
627 Aquarius Road
RR2 Victoria
British Columbia
Canada V9B 5B4

Dr Rory Putman
Deer Management Research Group
Department of Biology
Southampton University
Bassett Crescent East
Southampton SO16 7PX
UK

Mr Ian Redmond
Wildlife Consultant
PO Box 308
Bristol BS99 7LQ
UK

Dr Jonathan Reynolds
The Game Conservancy Trust
Fordingbridge
Hampshire SP6 1EF
UK

Professor John D. Skinner
Mammal Research Institute
University of Pretoria
Pretoria 0002
South Africa

Mr Richard J. Smith
Department of Zoology
University of Guelph
Guelph
Ontario
Canada N1G 2W1

Dr Stephen Tapper*
The Game Conservancy Trust
Fordingbridge
Hampshire SP6 1EF
UK

Contributors

Miss Victoria J. Taylor*
UFAW
8 Hamilton Close
South Mimms
Potters Bar
Herts EN6 3QD
UK

Professor James G. Teer
Welder Wildlife Foundation
PO Box 1400
Sinton
Texas 78387
USA

Mr Valmik Thapar
Ranthambhore Foundation
19 Kautilya Marg
Chanakyapuri
New Delhi 110 021
India

Dr Derek W. Yalden
School of Biological Sciences
3.239 Stopford Building
The Victoria University of
Manchester
Manchester M13 9PL
UK

Dr Lir V. Zhirnov
Institute of Evolutionary
Morphology and Animal Ecology
Russian Academy of Sciences
13 Fersman Street
Moscow 117312
Russia

*Authors to whom correspondence should be addressed.

Preface and Acknowledgements

This book is based, in part, on a joint symposium held by two scientific animal charities – the Universities Federation for Animal Welfare (UFAW) and the Mammal Society – on 25 and 26 November 1994 at the Meeting Rooms of the Zoological Society of London.

The symposium, on *The Exploitation of Mammals*, explored the various ways in which populations of mammals are currently or have been exploited world-wide; the concept and the practice of their sustainable use; and the welfare considerations for the animals involved. A wide range of uses of mammals were considered: harvesting, game ranching and domestication; hunting and trapping of animals for food, profit or recreation; the wildlife trade and conservation; and the growing industry of mammal oriented ecotourism. A further session on the welfare of mammals in zoos and captivity was included in the symposium. Whilst self-sustaining populations of animals in zoos are relevant in considerations of sustainable use and the survival of many species are already dependent on active management in zoos, the subject is not included in this volume as it could certainly constitute a book in its own right.

In the present book, various case studies and overviews of the status, conservation and welfare of exploited mammal populations are considered. The individual contributors were chosen because of their acknowledged expertise in an area or sometimes for their particular views on a subject, to try to achieve balanced coverage of this highly complex and controversial area. Some of the chapters portray current situations that are particularly damaging to the survival of species because of over-exploitation and their content simply reflects the consensus of those closely involved with the situation which, by its nature, does not permit more than informed impressions as well as anecdotal information. Individual chapters serve to highlight many of the often conflicting issues involved in the sustainable use debate; at times they are opinionated, but together they contribute towards the rational debate that is required on the effective and humane management of the exploitation of wild mammals. The views of individual contributors vary, but they are unanimous in encouraging responsible management of wildlife whilst maintaining the highest possible animal welfare standards. It must also be made clear that the views expressed in this volume do not necessarily reflect the views of the two organizing societies.

This book certainly does not attempt to resolve all the conflicts that arise in the exploitation and sustainable use of wild mammal populations. It is hoped that it will facilitate discussion and raise many of the questions and dilemmas that need to be addressed and incorporated into assessment and management procedures for the effective yet ethical use of exploited mammal populations throughout the world.

Acknowledgements

Thanks are due to all those who took part in the symposium, particularly all the speakers (some of whom are not represented in this book), the chairmen (Dr Pat Morris, Mr Harry Thompson, Dr Robert Hubrecht and Dr Martyn Gorman) and all the delegates, who helped to make the meeting such a success. In the preparation of this book, many individuals have contributed their time and expertise by refereeing manuscripts (anonymity precludes specific thanks), providing advice and support (Dr Rory Putman, Mr Roger Ewbank, Ms Lesley Diver) and throughout the publication process at Chapman & Hall (Dr Bob Carling, Mrs Helen Sharples, Ms Kim Worham). Acknowledgements are also due to the two organizing societies for their financial and administrative support of the meeting, to the Zoological Society of London for their generous help and to *BBC Wildlife Magazine* and Tusk Force for each sponsoring the participation of a contributor.

Abbreviations

AD
Anno Domini

BC
Before Christ

BP
Before the Present

CAMPFIRE
Communal Area Management Programme for Indigenous Resources

CITES
Convention on the International Trade in Endangered Species of Wild Fauna and Flora

ICRW
International Convention for the Regulation of Whaling (1946)

IFAW
International Fund for Animal Welfare

IPZ
Intensive Protection Zones

IUCN
International Union for Conservation of Nature and Natural Resources – commonly known as The World Conservation Union

IWC
International Whaling Commission

NGO
Non-governmental organization

RMP
Revised Management Procedure

RMS
Revised Management Scheme

SSC of IUCN
Species Survival Commission of the IUCN

UFAW
Universities Federation for Animal Welfare

UNCED
United Nations Conference on the Environment and Development

UNCLoS
United Nations Convention on the Law of the Sea

WWF
World Wide Fund for Nature

Part One

Exploitation of Mammal Populations: Past, Present and Future

1

The exploitation, sustainable use and welfare of wild mammals

Victoria J. Taylor and Nigel Dunstone

1.1 INTRODUCTION

It is difficult to think of any wildlife that is unaffected by humans or their actions. Humans exploit animals big and small, furred or feathered, mainly for subsistence or profit but also for sport and entertainment. If animals are not exploited directly, they may still be affected by human activities – either gaining or losing in terms of protection, food and available habitat. The human population is steadily increasing, requiring more space and demanding more resources. It is apparent that more extensive wildlife management is required to slow the inevitable loss of biodiversity, and that to be effective it must confine any exploitation within sustainable limits.

The successful management of biological resources has considerable conservation, political and economic significance for both the developing and developed world. The management of wild animals involves human intervention which inevitably has welfare implications, whether the animals are being protected, ranched, captured, culled or simply disturbed in their natural habitat. Consideration of welfare is important to many on ethical grounds because of the suffering inflicted on wild animals but it also has considerable implications for the success of any sustainable use programme. The responsible use of animals will reduce the numbers required if they receive appropriate care, or the numbers captured if the methods are specific to requirements, and as a result will improve the overall acceptability and 'profitability' of the enterprise.

1.2 THE EXPLOITATION OF MAMMALS

Exploitation is a broad term which covers the utilization of animals for reasons such as pure commercial gain, subsistence or in the interests of conservation or control. The means by which it is carried out can be consumptive, either permanently removing animals from the population by hunting or live-trapping or harvesting products from wild individuals under management regimes, such as musk from captive wild Indian civets. Exploitation also involves non-consumptive (non-lethal) enterprises such as ecotourism. The effects of exploitation on animals in terms of their welfare can be marginal or extreme, but in most cases improvements are possible and further research is essential.

The exploitation of wild mammals or their products is emotive and has stimulated extensive debate on moral, pragmatic and economic grounds (Robinson and Redford, 1991; Barbier, 1992; Swanson and Barbier, 1992; IIED, 1994). So when is exploitation morally defensible? Can exploitation help to conserve species, particularly in the long term? Does assigning a monetary value to wildlife protect it or hasten its demise? These considerations are of particular relevance but meanwhile the fact remains that people throughout the world use a great number of wild species for a wide variety of benefits ranging from food to recreation (Prescott-Allen and Prescott-Allen, 1982) and it is likely that they will continue to do so, even if this results in the eventual extinction of species. In **Chapter 2**, **Derek Yalden** documents how, historically, man's three eras of exploitation of mammals have numerically benefited numerous species of domesticated or pest animals. But this has been at the expense of many others, and reinforces the concern that this situation will probably continue unless biodiversity is both preserved and harnessed for human benefit as we move into a conscious new era of exploitation.

1.3   THE SUSTAINABLE USE OF WILDLIFE

The sustainable use of wildlife is a relatively new conservation concept. It promotes the use of biological resources within their capacity for renewal, so that the potential for any future use is not impaired (WCED, 1987; IUCN/UNEP/WWF, 1980, 1991). It is also based on the principle that, wherever possible, local communities should be granted property rights over the wildlife populations and should manage them for their own long-term interests (IIED, 1994). There is no generally accepted definition of sustainable use and this has led to a diversity of interpretations of its meaning (Conway, 1993; Robinson, 1993). Sustainable use has its advocates and opponents (Willers, 1994) arguing from a variety of positions including on conservation, welfare, socio-economic and purely ethical grounds. Clearly, there is a need for some scientific definition or practicable framework to encompass the wide range of situations and approaches involved, to avoid so-called benefits in one area being obtained at the expense of less obvious costs in other areas (Conway,

1993). In particular, uses can be economically sustainable without being ecologically sustainable, leading to rapid over-exploitation. Economic considerations of resource uses have not paid sufficient attention to the value (be it commercial or social) of wildlife in development, and as a consequence they have been overlooked in policy and strategy processes (Swanson and Barbier, 1992; Barbier, 1992) to the detriment of humans and natural resources.

A comprehensive assessment of all the values and issues important to people and the environment needs to be evaluated, otherwise unsustainable uses can superficially appear to be sustainable. For instance, uses related to ecotourism are often promoted as being more sustainable than consumptive uses of wildlife, on the basis of the revenue generated and a dislike of killing animals, but on examination this is not always clearcut: factors such as disturbance, the potential for disease transmission, flow of profits out of the area and detrimental habitat alteration need to be taken into account (**Chapters 4, 18, 19, 21**; Goodwin, 1995). In contrast, the extent of the historical exploitation of furbearers in North America reviewed by **Stephen Tapper** and **Jonathan Reynolds (Chapter 3)** is an example of how an economically and ecologically sustainable consumptive industry has eventually developed from an over-exploited situation. But there are still ethical and welfare problems associated with it because of the trapping methods used.

Sustainable use, which has been promoted by some as the only long-term possibility for the preservation of many wild species, advocates the utilitarian view that species must pay for their continued existence. Others express the concern that sustainable use might never be achievable for a variety of reasons, mainly socio-economic, many of which are highlighted throughout this volume. Some species may never have an economic value, although in some instances they may benefit from the commercial harvest of others. Some opponents of the sustainable use ethic believe that animals should not be used at all which is ultimately a preservationist ideal. In some cases, strict preservation may be the only way to save species from impending extinction. Thus while some species may be able to pay their way, the costs of conserving species that will never have an economic value, or one that can be regulated, still need to be borne. To be feasible, it should not be by those who can least afford it (Eltringham, 1994) (Plate 1).

Sustainable use has to be seen for what it really is – one conservation tool of many (Robinson, 1993), with potential for associated social and economic benefits. It can only be applicable under specific circumstances and it is not a panacea which will halt the serious threats to biodiversity which still urgently need to be addressed. **David Lavigne** *et al.* **(Chapter 14)** argue forcibly, with reference to the North American situation, that indiscriminate promotion of the general sustainable use ethic, particularly with regard to international trade in dead wildlife or their derivatives, is consistent with a return to nineteenth century practices which resulted in the over-exploitation of wildlife. However, they do consider that some utilization, when regulated closely and supported by finance derived from related services, can be sustainable. As most developing countries cannot afford the control required to fulfil sustainable use criteria, the possibility of gaining the revenue to do so by exploiting wildlife is compelling. The African perspective is contributed by **Michael Kock** (**Chapter 13**) who demonstrates how the value of wildlife and changed land management practices have benefited local communities as well as contributed to the conservation of many species in Zimbabwe.

How can the sustainability of a use be judged? Who should make and can enforce the decisions? Are there reasons to believe that future uses can be sustainable? The World Conservation Union (IUCN) is developing sustainable use guidelines. These have yet to be finalized but the IUCN has concluded that exploitation in the service of conservation should be considered as one option where sustainability can be proved, where any harvest taken is by humane methods and where the proceeds directly benefit local communities as well as demonstrably improve conservation of the species and ecosystems concerned (R. Putman, personal communication). **Robert and Christine Prescott-Allen**, former co-chairs of the IUCN SSC on Sustainable Use of Wild Species (**Chapter 4**), propose a framework developed to assess the impact of uses on both the ecosystem and the human system; using limited information, they draw some revealing conclusions on the impact of various types of exploitation on mammal populations. This type of approach is a valuable first step in ranking current uses and prioritizing areas that require attention for remedial action.

The wildlife conservation movement grew from a concern, often from the exploiters themselves (Eltringham, 1994), that wild animals were being over-exploited. Unsuccessfully controlled or indiscriminate exploitation of mammals has reduced many populations of economically important animals, such as rhinos and whales, to levels where they cannot effect their own recovery. Exploitation was once thought to be self-controlling, but the greater costs in time and effort of hunting a dwindling population cannot be guaranteed to reduce the offtake, particularly where the market value of the resource increases with its rarity. This emphasizes the double-edged sword of sustainable use and assigning a value to wildlife: the commercial value of wildlife does not necessarily ensure its preservation, as market forces can work both for and against conservation.

Most of the previous strategies to limit exploitation have been based on incomplete objective evidence, such as population estimates, and little is known why some ecosystems are easily pushed into alternative stable states whereas others are not (Orians, 1990). This situation is unlikely to change rapidly enough to satisfy all the criteria required to guarantee sustainable use. Any relaxation of conservation safeguards to facilitate the exploitation of mammals, without responsive adaptive management and continuous monitoring, is likely to contribute towards over-exploitation and lower animal welfare standards. It is for these reasons that any exploitation of wildlife must be rigorously regulated and managed, preferably at international, government and local community level. In the case of commercial whaling, as outlined by **Vassili Papastavrou** in **Chapter 8**, management strategies are now being developed to be flexible enough or precautionary enough to prevent any future over-use; there are important lessons to be learnt from such approaches. The paucity of basic biological data available to wildlife managers means that there is a need to continually re-assess the sustainability of any uses of wild mammals.

1.4   THE ASSESSMENT OF SUSTAINABLE USE AND ANIMAL WELFARE

Current uses of animals need to be assessed and brought into line with management and conservation goals and where possible the detrimental effects must be minimized to achieve sustainable use (**Chapter 20**; Robinson, 1993). From a conservation point of view, it is the lack of management of utilized wildlife that is thought to be one of the main causes of depletion of animal populations, and it is advocated that unmanaged systems should be avoided (Redford and Bodmer, 1995). From a welfare point of view, it is the unmanaged systems that often result in inhumane practices causing pain and suffering (for example, the kangaroo case study in **Chapter 4**). Animals without value or considered as pests usually suffer at human hands (Putman, 1989). If sustainable use programmes can change the concept of wild animals from an inconvenience to a valued resource and provide the necessary expertise, a concomitant improvement in animal welfare standards should result.

Even when wildlife is managed, welfare considerations are often insufficient. The welfare of wild animals has only occasionally been included in the formal consideration of sustainability of wildlife. This is probably because of the inherent difficulties associated with its measurement and application, and due to the extreme views that can be held on the subject, but some form of compromise is obviously required. A methodology for wildlife welfare assessment has been developed (Kirkwood *et al.*, 1994; A. Lindley, personal communication). There are differences in opinion regarding the extent to which humans should intervene on behalf of the welfare of free-living but partially managed wild animals (Kirkwood, 1994) but, regardless of this, the obligation to animals under human stewardship remains. In this volume, the term animal welfare is used in the context that wherever humans affect animals it is their responsibility to minimize the negative effects – an ethic which is consistent with responsible use.

1.4.1 Considerations and limitations of sustainable use

(a) Use of the resource: renewable resources and harvesting methods

Sustainability requires, as a prerequisite, careful conservation of renewable biological resources and any strategies must take account of certain key biological and ecological principles (Orians, 1990). The ability of populations and ecosystems to cope with unexpected natural events, such as disease, and the prolonged stress of harvesting has not always been considered in the past, particularly when offtake is biased to a particular sex or size of animal. If the potential of particular species for sustainable use is to be achieved, strategies for their management must now take a long-term, overall ecological view and be based on modern population biology and knowledge.

Wildlife is a renewable resource if managed correctly. Even those living organisms with slow reproductive rates (*K*-selected species, such as elephants and rhinos), can increase in number despite closely regulated offtake as in trophy hunting. In other instances such as whaling, it may be more appropriate to allow the population to recover before exploitation resumes. Generally, the *r*-strategists (organisms that have a rapid reproductive, growth or immigration rate) are more resistant to substantial consumptive use. However, limits are still essential, often because of the inappropriate harvesting methods used. **José Roberto Moreira** and **David Macdonald** give an overview in **Chapter 7** of the exploitation of capybara, which are seen as the South American mammal with the greatest potential for sustainable use, not only because of their productivity but also because of the expanding areas of cleared forest which provide suitable habitat.

Indiscriminate harvesting techniques may cause dispersal and fragmentation of groups of animals or the animals may alter their behaviour to become more secretive. To minimize disruption of the population structure and social behaviour of animals, selective harvesting methods are required so that optimum proportions of a particular sex or size of individuals are maintained. For example, one of the methods of hunting capybara tends to take a higher number of pregnant females (**Chapter 7**) and computer modelling shows that this in time will make the harvest unsustainable. In **Chapter 6**, **James Teer** *et al.* document the history of the traditionally exploited saiga antelope of Kazakhstan, a species robust enough to sustain a substantial harvest, but the effects of habitat degradation as well as uncontrolled poaching for males have jeopardized any sustainable potential and threatened its conservation, so that it is now listed on CITES Appendix II.

The potential for domesticating wild African ungulate species, and the productivity of multi-species systems to minimize habitat degradation by current agricultural domesticated mammals, is considered by **John Skinner** in **Chapter 5** on the development of game ranching in South Africa (Plate 2). It is apparent that more behavioural research is required to be able to manage wild animals optimally. The welfare aspects of consumptive and non-consumptive uses include the methods of capture, holding, transportation, killing and disturbance in their natural habitat (A. Lindley, personal communication). Additionally, for the harvesting of products from live wild animals, the permanent keeping or holding of them and the resource harvesting methods are welfare concerns. There is currently limited information available to define optimum post-capture management strategies to ensure good welfare and maximize survival and productivity of wild-caught animals.

In wildlife harvesting (which includes live and dead uses) or game ranching, monitoring procedures need to be further developed to determine the least stressful methods of wildlife husbandry and to improve the survival rates. Such requirements are similar in principle to those required for domesticated species but are invariably harder to implement due to the practicalities involved in dealing with wild animals (Plate 3).

To minimize the stress caused in capturing wild game species, the use of drug immobilization, habituation to handling and the administration of long-acting neuroleptics can have beneficial effects in a number of species (Morton *et al.*, 1995). Manipulation of behaviour, such as integrating wild with farmed animals (red deer: Leonard *et al.*, 1994) have great potential for humane handling procedures in wild animals because they help to minimize the negative effects associated with removing them from the wild. Humane capturing and killing methods have not been developed for all animals but where knowledge exists it should be implemented as part of a sustainable use programme. **Vassili Papastavrou** (**Chapter 8**) mentions how the International Whaling Commission has recognized the importance of animal welfare and banned various devices for killing whales; it is also continuing discussions on humane killing. Techniques for killing wild animals need to be further developed where possible, especially for use in the field, and the technology must be passed on to the users (Gardner, 1991; UFAW, 1996).

Hunting and trapping are common methods of harvesting mammals for subsistence, control and profit. Subsistence hunting is particularly relevant in discussions of conservation and sustainable use, mainly because of the total numbers of animals involved and the lack of legislative enforcement where such persistent activities are illegal. **Clare FitzGibbon** *et al.* compare two different methods of hunting in **Chapter 10** and find that they have different effects on the wildlife populations – the authors advocate that it would be preferable to encourage the more sustainable method of subsistence hunting. Commercial hunting for bushmeat is much harder to control (**Heribert Hofer** *et al.*, **Chapter 9**): the impact of illegal hunting by trapping on both target and non-target species in the Serengeti is such that, in the long term, unchecked activities are unlikely to be sustainable unless socio-economic improvements for the local communities are developed. There is little doubt that most forms of trapping cause pain and suffering, especially those that restrain an animal by a limb. Where trapping occurs, particularly in a system claiming to be sustainable, every effort must be made to reduce the suffering inflicted through the development of more humane, species-specific traps and increased training of trappers. In addition to the wastage from injury and mortality in any harvesting programme, accidental take of unrequired or surplus animals will affect the overall sustainability of any management programme (Plate 4). Methods of harvest should be assessed for their selectivity to minimize effects on non-target individuals or species.

Hunting is also carried out for sport and entertainment. The chapters by **David Macdonald** and **Paul Johnson** on foxhunting (**Chapter 11**) and **Jochen Langbein** and **Rory Putman** on deer hunting (**Chapter 12**) show how current levels of these sports in Britain do not affect the overall sustainability of the use and have been instrumental in conserving habitat. These are further examples of the ethical dilemmas of currently sustainable uses – welfare is still a major problem because suffering is undoubtedly caused to the hunted animals. The authors of **Chapter 11** suggest an interesting compromise: to encourage and promote more the skill of the hunt instead of killing animals – and they further suggest certain aspects of the hunt that could be changed on welfare grounds to make it more acceptable.

A growing non-consumptive use of mammals is nature-viewing or ecotourism (Plate 5) which, whilst not a harvest as such, can still have detrimental disturbance effects. Animals can become a nuisance or more secretive in their behaviour to avoid humans, and sensitive keystone resource sites for breeding or nutrition require adequate protection (**Chapter 18**). Guidelines and codes of conduct are needed to minimize disruption. **Peter Evans** reviews the potential disturbance of cetaceans in the marine environment and suggests guidelines for managing whale watching in **Chapter 22**. People will often pay large amounts to be taken in small exclusive groups to see wild animals such as mountain gorillas (see **Chapter 19** by **Alastair McNeilage**). This has the added advantage of maximizing profits and reducing the environmental impact although, to put this in a world-wide context, ecotourism only constitutes 5% of the growing tourism industry (Goodwin, 1995). Additionally, to fulfil sustainable use criteria the profits need to reach local conservation projects and communities, but in the worst cases up to 80–90% of gross tourism revenues go out of the area (Goodwin, 1995).

(b) Economic considerations

The commercial value of wildlife does not necessarily ensure its continued preservation in the wild, particularly when trade is unregulated. When profit is the primary goal, less valuable species can be destroyed at the expense of others and highly valued animals may be over-exploited even to extinction, because the returns are greater than those from a sustainable use project (**Chapters 14, 15, 16**; Lander *et al.*, 1994). A number of economic issues and control avoidance are considered in **Chapter 15**.

If managed correctly for conservation, market values can be used to increase the probability of preservation, but this requires that the flow of profits remains in the area, particularly to benefit local communities (**Chapters 3, 13, 19**). As highlighted in **Chapter 9**, with commercial uses of wildlife, people may move to areas where market hunting is more profitable once they have depleted their current area, and more people move into the area as a response to opportunities, leaving no room for regeneration or recovery of the mammal populations and supporting ecosystems.

(c) Monitoring and legal protection

Wildlife management has become more sophisticated with evaluations and monitoring schemes but the development and implementation of effective enforcement procedures are ultimately the key to successful sustainable use. More consideration has been given to developing methods of taking a sustainable harvest from wild animals than has been paid to enforcing such activities. Illegal activities will always compromise conservation, animal welfare and sustainable use programmes. Any legal trade in wildlife always carries the risk that illegal trade will also occur, but regulation and monitoring will always be required whether uses are legal or not. Since the ban on the ivory trade, some African conservation budgets have been cut substantially (Stuart, 1995).

The international wildlife trade of CITES listed species is examined in **Chapter 15** by **David Bowles** who concludes that only small-scale international trade in valuable threatened wild mammals has any potential to be sustainable in the long term. Small-scale regulated trade would have the most potential for maintaining good enforcement procedures and the implementation of high animal welfare standards. The fur trade in

Chapter 3, the case study of kangaroos used in **Chapter 4** and the continuing development of the Revised Management Scheme for whaling in **Chapter 8** all highlight the extent of control required to help to ensure future sustainable use.

(d) Community involvement

The overall success of any management regime depends on the quality of the underlying science (Singer, 1994) but science alone cannot provide all of the solutions, particularly when some of the considerations are of a socio-economic or cultural nature or due to conflicts between humans and wildlife, such as Asian elephants (detailed in **Chapter 17** by **Jacob Cheeran** and **Trevor Poole**). Often, the exploitation of wild areas for conservation has precluded the involvement of local communities who have relied upon them for resources. Community-based projects help to minimize the conflicts between agriculture and wildlife by halting environmental degradation, changing pest or competing wildlife status to a resource worthy of management, reducing poaching and providing a monitoring system of local wildlife.

(e) Human values and ethics

The concept of sustainable use involves value judgements and these will vary between cultures and religions or with personal moral beliefs regarding the welfare and use of animals. This makes the importance of decisions at a local community level particularly relevant to sustainability. To what extent can we afford to adopt the preservationist ideal? Ethically, the criteria for sustainability are more difficult to justify with such uses as sport-hunting than they are for activities like ecotourism (**Chapter 21** by **Ian Redmond**). On the other hand, economic arguments involving sport-hunting as a means to finance conservation can be equally compelling when alternative means of sufficient finance are not available (**Chapter 13**). Value judgements and ethics have their place in deciding whether a use is acceptable or not but in terms of animal welfare they also have a practical value when the development of sustainable use programmes can lead to improvements in the way animals are treated.

1.5 THE WAY FORWARD

Extensive wildlife management, with adaptive regulations and monitoring procedures, is crucial with regard to achieving sustainable use. Mammal populations are used either consumptively or non-consumptively but this distinction does not necessarily reflect whether a use will

Plate 1 Dehorning black rhino in Zimbabwe. The protection of rhinos from poachers involves significant expense. (Photo copyright A. Allan.)

Plate 2 Game ranching of tsessebe in Zimbabwe. (Photo copyright A. Allan.)

Plate 3 Immobilizing a bull elephant in Kruger National Park, South Africa. (Photo copyright A. Allan.)

Plate 4 Accidental injury: young male lion in the Serengeti with injuries from a snare. (Photo copyright M. East/H. Hofer.)

Plate 5 Long-distance viewing of elephants by tourists, Parc des Virungas, Eastern Zaire. (Photo copyright I. Redmond.)

Plate 6 Spotted hyaena suffering severe constriction and lacerations from the hunter's snare still encircling its neck. (Photo copyright M. East/H. Hofer.)

Plate 7 African elephants performing in a circus, Battersea, London, 1990. (Photo copyright I. Redmond.)

Plate 8 Local people enjoying the company of a tame bull elephant, Akagera National Park, Rwanda. (Photo copyright I. Redmond.)

be sustainable or not. Current uses of mammal populations need to be assessed to gauge their impacts on the humans and ecosystems concerned; unsustainable uses need to be improved or stopped. New ventures need to be rigorously planned, taking objective biological, ecological and welfare information into consideration along with economic and social factors. The priorities for action and research in this area include:

- understanding the natural history of species to be able to develop sustainable use strategies with the least negative effect;
- developing suitable capture and cropping techniques;
- transferring that technology to the users;
- the development of management strategies which take into account the market for such resources, local community requirements and how competing land uses will affect sustainability.

If they are to be achieved and maintained, sustainable use programmes require considerable social and political will and rigorous, dynamic and innovative management to be able to provide all the benefits required of them, including the use and conservation of a resource whilst contributing towards the development of local communities and economies.

There is a lot of rhetoric in the debate on sustainable use; there should be no illusions. Exploiting wildlife solely for profit will always carry a serious risk of extinction – any use of living organisms depletes wildlife resources, is often associated with a degradation of the supporting ecosystems and sometimes eventually results in a loss of populations or species (Robinson, 1993; Willers, 1994). A sustainable use programme for a particular species in isolation is unlikely to guarantee the conservation of that particular species or its habitat. Wildlife conservation in the long term requires conservation of genetic diversity, natural selective forces and the maintenance of the whole range of species interactions (McNab, 1991).

Efforts should also be concentrated on developing uses that are more likely to be sustainable than others. Some categories of use do have limited positive conservation effects in that they have a lesser detrimental effect than others, for instance game ranching when compared with domestic livestock or well managed ecotourism ventures compared with the trade in endangered wildlife. Most of the uses, particularly those removing a substantial harvest, have potential deleterious side-effects that could threaten long-term conservation of biodiversity. The drawbacks of game ranching and cropping include the elimination of local predators, genetic hybridization and selection for desirable production traits (McNab, 1991, citing Geist, 1988); and in the process of making such ventures economically viable, a degree of habitat modification will often occur. At the same time the preservation of wild populations with minimal human interference is required for genetic material to maintain such ranching endeavours (Orians, 1990).

The limited information available regarding the uses of wild mammals and their overall effects suggests that broadly based, integrated programmes of wildlife utilization have the most potential to be sustainable. Some of the uses may be preferable because they are benign but other limited lethal uses may be profitable enough to justify their employment when they can subsidize conservation, though it is probable such examples will not be numerous. In this way, some species can be harvested to pay for their continued survival and in so doing contribute towards the conservation of others less suited for sustainable use, but other forms of conservation, protection and regulation still need to be maintained. In this respect multi-species and multi-use projects can provide an integrated approach to some of the problems of sustainable use, conservation and the welfare of wild mammals. Such work has only just begun and this book is a collective contribution to the ongoing debate.

REFERENCES

Barbier, E.B. (1992) Economics for the wilds, in *Economics for the Wilds: wildlife, wildlands, diversity and development*, (eds T. M. Swanson and E. B. Barbier), Earthscan Publications, London, pp. 15–34.

Conway, G.R. (1993) Sustainable agriculture: the trade-offs with productivity, stability and equitability, in *Economics and Ecology*, (ed. E.B. Barbier), Chapman & Hall, London, pp. 46–65.

Eltringham, S.K. (1994) Can wildlife pay its way? *Oryx*, **28**, 163–168.

Gardner, A.L. (1991) Foreword, in *Neotropical Wildlife Use and Conservation*, (eds J.G. Robinson and K.H. Redford), University of Chicago Press, Chicago and London.

Goodwin, H. (1995) Tourism and the environment. *Biologist*, **42**(3), 129–133

IIED (1994) *Whose Eden? An Overview of Community Approaches to Wildlife Management*, International Institute for Environment and Development, London.

IUCN/UNEP/WWF (1980) *World Conservation Strategy. Living resource conservation for sustainable development*, IUCN, Gland, Switzerland.

IUCN/UNEP/WWF (1991) *Caring for the Earth: a strategy for sustainable living*, IUCN, Gland, Switzerland.

Kirkwood, J.K. (1994) Veterinary education for wildlife conservation, health and welfare. *Veterinary Record*, **135**, 148–151.

Kirkwood, J.K., Sainsbury, A.W. and Bennett, P.M. (1994) The welfare of free-living wild animals: methods of assessment. *Animal Welfare* **3**, 257–273.

Lander, R., Engen, S. and Snether, B.-E. (1994) Optimal harvesting, economic discounting and extinction risk in fluctuating populations. *Nature*, **372**, 88–89.

Leonard, F., Goddard, P.J. and Gordon, I.J. (1994) The effect of the presence of farmed red deer (*Cervus elaphus*) hinds on the mother–offspring behaviour of captive wild red deer. *Applied Animal Behaviour Science*, **40**, 179–185.

Lindley, A. Sustainable use of wildlife: assessment of animal welfare. (Unpublished.)

MacNab, J. (1991) Does game cropping serve conservation? A re-examination of the African data. *Can. J. Zool.*, **69**, 2283–2290.

Morton, D.J., Anderson, E., Foggin, C.M., Kock, M.D. and Tiran, E.P. (1995) Plasma cortisol as an indicator of stress due to capture and translocation in wildlife species. *Veterinary Record*, **136**, 60–63.

Orians, G.H. (1990) Ecological concepts of sustainability. *Environment*, **32**, 10–39.

Prescott-Allen, R. and Prescott-Allen, C. (1982) *What's Wildlife Worth?*, Earthscan Publications, London.

Putman, R.J. (1989) Introduction: mammals as pests, in *Mammals as Pests*, (ed. R.J. Putman), Chapman & Hall, London.

Redford, K. and Bodmer, R.E. (1995) Wildlife use in the Neotropics. Guest editorial. *Oryx*, **29**, 1.

Robinson, J.G. and Redford, K.H. (eds) (1991) *Neotropical Wildlife Use and Conservation*, University of Chicago Press, Chicago and London.

Robinson, J.G. (1993) The limits to caring: sustainable living and the loss of biodiversity. *Conservation Biology*, **7**, 20–28.

Singer, S.F. (1994) Problems and strategies in the scientific management of fisheries and marine mammals: from 'the tragedy of the commons' to an era of sustainable development. *Environmental Conservation*, **21**, 184–185.

Stuart, S.N. (1995) High profile species issues, SSC Office Report. *Species*, **23**, 5.

Swanson, T.M. and Barbier, E.B. (eds) (1992) *Economics for the Wilds: wildlife, wildlands, diversity and development*, Earthscan Publications, London.

UFAW (1996) *Humane Killing*, 5th edn, Universities Federation for Animal Welfare, Potters Bar.

Willers, B. (1994) Sustainable development: a new world deception. *Conservation Biology*, **8**, 1146–-1148

World Commission on Environment and Development (1987) *Our Common Future*, Oxford University Press, Oxford, England.

2

Historical dichotomies in the exploitation of mammals

Derek W. Yalden

SYNOPSIS

The initial success (population increase and range expansion) of humans relative to the other hominoids seems to stem at least in part from a change in diet. Hunting, particularly of ungulates, was a key evolutionary initiative. Ungulates are scarce in forested areas but much more abundant in grasslands, so a habitat shift was inevitable. Hunting was thus the earliest form of exploitation; it seems to have led to the extinction of prey in many local situations (such as Madagascar and New Zealand) and, if Martin's 'overkill hypothesis' for mega-faunal extinction is accepted, throughout the Americas.

At about the time that humans entered North America, other humans in the Middle East were beginning the next great evolutionary innovation: the agricultural revolution. Both animals and plants were taken into management, and the domesticated species are now, with *Homo sapiens*, the most abundant and widespread mammals in the world. The wolf was the first to be domesticated, apparently to assist in hunting. Subsequently sheep, goats, pigs and cattle were domesticated primarily as food; asses, horses and camels (partly as transport) and cats (perhaps for pest control or just as pets) have been spread from the Middle East around much of the world. In a few cases, local species have also been domesticated, quite independently.

These and species subsequently domesticated for other purposes (notably rats, mice and rabbits as laboratory animals) are now collectively a greater biomass of mammals than remain in the wild. Clearly, in evolutionary terms, they seem very successful species.

Along with the huge expansion in range and numbers of *Homo*, sheep, cattle and other species, a new phase of hunting took place by way of exploitation of many of the larger and valuable mammals. Fur seals, large whales, sea otters and many other species declined severely, to the point of actual extinction in some cases. What, from an ethical viewpoint, should we argue? Should we rejoice that horses or cattle are such widespread and numerous mammals or regret that they are extinct as wild mammals? Is it possible to have both?

2.1 INTRODUCTION

Human exploitation of other mammals has passed through three historical phases, distinct in their ecological significance though overlapping in
time. Initially, *Homo sapiens* was a predator, particularly of herbivores but
also of fur-bearing predators. From about 11 000 years ago, goats and
sheep were domesticated in the Middle East, rapidly replacing gazelles
and other game as the principal source of meat. The principal crops,
including wheat and barley, were taken into agriculture at about the
same time, and the resulting Neolithic farming culture spread slowly
from there over the subsequent 10 500 years. In a few places such as
Mexico, Peru and China, this Middle Eastern culture met and merged
with agricultural traditions that had made a similar but independent
transition. These agricultural traditions provided the essential support
for the industrial revolution, and for a third phase of industrial exploitation of mammals.

In this chapter, these themes are drawn out and their ecological significance is investigated. Some of the impacts of humans on other mammals
require consideration on a world-wide basis, but the chapter concentrates, parochially, on Great Britain. What have been the ecological
consequences of our exploitation of other mammals?

2.2 HISTORICAL PHASES OF EXPLOITATION

2.2.1 Predatory man

Our nearest relatives – chimpanzees, orang utans and gorillas – are
essentially forest species, deriving most of their diet from the fruits of
forest trees and the shoots and leaves of plants. While chimpanzees certainly eat termites and ants fairly regularly, kill larger animals for meat
occasionally, and even organize hunting of monkeys in the Côte d'Ivoire
study area (Boesch, 1994), meat is still a minor item of diet. This is clearly
not true of *Homo* species and most early archaeological sites are recogniz-
able both from the stone tools that they contain and from the bones of the
prey that were hunted. Mammals of a suitable size yet not too dangerous
to be worthwhile prey, such as the smaller ungulates of gazelle size, are
not common in rainforest but much more abundant in savannah. For
example, bushbuck (*Tragelaphus scriptus*) live at densities of 4 per km^2
whereas Thomson's gazelle (*Gazella thomsoni*) occur on the Serengeti
Plain at 39 per km^2 (Houston, 1979; Nowak and Paradiso, 1983). Thus a
transition from forest to plains habitat, along with the transition to
bipedal walking and becoming a hunter, was surely a fundamental part
of the evolutionary divergence and subsequent success of *Homo*.

Towards the end of the last glaciation, 15 000 to 12 000 years ago, people in the Middle East were hunting fallow deer (*Dama dama* and *D. mesopotamica*), gazelles (*Gazella subgutturosa*) and other species (Uerpmann, 1987). In France, reindeer (*Rangifer tarandus*), horse (*Equus ferus*), ibex (*Capra ibex*) and chamois (*Rupicapra rupicapra*) were the main prey, while south of the Pyrenees red deer (*Cervus elaphus*) predominated (Straus, 1987). In Britain, reindeer were the principal prey of Palaeolithic hunters at this time.

The impact of predatory man on his ecosystem was surely insignificant by the standards of what followed. McEvedy and Jones (1979) make some interesting ball-park comparisons. Gorillas occur at about 1 per km^2, in a limited African range, giving a total population of about 70 000 and a total biomass of about 10 500 tonnes (t). The smaller chimpanzees live at a higher density, 3–4 per km^2, across a larger range and total over 1 million (biomass about 62 000 t). Our australopithecine ancestors of 2–3 million years ago, confined to Africa across a similar range to that of the chimpanzees (so far as we know), probably lived at densities between these – about 2 per km^2. This suggests a population of perhaps a half million and a total biomass of 25 000 t. Our ancestors therefore had less ecological impact than chimpanzees.

Expansion of range by *Homo erectus* to much of the Old World might have produced a population of 1.7 million. Expansion into Australia and the Americas, as well as into northern latitudes with the retreat of the ice caps, might have brought the population near 4 million, but this is on the assumption of a density of only 1 per 10 km^2 over habitable terrain. Even so, there is at least an argument that the extinction of large mammals in America was the result of over-hunting by humans (Martin and Wright, 1967). This argument seems unconvincing: it is hard to believe that a population of predatory humans would have supported itself on a declining prey base, and the coincidence of human invasion with rapid climatic change makes it extremely difficult to disentangle the evidence (Stuart, 1991). Perhaps, in Britain, moose (*Alces alces*), wild horse and lynx (*Lynx lynx*) were hunted to extinction during this phase of our population development, but the evidence is negligible. It seems more likely that predatory man was regulated by the populations of his prey (*cf.* Jarman, 1972).

2.2.2 Agricultural man

The places and dates of animal domestication have been well reviewed by Clutton-Brock (1981) and Davis (1987); most of what follows is taken from their accounts. The earliest domesticated species seems to be the wolf (*Canis lupus*), remoulded as the dog (*C. familiaris*), with the earliest archaeological examples from Ein Mallaha, Israel, 11 600 years before the present (BP), Palegawra, Iraq (12 000 BP) and, more surprisingly, Japan (11 000 BP) (Reed, 1974). The dog was the only domesticated animal to accompany the human invasion of both North America and Australia, and must have been domesticated very early, perhaps on several independent occasions, as a hunting companion.

The earliest agricultural domestications concerned sheep (*Ovis orientalis* – *O. aries*) and goats (*Capra aegagrus* – *C. hircus*) in the Middle East, around 11 000 to 10 000 BP; the earliest domestic sheep are probably from Zawi Chemi Shanidar, Iraq (10 870 BP), and the earliest goats from Jericho, where the change from hunting gazelles to harvesting goats is documented between 10 000 and 9000 BP. The pig (*Sus scrofa*) was domesticated a little later perhaps, but it too was present at Jericho by 9000 BP; possibly its domestication also took place independently but at an early date in South-east Asia or China. Cattle were certainly domesticated a little later, around 8500 BP at Catal Huyuk, Turkey, but the decorated altars there, dated around 8000 BP, suggest that they may have been initially kept for ceremonial purposes, not for food. However, by 8500 BP, all four domestic ungulates were present in Greece, well outside the natural range of sheep and goats.

These four became the dominant food animals throughout the Middle East by about 7000 BP, and subsequently throughout the world. Other species have since been domesticated, but primarily as transport animals – in particular, horse (*Equus caballus*), ass (*Equus asinus*) and the two camels (*Camelus bactrianus* and *C. dromedarius*). In South America, the dominant food animals were guinea pigs (*Cavia porcellus*) and camelids (*Lama* spp.), domesticated locally by 3000 BP.

If predatory humans had an insignificant effect on their ecosystem, this is clearly not true for agricultural humans. Early agriculture sustained perhaps 10 times the density, that is 1 person per km^2, but by 5000 years before Christ (BC) the area under agriculture was still so small that the world human population was only around 5 million. Over the next 4000 years, however, it grew to around 50 million as agricultural knowledge spread (McEvedy and Jones, 1979). The beginning of the Iron Age in the Middle East at about 1000 BC might be regarded as the tentative start of the industrial age, but ecologically this is clearly not so. Sources of food and energy would remain, largely, those of agricultural man up to AD 1800 and human populations remained constrained by those resources. By medieval times, the human population of the world reached a peak of 360 million; after a decline caused largely by outbreaks of bubonic plague, it was still 'only' 425 million by AD 1500.

However, by this time the impact of agricultural man was evident. In Madagascar, the large diurnal lemurs and *Aepyornis* had been exterminated; in New Zealand the moas had been wiped out by moa-hunters, supported apparently by a limited agricultural system lacking domestic herbivores. In Britain, the aurochs (*Bos primigenius*), beaver (*Castor fiber*) and bear (*Ursus arctos*) had certainly been exterminated by then, and the wolf and wild boar did not survive much longer. Agricultural humans continued to hunt and the wild species they hunted could now be exterminated without cause for concern, because they no longer supported the human population. In addition, humans inflicted major changes of habitat on their fellow mammals.

2.2.3 Industrial man

From about the fifteenth century, and particularly in the nineteenth and twentieth centuries, an industrialized society was built on the exploitation of power from fossil fuels rather than muscles, wind or water. This allowed more intensive agriculture, more intensive exploitation of other resources and an increase in human population, from 425 million in 1500 to 900 million around 1800, 1625 million by 1900 and approaching 6000 million now (McEvedy and Jones, 1979). From the viewpoint of other mammals, both the enormous biomass of the human population and its freedom from the conventional ecological constraints of predator–prey relationships and, largely, epidemic diseases, have had enormous consequences. The extinction of Steller's sea cow (*Hydrodamalis gigas*), bluebuck (*Hippotragus leucophaeus*) and quagga (*Equus quagga*) could not have occurred if the human populations which wiped out those species were actually dependent upon them for their basic food supply. Even more obviously, the extent to which furbearers such as sea otters (*Enhydra lutris*) or, in Europe, beavers were reduced, and the impact of industrialized harvesting of fur seals and large whales, depended on an industrial base to fit out the expeditions which damaged them, and on the wealth that an industrial society could generate to give them their fatal value. Though international treaties have done much to regulate some of these pressures, this has often been only when (as with sea otters, whales and fur seals) the resource has been so damaged that commercial exploitation has itself become threatened. The recent experiences with the ivory trade and African elephants, with rhino horn and with tiger bones, do not suggest much optimism for the future of most of our fellow larger mammals. The global picture is too large to examine further, and in any case many of these examples form the subject of later chapters. This chapter therefore turns to the parochial case and examines what has happened in Great Britain (Ireland is deliberately excluded simply because there is less certainty about its fauna 10 000 years ago).

2.3   GREAT BRITAIN AS A CASE HISTORY

Great Britain has a land area of 230 367 km^2. Between 9000 and 5000 years ago (the Mesolithic period), it was largely covered in deciduous woodland, though there was some pine woodland in northern Scotland, open country on the higher mountains and perhaps on exposed islands, and much more fenland than we can envisage now. The human population of hunter-gatherers was perhaps only 2000–3000 but they fed on moose, red and roe deer, aurochs and wild boar, among other prey. They also hunted beaver and red fox (*Vulpes vulpes*), presumably for their fur. Their only domestic animal was the dog.

To assess the impact of ourselves, let us try to estimate the size of the prey-base available to our Mesolithic forbears. Jedrzejewski *et al.* (1993) have recently published a fascinating account of the lynx and its prey in the Bialowieza Forest. It is a long way east, a relatively small area (1250 km^2) to use as a comparison and itself not totally pristine, but it is the nearest fragment available to tell us what the western European mammal fauna might have looked like in a functioning ecosystem. Their paper gives the results of prey censuses undertaken in March (i.e. numbers at the end of winter/beginning of the breeding season). It includes a population of 250 European bison (*Bison bonasus*), maintained artificially at that level by export and selective culling. Let it be assumed that Britain, which lacked bison, had a similar biomass of aurochs (*Bos primigenius*); and further that, if the bison population is lower than the habitat would sustain, then the other ungulate populations have expanded to equilibrium with the available food supplies.

Great Britain is 397 times larger than the 580 km^2 pristine part of Bialowieza; the populations have therefore been taken in the latter, multiplied by this number to estimate the potential native herbivore population of Great Britain and, taking likely individual mean masses, further extrapolated to biomass (Table 2.1). This suggests that Mesolithic Britain held a native herbivore population of over 4 million animals, with a collective biomass of nearly half a million tonnes. Red deer, wild boar and roe deer would each have numbered over 1 million, and the two larger species about 50 000–100 000. Biomass would have been dominated by red deer and wild boar, each contributing over 100 000 tonnes, respectively, and the other three species would have made similar contributions of about 14 000–40 000 tonnes each.

Interestingly, the Mesolithic hunters of Star Carr, Yorkshire, seem not to have taken prey in proportion to their availability, if these calculations of 'availability' are correct. Red deer were by far their most numerous prey, with 80 individuals represented. Moose (11) and aurochs (9) were also well represented, but roe deer (33) and wild boar (5) seem much under-utilized. It is, of course, possible that this was a local environmental effect; at Thatcham, Berkshire, the same five ungulates were hunted, and red deer (8) again were the most numerous prey, but roe deer (6) and wild boar (7) were also numerous, while moose (1) and aurochs (2) were scarce (sources in Yalden, 1982, Table 4). There is a big difference between the Star Carr fauna and the likely availability ($\chi^2_4 = 91.2$,

Table 2.1 Populations and biomass of large mammals in Bialowieza Primeval Forest (580 km^2) (data from Jedrzejewski *et al.*, 1993) extrapolated to give estimate of Mesolithic fauna of Great Britain (230 367 km^2) (individual masses from Jedrzejewska *et al.*, 1994)

Species	Individual mass (kg)	Bialowieza Forest		Great Britain	
		n	Biomass (kg)	n	Biomass (kg)
Bison *Bison bonasus* (≡Aurochs *Bos primigenius*)	400	250	100 000	99 250	39 700 000
Red deer *Cervus elaphus*	100	3 710	371 000	1 472 870	147 287 000
Roe deer *Capreolus capreolus*	20	2 730	54 600	1 083 810	21 676 200
Moose *Alces alces*	200	170	34 000	67 490	13 498 000
Wild boar *Sus scrofa*	80	3 420	273 600	1 357 740	108 619 200
Herbivore total		**10 280**	**833 200**	**4 081 160**	**330 780 400**
Lynx *Lynx lynx*	15	19	285	7 543	113 145
Wolf *Canis lupus*	32	20	640	7 940	254 080

$P < 0.0001$), whereas the Thatcham fauna is closer to availability ($\chi^2_4 = 4.49$, $P = 0.03$). In preferring red deer and avoiding wild boar, the Mesolithic hunters of Star Carr seem to have behaved like the wolves of Bialowieza (Jedrzejewski *et al.*, 1993); it seems likely that red deer were both abundant and large enough to be worthwhile prey, yet not so dangerous as wild boar.

On the basis of the Bialowieza figures (Jedrzejewska *et al.*, 1994), there should have also been about 7500 lynx and 8000 wolves preying on these herbivores, but these figures seem rather low. Numerically, there are 264 herbivores per large carnivore at Bialowieza; in terms of biomass, 900 kg herbivore/kg carnivore. On Isle Royale, in Lake Superior, as studied by Mech (1966) and Peterson (1977), there were about 800 moose supporting 22 wolves, or 351 kg herbivore/kg carnivore. Similarly, in the Serengeti ecosystem, there are over 2 million herbivores sustaining 7000 large carnivores; that is, 336 herbivores per carnivore or 431 kg herbivore/kg carnivore (Houston, 1979, Tables 11.1 and 11.2). At 350 kg herbivore/kg carnivore, one might expect 75 wolf-sized predators in the Bialowieza ecosystem at present, and on the same basis 29 500 wolves (or equivalents) in Mesolithic Britain.

Brown bears (*Ursus arctos*), now absent from Bialowieza, would have contributed some of the 'missing' predator biomass, both there and in Britain, but one must suspect that poaching or emigration contrive to constrain the predator population of Bialowieza well below its potential. Apparently wolves are still persecuted in the Byelorussian part of Bialowieza, and their density in the Polish section is indeed depressed; there were 39 in the 1950s (W. Jedrzejewski, personal communication).

Returning to the impact of humans on the mammals of Britain, what has replaced the former ungulate biomass? With forest clearance, much more of the landscape is now grassland (or cropland), so that forest, much of it coniferous, now covers only 8% of the landscape. Grassland – even unfertilized natural grassland – has a much higher and more available productivity and can therefore support a higher biomass of herbivores; in the New Forest, Putman (1986, Table 7.3) suggests a productivity over the year of 105.4 g/m^2 in wooded glades, compared with 656.6 g/m^2 for streamside lawns. Aurochs, wild boar and moose are long extinct, along with the large predators; Harris *et al.* (1995) suggest that 360 000 red deer and 500 000 roe deer (perhaps 24% and 46% of their native populations, respectively) remain in Great Britain. Instead, there are 3.9 million breeding cattle, 20.3 million breeding sheep and 0.8 million pigs, according to the June census returns of the various agriculture departments (Government Statistical Service, 1992). (The full census returns indicate 10.3 million cattle, 41 million sheep and 7 million pigs, but these include the young of the year and are not comparable with the end of winter/beginning of breeding season population estimates being used for other species.) In the absence of wolves, there are now some 7 million dogs, and also 55 million humans preying on these herbivores. Converting to biomass (Table 2.2) there are now about 3.2 million tonnes of herbivore in Great Britain, and the addition (or subtraction) of 70 000 tonnes of wild ungulates makes little difference to this total.

The estimate of the Mesolithic ungulate biomass works out as a crude density of 1436 kg/km^2, whereas the density of domestic plus wild ungulates is now about 14 169 kg/km^2 – a 10-fold increase. This is a modest overall density of livestock, comparable with the figures quoted by Sharkey (1970) for cattle on native pastures in temperate New Zealand and the United States. For *Lolium/Trifolium* reseeds, with added phosphate fertilizer, he suggests that English pasture can support 135 000 kg cattle/km^2. However, arable land, which directly sustains few ungulates, 'dilutes' the area of pasture available to livestock; put another way, it either feeds humans directly, or provides supplementary food for livestock in winter. Thus this modest overall density is probably the appropriate figure for this comparison.

Table 2.2 Current large mammal population of Great Britain, for comparison with the 'wild fauna' in Table 2.1 (domestic mammals from June census, Government Statistical Service, 1992; wild mammals from Harris *et al.*, 1995)

Mammal group	Individual mass (kg)	n	Biomass (kg)
Domestic (breeding adults)			
Pigs [7 000 800]	150	853 000	127 950 000
Sheep [41 047 000]	45	20 364 600	916 407 000
Cattle [10 332 800]	550	3 908 900	2 149 895 000
Total domestic [58 380 600]		25 126 500	3 194 252 000
Wild herbivores (pre-breeding season)			
Red deer *Cervus elaphus*	150	360 000	54 000 000
Fallow deer *Dama dama*	45	100 000	4 500 000
Sika deer *Cervus nippon*	45	11 500	517 500
Roe deer *Capreolus capreolus*	21	500 000	10 500 000
Feral goats *Capra hircus*	45	3 600	162 000
Feral sheep *Ovis aries*	45	2 100	94 500
Total wild herbivores		977 200	69 774.000
Total herbivores			3 264 026 000
Humans (all ages)	50	55 000 000	2 750 000 000

Figures in square brackets are total populations (June census).

2.4 DISCUSSION

How do we evaluate the changes wrought in Great Britain or in the world as a whole? If these calculations are at all credible, there are now in Britain 40 times more cattle than there were aurochs, and 15–20 times more sheep than there were red deer or roe deer. Clearly, cattle and sheep are very successful mammals – far more populous than they could ever have been before domestication. Similarly, the domestic *Canis familiaris* is over 200 times more numerous than *C. lupus* ever was. If the measure of evolutionary fitness is the quantity of surviving genotypes, then aurochsen, wolves and urials have done very well by becoming domestic and we now have far more large mammals in Britain than we had 7000 years ago.

Among other domesticated species world-wide, the two camels are also very successful, numbering some 17 million, yet extinct in the wild; the horse (*Equus caballus*) numbers 65 million world-wide, yet it is the subject of a 'survival' plan and reintroduction scheme in its ancestral guise as *E. przewalskii*; the ass world-wide numbers some 40 million, yet wild *E. asinus* number fewer than 5000 (FAO, 1987; Duncan, 1992). Clearly these too, are species which have done very well out of human exploitation (Table 2.3), and not only in captivity: substantial feral populations of asses, horses and camels occur on continents well beyond their native ranges.

Table 2.3 World populations of large domestic mammals in 1986 (from FAO Production Yearbook, 1987)

Species	n	Species	n
Horses	65 064 000	Cattle	1 271 810 000
Mules	15 142 000	Buffalo	138 352 000
Asses	40 477 000	Sheep	1 145 690 000
Camels	17 446 000	Goats	492 192 000
Pigs	822 443 000		

Neither are food and transport species the only ones to thrive under human exploitation. Not only have mice (*Mus domesticus*), rats (*Rattus norvegicus*) and rabbits (*Oryctolagus cuniculus*) successfully spread extensively round the world as wild mammals, but they also number millions as laboratory and pet animals, along with guinea pigs (*Cavia porcellus*), hamsters (*Mesocricetus auratus*) – virtually unknown in the wild – and gerbils (*Meriones unguiculatus*). Among the furbearers, chinchilla (*Chinchilla lanigera*) is barely surviving as a wild mammal; and while mink (*Mustela vison*) and fox (*Vulpes vulpes*) are hardly endangered as wild mammals, their populations also have benefited substantially from exploitation.

It seems that human exploitation is not necessarily a bad thing from the mammal's perspective – hence the dichotomy of the chapter title. Those species which have been completely exterminated are a major entry in the debit side of the evaluation: Steller's sea cow, the sea mink, bluebuck and the like. Far too many other large mammals – obviously large predators such as lions (*Panthera leo*) and tigers (*P. tigris*), but also such important distinctive large herbivores as elephants (*Loxodonta africana*, *Elephas maximus*) and the rhinos (Rhinocerotidae) – are too close to extinction, largely due to over-exploitation though habitat loss has also played a major role. What is wrong, of course, is that the calculations in this chapter ignore the loss of diversity (Table 2.4); numbers alone are no substitute, and the species under threat are frequently the most distinctive. Another million cattle are no substitute for a few thousand Steller's sea cows.

This is, of course, a serious loss from our world; to paraphrase John Donne, every species lost diminishes me, for I am part of the natural world. If exploitation is not necessarily a bad thing from certain mammals' viewpoints, it is clearly a bad thing from a human perspective. Yet there would be far fewer human perspectives if it had not happened. How many human perspectives should there be? Most of the world's 6000 million people would not exist but for their successful exploitation of domestic ungulates and, if asked, most of them would be very happy with that success. There lies another dichotomy: their happiness is at the same time our sadness.

Table 2.4 Diversity (Shannon's Index H') calculated for modern ungulate fauna of Britain, for comparison with estimated Mesolithic British fauna and with modern Serengeti fauna (from Houston, 1979; more recent figures for the Serengeti fauna in Chapter 9)

Mesolithic Britain		Modern Britain		Serengeti	
Aurochs	99 250	Cattle	3 908 900	Wildebeest	720 000
Bos primigenius		*Bos taurus*		*Connochaetes taurinus*	
Red deer	1 472 870	Sheep	20 364 600	Zebra	240 000
Cervus elaphus		*Ovis aries*		*Equus burchellii*	
Wild boar	1 357 740	Pigs	853 000	Buffalo	108 000
Sus scrofa		*Sus scrofa*		*Syncerus caffer*	
Moose	67 490	Red deer	360 000	Th. gazelle	981 000
Alces alces		*Cervus elaphus*		*Gazella thomsoni*	
Roe deer	1 083 810	Roe deer	500 000	Giraffe	17 400
Capreolus capreolus		*Capreolus capreolus*		*Giraffa camelopardalis*	
		Fallow deer	100 000	Eland	24 000
		Dama dama		*Taurotragus oryx*	
		Sika	11 500	Topi	55 500
		Cervus nippon		*Damaliscus lunatus*	
		Goats	3 600	Impala	119 100
		Capra hircus		*Aepyceros melampus*	
				Elephant	4 500
				Loxodonta africana	
				Kongoni	20 700
				Alcelaphus buselaphus	
				Warthog	34 200
				Phacochoerus aethiopicus	
				Black rhino	900
				Diceros bicornis	
				Gr. gazelle	6 000
				Gazella granti	
				Hippopotamus	2 400
				Hippopotamus amphibius	
				Waterbuck	3 000
				Kobus ellipsiprymnus	
$n = 4\ 081\ 160$		$n = 26\ 101\ 600$		$n = 2\ 336\ 700$	
$s = 5$		$s = 8$		$s = 15$	
H' = 1.2243		H' = 0.7506		H' = 1.5767	
var H' = 0.0001		var H' = 0.0001		var H' < 0.0001	

REFERENCES

Boesch, C. (1994) Chimpanzees–red colobus: a predator–prey system. *Animal Behaviour*, **47**, 1135–1148.

Clutton-Brock, J. (1981) *Domesticated Animals from Early Times*, British Museum (Natural History) and Heineman, London.

Davis, S.J.M. (1987) *The Archaeology of Animals*, Batsford, London.

Duncan, R. (1992) *Zebra, Asses and Horses, An action plan for the conservation of wild equids*, IUCN, Gland, Switzerland.

FAO (1987) *Production Yearbook 40 (for 1986)*, FAO/UN, Rome.

Government Statistical Service (1992) *The Digest of Agricultural Census Statistics; United Kingdom 1991*, HMSO, London.

Harris, S., Morris, P., Wray, S. and Yalden, D.W. (1995) *A Review of British Mammals: population estimates and conservation status of British mammals other than cetaceans*, JNCC, Peterborough.

Houston, D.C. (1979) The adaptations of scavengers, in *Serengeti: Dynamics of an Ecosystem*, (eds A.R.E. Sinclair and M. Norton-Griffiths), Chicago University Press, Chicago and London, pp. 263–286.

Jarman, M.R. (1972) European deer economies and the advent of the Neolithic, in *Papers on Economic Prehistory*, (ed. E.S. Higgs), Cambridge University Press, Cambridge, UK, pp. 125–149.

Jedrzejewska, B., Okarma, H., Jedrzejewski, W. and Milkowski, L. (1994) Effects of exploitation and protection on forest structure, ungulate density and wolf predation in Bialowieza Primeval Forest, Poland. *Journal of Applied Ecology*, **31**, 664–676.

Jedrzejewski, W., Schmidt, K., Milkowski, L., Jedrzewska, B. and Okarma, H. (1993) Foraging by lynx and its role in ungulate mortality: the local (Bialowieza Forest) and the Palaearctic viewpoints. *Acta Theriologica*, **38**, 385–403.

Martin, R.S. and Wright, H.E. (eds) (1967) *Pleistocene Extinctions: the search for a cause*, University Press, Yale.

McEvedy, C. and Jones, R. (1979) *Atlas of World Population History*, Penguin, London.

Mech, D. (1966) *The Wolves of Isle Royale*, Fauna of the National Parks of the United States, Fauna Series 7, Washington.

Nowak, R.M. and Paradiso, S.L. (1983) *Walker's Mammals of the World*, 4th edn, John Hopkins University Press.

Peterson, R. (1977) *Wolf Ecology and Prey Relationships on Isle Royale*, National Park Service Monograph Series, Number 11, Washington.

Putman, R.J. (1986) *Grazing in Temperate Ecosystems: large herbivores and the ecology of the New Forest*, Croom Helm, London and Sydney.

Reed, C.A. (1974) The beginnings of animal domestication, in *Animal Agriculture: the biology of domestic animals and their use by man*, (eds H.H. Cole and M.Ronning), W.H. Freeman, San Francisco, pp. 5–17.

Sharkey, M.J. (1970) The carrying capacity of natural and improved land in different climatic zones. *Mammalia*, **34**, 564–572.

Straus, L.G. (1987) Upper Palaeolithic Ibex hunting in the southwest Europe. *Journal of Archaeological Science*, **14**, 163–178.

Stuart, A.J. (1991) Mammalian extinctions in the late Pleistocene of Northern Eurasia and North America. *Biological Reviews*, **66**, 453–562.

Uerpmann, H.P. (1987) *The Ancient Distribution of Ungulate Mammals in the Middle East*, Tubinger Atlas des Vorderen Orients, Reihe A (Naturwissenschaften) Nr 27, Reichert, Wiesbaden.

Yalden, D.W. (1982) When did the mammal fauna of the British Isles arrive? *Mammal Review*, **12**, 1–57.

3

The wild fur trade: historical and ecological perspectives

Stephen Tapper and Jonathan Reynolds

SYNOPSIS

The exploitation of mammals for fur has existed since prehistoric times and in the Old World it is recorded in a fragmentary way from the sixth century. It has always been an important export from boreal regions and populations were probably not over-exploited until the nineteenth century.

In the New World, systematic exploitation began in the sixteenth century as an adjunct to high seas fishing in the western Atlantic. French and English companies exploited large areas of the continental interior by trading useful commodities such as steel tools and muskets to native Americans who in turn acted as middlemen to more inaccessible tribes. After American independence and the withdrawal of France, the trade was dominated by several independent companies. During periods of competition many furbearer species were over-exploited and some were extirpated from their native range. The end of the nineteenth century marked a nadir for most furbearer species. In the twentieth century, fur trapping is highly regulated in the United States and Canada and the trapline system encourages individual trappers to conserve furbearer stocks and take only a sustainable harvest. Populations of most species have now wholly or partially recovered. Indeed for some species like the beaver the current harvest is apparently larger than at any time in history.

The continued trapping and trading of furs is highly contentious and many argue that, although sustainable harvests are possible, they will always be difficult to regulate except in the best-ordered western societies. History shows that furbearers are very easily over-exploited and, even with a sustained harvest, furbearer numbers are maintained at much lower population levels than would otherwise be the case. Legal trade in furs also makes the illegal trade easier. However, advocates of the trade argue that in wilderness areas fur trapping is less environmentally damaging than other forms of use (e.g. logging or even tourism). Income from the trade can be used to manage species which in many cases would be considered pests in non-wilderness areas. Finally, fur trapping allows native peoples to retain their culture better than any other way of life.

3.1 INTRODUCTION

Furs and skins leave little archaeological evidence, but it is clear that pre-historic societies traded industriously in such products (Wright, 1987). In northern communities, furs would have been one of the few exports.

In the Old World, a fur trade was widespread by the time written records began. By contrast, large-scale exploitation of furbearers in the New World is more recent and better documented. This chapter draws primarily on the North American story, in which the established Old World fur trading and manufacturing system was suddenly put in touch with totally new sources of furs. It avoids ethical issues and concentrates on the ecological history and the current basis of harvesting wild furs.

3.2 THE OLD WORLD FUR TRADE UP TO THE DISCOVERY OF AMERICA

Since at least the sixth century, the Svea in Scandinavia collected tributes from the Saami (Lapps) in the form of furs, which were traded as far afield as England and Rome (Nockert, 1991; Wheelersberg, 1991), and by the ninth century this trade was an important part of Sweden's economy. There was also a long-established trade in furs centred on Bulgar-on-the-Volga, linking northern Europe via the Levant to the Middle East and Central Asia (Martin, 1986).

Later, safe trade routes between northern tribes and powerful southern societies were often the cause of wars. In some cases, fur supplies came as tribute from subjugated tribes rather than through exchange of goods, but the distinction is often unclear. Trade connections became ever more extensive until furs draining from the boreal region into merchant cities were being traded as far afield as Spain (via the Middle East and North Africa) and China. The centre of distribution shifted between Bulgar, Kiev and Novgorod, and finally to Moscow. The exploitation of fur production from outlying districts became more organized, with wealthy Russian 'boyars' extracting annual levies in furs from huge distant estates (Veale, 1966).

The ecological impact of the medieval Old World fur trade is unclear, because its effects are indistinguishable from human settlement and habitat change (particularly forest clearance). However, it is not correct to infer – as Ponting (1991) has done – that supply shortages, restrictions on the use of certain furs, the imposition of trade restrictions, or a final collapse in some aspect of fur trade, indicate the over-exploitation of furbearers. Ponting (1991) depicts the entire Old World fur trade as a 'one-off' use of a natural resource, even though key components of it – such as the boreal populations of squirrel (*Sciurus vulgaris*), marten (*Martes*

martes), sable (*Martes zibellina*) and fox (*Vulpes vulpes*) – have been commercially exploited for at least a thousand years. Although marten and sable populations did suffer over-exploitation in the nineteenth century, this was followed in the present century by an era of research and management across the Soviet Union which restored them to most of their former range (Bakeyev and Sinitsyn, 1994). In more southern Old World countries, it is impossible to distinguish the relative historical importance of exploitation, habitat destruction and climatic change in the extermination of indigenous fur-bearers like beaver (*Castor fiber*) and marten, as it is for other species like elk (*Alces alces*) or reindeer (*Rangifer tarandus*) (Ritchie, 1930; Dahl, 1986).

3.3   FUR TRADE HISTORY OF NORTH AMERICA

The history of the North American fur trade has been exhaustively described by Innis (1956), Newman (1985, 1988) and Wishart (1979), and briefer descriptions can be found in Obbard *et al.* (1987) and Ray (1987). What follows is based on these accounts.

3.3.1 The early trade

Early fur trading began some 450 years ago as an adjunct to North Atlantic high seas fishing. In the sixteenth century European vessels regularly fished the cod banks off Newfoundland, returning with salted fish and fur traded with coastal Indians. Ships bound for North America, which would otherwise be empty, could be loaded with heavy trade goods such as metal kettles and hatchets, while the pelts for the return trip were light and easily carried in addition to the salted fish.

By 1600 the French and later the Dutch were making a more serious effort to exploit the interior. The French pushed up the St Lawrence to trade with the Huron. Though agricultural people, the Huron trapped furs themselves and acted as intermediary traders with other tribes. Thus the fur sold to Europeans came from the regions which drained into the rivers Saguenay, St Maurice and Ottawa. The Dutch concentrated on the Hudson River, the headwaters of which pushed into land inhabited by the Iroquois. This early trade included skins of moose (*Alces alces*), lynx (*Lynx canadensis*), otter (*Lutra canadensis*), marten (*Martes americana*) and badger (*Taxidea taxus*) as well as beaver (*Castor canadensis*). By the 1620s, in June each year two ships laden with trade goods would arrive from France and return with 5000–12 000 beaver skins. In one year over 22 000 beaver pelts were carried. Even at this early stage beaver were being trapped at a faster rate than local populations could replenish and by 1623 most of the beaver had gone from the St Lawrence river basin.

The beaver soon became the most important furbearer because its very dense underfur could be used to make a high quality waterproof felt much in demand for the fashionable wide-brimmed and tri-cornered hats of the period. Beaver underfur is protected by long guard hairs and during felt-making these have to be stripped out and the underfur cut off the hide. This process is technically difficult and much of the European expertise came from the old fur-producing countries like Russia. Because of this, coat beaver was preferred. These were hides that had already been made into robes by the Indians and worn fur side in, causing the guard hairs to slip out. Later as coat beaver became scarce the trade shifted to freshly trapped animals whose pelts had been stretched and dried: the so-called parchment beaver. Quality was important and varied regionally; for example, roughly twice as many hats could be manufactured from beaver from a northern district as from a southern one.

Following the French example the British set up the Hudson's Bay Company (HBC) to establish its own corner of the growing trade. By Royal Charter in 1670, it was given the sole rights to 'Rupert's land' or all land that drained northwards into Hudson Bay. From the outset HBC was conservative, commercial, apolitical and controlled from London. Furs were brought by canoe to stations on the Hudson Bay coast largely by Cree Indians, who had in turn traded them with other tribes in the interior. This commerce remained essentially unchanged for almost a hundred years, interrupted only by natural disasters and raids by the French.

By contrast the French were aggressive, entrepreneurial and very political. During the hundred years from 1670 they had pushed westwards, largely cutting off HBC trade from the south. By 1682 they had reached down the Mississippi to New Orleans and as far west as Lake Manitoba. Expansion was easier for the French as the native people of the St Lawrence were agricultural and could provision expeditions with Indian corn and animal grease. Resources were scarce on Hudson Bay and expeditions had to be largely provisioned from London.

This early trade is often seen as unscrupulous exploitation of native peoples by Europeans – the exchange of cheap, worthless items for fur at immense profits. In fact most trade goods were of real value for the Indians. Steel axes, knives, awls, needles, metal cooking pots, steel traps, guns and blankets significantly improved their way of life. Furthermore the costs of the trade were very high and HBC investors received dividends in only four out of the first 47 years of the company's history (Newman, 1985).

Within North America, Indians proved to be sophisticated entrepreneurs, jealously guarding their contacts with European sources of trade goods, and selling on within or between tribes at substantial (250% to 1200%) profits (Innis, 1956). For this reason they discouraged Europeans from penetrating inland, and it was many years before European traders managed to circumvent these inflationary trade links.

By the middle of the eighteenth century much of central and eastern Canada and the American Midwest down to the Gulf of Mexico had come under the influence of the fur trade (Figure 3.1).

3.3.2 The expansion west

Following their defeat at Quebec in 1758, France withdrew and their operation became the North West Company (NWC), run mainly by Canadians of Scottish descent out of Montreal, but with a French Canadian backbone of *voyageurs* and *coureurs de bois*.

The late eighteenth century was a period of intense competition between HBC and NWC which was ultimately detrimental to the companies, the furbearers and the native people. Competing traders deliberately encouraged fur to be depleted to prevent access by the competition; Indian long-term interests were pushed aside and drink was often traded instead of useful goods. Furthermore at this time many Indian tribes were unintentionally decimated by European diseases.

In the end the two companies merged in 1821, and for four decades there was a long period of stability in which trading posts were rationalized and manning levels reduced by one third (Figure 3.2). Fur stocks began to be conserved and quotas, closed seasons and differential pricing were used to reduce the harvest of beaver and increase exploitation of alternate species such as muskrat (*Ondatra zibethicus*). Although not always popular, these measures worked and beaver stocks in many areas had recovered by the 1840s. However, subsequently the beaver price collapsed as fashions turned to silk rather than felt hats.

Over the same period in the American west, south of the 49th parallel, trading was conducted without native Indians or trading posts. Instead white trappers (mountain men) were sent into these regions to trap furs which were traded at an annual rendezvous. This was effective and cheap, but it rapidly depleted fur resources and alienated the Indians. Much of the American trade relied on buffalo (*Bison bison*) robes and these mammals, like the beaver, were over-exploited. To a large extent it was over-exploitation and competition with HBC in the Rocky Mountains that caused the collapse of most of this American fur trade by the 1840s.

During the latter half of the nineteenth century the situation once again became destabilized. The British North America Act 1867 founded

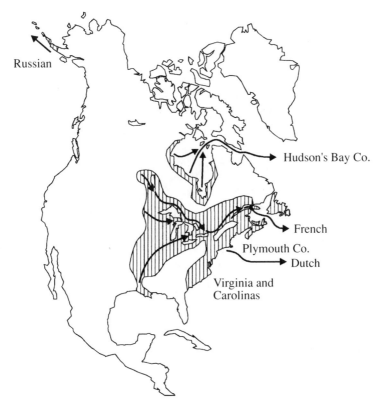

Figure 3.1 Extent of the North American fur trade by the middle of the eighteenth century. Hatched area indicates main regions of wild fur production; arrows are main trade routes. Names of companies or national enterprises are shown. Simplified from Obbard *et al.* (1987).

Canada, and furbearer management was given to the new provinces. In 1885 the completion of the Canadian Pacific Railway created new trade routes along the southern margins of the boreal forest. Free traders could keep overhead costs low by offering cash for fur instead of goods, and they could rely on the railway for transport and keep in touch with market prices by telegraph. In the United States, after the disruption of the Civil War and large-scale settlement of the west, most furbearer populations were depleted. However, an increasing urban population was building a strong home market and by 1900 the United States was importing three times as many furs as it was exporting. Many of these furs came from the HBC via London, but later the cash fur traders of southern Canada sold directly to auction houses in St Louis and New York.

Competition and lack of either government or company control

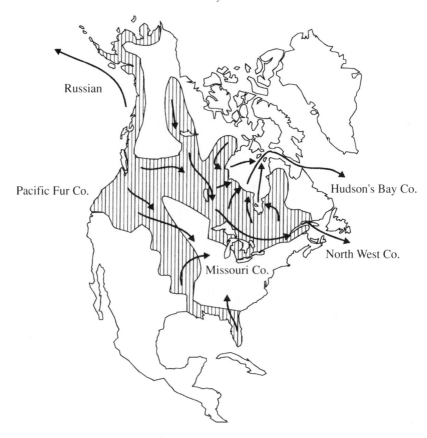

Figure 3.2 Extent of the North American fur trade by the middle of the nineteenth century. Hatched area indicates main regions of wild fur production; arrows are main trade routes. Names of companies or national enterprises are shown. Simplified from Obbard *et al.* (1987).

inevitably led to a depletion of most furbearers. The turn of the century marked a nadir in wild fur production that resulted, amongst other things, in the ranching of captive-reared animals instead.

3.3.3 Modern wild fur production and marketing

During the twentieth century, governments in Canada and the United States have gradually implemented conservation programmes and re-organized trapping regulations so that fur harvests are adjusted to the abundance and reproductive success of each species.

Most schemes are organized at the state or province level and Ontario is a well documented and appropriate example as it covers the core area of fur production since the seventeenth century (Novak, 1987b). Here the regulations are sufficiently watertight to discourage abuse, and are designed to encourage the trappers themselves to conserve the species upon which their livelihood depends. Each trapper needs a licence and formal training. He also needs trapping rights to an individual area (a trapline) where he must conform to annual quotas set by the province. All his pelts must be tagged (sealed) by a wildlife officer before sale, and he must pay a royalty on each skin. Revenues from royalties finance the cost of administering the fur regulations. The key to the system has been the development of the trapline. This gives each trapper (or head trapper, if he has helpers) a well defined and restricted area within which to trap. The trapper will keep this area from year to year so it is always in his interest to ensure adequate breeding stock. In addition the quotas set for him will be appropriate for the extent and quality of his trapline. In Ontario there are some 3000 traplines on crown land and these fall within 48 fur administration districts. Traplines vary in size but the average is 109 km^2 and most are configured around watersheds to encourage the management of beaver, muskrat and mink (*Mustela vison*).

Furs from a winter's trapping are sold to fur traders, trapping associations, or directly to auction houses. Sheiff and Baker (1987) give a detailed account of recent fur marketing operations. Ultimately most furs pass through one of about a dozen auction houses world-wide and are graded into individual lots of similar quality and colour. Quantities and values of wild fur produced in a typical recent year are shown in Table 3.1. These are ranked according to their economic value and in total this amounts to 17.8 million pelts with a value of nearly Can\$250 million (£120 million). Many species in Table 3.1 may also be killed as pests, predators or game species and do not appear in this table. Furs sold at auction go to garment manufacturers and eventually to retailers. This increases their value. Sheiff and Baker (1987) show that \$500 paid for 10 beaver pelts end up in a coat for which the consumer pays \$3100. Applying this same ratio to all furs suggests that the retail value from North American wild furs is some £744 million annually.

Trends in the market of North American furs to the mid 1980s suggest a large reduction in European sales, but markets such as Asia have increased and greater affluence in developed countries has increased demand (Sheiff and Baker, 1987).

3.4 THE IMPACT ON FURBEARER NUMBERS

The wide range of species trapped means that it is not always easy to determine the true impact of the trade for all of them in a historical sense, particularly if, as with wolves (*Canis lupus*) and coyotes (*Canis latrans*), records have grouped them together. It is clear that impacts have

Table 3.1 Combined wild fur harvest of United States and Canada ranked in order of commercial value; values are in Canadian dollars and based on average pelt price paid to the trapper (data from 1982/83, derived from Sheiff and Baker, 1987)

Species		Total value	Total pelts	Historical notes
Raccoon	*Procyon lotor*	90 670 494	4 917 055	Current harvests about 10 times earlier times, due to fashion and increased population.
Muskrat	*Ondatra zibethicus*	32 861 813	7 351 636	Largely ignored until mid 19th century. Never apparently over-harvested.
Red fox	*Vulpes vulpes*	28 378 233	534 430	Poorly recorded in the past. Many others currently killed to protect livestock.
Coyote	*Canis latrans*	19 953 490	612 446	Mainly controlled as a livestock predator. Probably more common than previously.
Beaver	*Castor canadensis*	13 700 050	641 088	Historically the most important furbearer. Populations now more than recovered from over-trapping.
Gray fox	*Urocyon cinereoargenteus*	13 589 844	315 456	Historically records not separated from other foxes. Recent USA figures suggest increasing numbers.
Mink	*Mustela vison*	11 101 126	430 610	Unimportant until 1820s but very fashionable by 1930s. Ranched animals now far exceed wild harvest.
Lynx	*Lynx canadensis*	9 493 793	35 268	Very cyclic species. Apart from early 20th century when over-trapped average harvests similar since 1800.
Marten	*Martes americana*	6 815 273	178 457	Over-trapped in 19th century – now recovered and high harvests sustainable.
Bobcat	*Lynx rufus*	6 624 902	81 387	Little data until 20th century. Harvest increased since but impact on populations is unclear. CITES listed.
Fisher	*Martes pennanti*	4 004 341	25 082	Over-trapped in 19th century. Now recovered in most areas.
Virginia opossum	*Didelphis virginiana*	3 099 670	990 310	Low value fur with maximum harvests in the 1920s.
River otter	*Lutra canadensis*	2 032 106	34 542	Over-trapped in 19th century. Populations and harvests increased this century.

Common name	Scientific name			Notes
Nutria	*Myocaster coypus*	1 399 213	799 550	Introduced to USA in 1899. Agricultural pest and furbearer of southern states.
Black bear	*Ursus americanus*	1 315 531	26 919	Pre-1900 data combined with other bears. A game animal and most trophy hunters retain the pelt.
Striped skunk	*Mephitis mephitis*	869 616	271 715	Cheap fur often sold as 'Alaska sable'. New laws compel them to be identified as 'skunk' – sales have dropped.
Polar bear	*Ursus maritimus*	629 582	727	No historical data. Also a trophy species. Quotas for 1980s higher than earlier this century.
Brown bear	*Ursus horribilis*	593 275	1 249	No historical data. Trophy animal. Harvest increasing slightly.
Arctic fox	*Alopex lagopus*	455 503	15 707	All foxes combined before 1900. Harvest has declined since 1930s due to fashion and ranched animals.
Badger	*Taxidea taxus*	434 630	43 463	Poor historical records. Popularity as fur fluctuated this century. Numbers subject to habitat and agricultural pressures.
Wolverine	*Gulo gulo*	350 283	1 774	Specialist fur. Loss of habitat may have caused decline in numbers. Current harvest about half 19th century levels.
Red squirrel	*Tamiasciurus hudsonicus*	338 218	348 678	Little historical data. Low value fur, with low exploitation rate. Biggest harvest this century during depression years.
Mountain lion	*Felis concolor*	263 200	1 316	No historical data. Now rare and infrequently trapped for fur. Trophy animal . Extensively reduced by predator control schemes.
Gray wolf	*Canis lupus*	261 371	4 133	Early data includes coyotes. Predator control programmes reduced numbers until 1960s. Numbers recovering.
Ringtail	*Bassariscus astutus*	220 799	67 938	Low value fur. No historical data. Protected in some US states.
Weasels	*Mustela* spp.	85 768	51 358	Includes three species. Number harvested has declined due to low value, changes in trapping and habitat loss.
Spotted skunk	*Spilogale putorius*	850	7 724	No historical data. Lower harvests now than in earlier decades. Populations have increased and spread.
Total		**249 542 974**	**17 790 018**	

The wild fur trade

ranged from little or none to extreme. The following examples illustrate this range.

3.4.1 Muskrat

Muskrat are perhaps the ideal furbearer in that they are widespread, common and prolific, and the pelts can be made into fur garments that are reasonably priced. Historically they were of little value until the latter half of the nineteenth century (Figure 3.3) when up to 2 million were taken annually. In the twentieth century more muskrat pelts have been taken than any other single furbearer, but there is little evidence that they have ever been seriously over-trapped. Indeed some authors regard them as more or less immune to over-harvesting, although studies are needed to develop more efficient management strategies (Boutin and Birkenholz, 1987).

3.4.2 Beaver

Beaver management has been reviewed by Novak (1987a). Beaver were formerly distributed throughout most of North America wherever riverine or lake communities existed (Seton, 1910; Hall, 1981). In the past exploitation rates were excessive and populations were trapped out successively as the fur trade pushed west (Wishart, 1979; Novak, 1987a; Obbard *et al.*, 1987). Today they have recovered nearly all of this former range of habitats (Allen, 1987). They have always been used by native

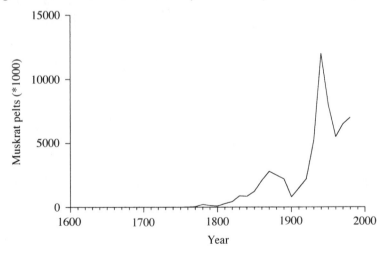

Figure 3.3 Estimated total commercial harvest of muskrat pelts from North America since 1770. Figures are averaged by decade. Simplified from Obbard *et al.* (1987).

Americans for food as much as for fur and today 90% of beavers trapped by Indians are used for meat as well as fur, as are 38% of trapped beavers overall in Ontario (Todd and Boggless, 1987). Being territorial and living in conspicuous family groups, beavers are a straightforward furbearer to manage. Today, harvesting rates in Ontario are set at 30% per annum and trappers receive quotas of 1–2.5 beavers per lodge within their trapline.

Today's beaver harvest appears to be larger than at any time in history (Figure 3.4) and probably beaver numbers are also higher.

Beavers are extreme habitat modifiers: they cause economic losses to forestry, and their dam-building causes flooding, road and rail washouts and blocked culverts. It is argued these losses even exceed the animals' economic value as a furbearer.

3.4.3 Marten and fisher

Martens and fishers (*Martes pennanti*) are furbearers of the coniferous forest, though fishers are rarer and their pelts more valuable. In Ontario in the 1920s, a single fisher pelt would cover the food costs for three trappers all winter, and in 1986 a prime pelt was worth US$450 (Douglas and Strickland, 1987). Dispersed and slow breeders, they are also vulnerable to over-exploitation (Quick, 1956). Although data are fragmentary for the seventeenth century, it is evident that there was an increasing harvest of both species until the mid nineteenth century, after which numbers steadily declined, suggesting over-trapping, and both species were extirpated from parts of their range (Douglas and Strickland, 1987; Strickland

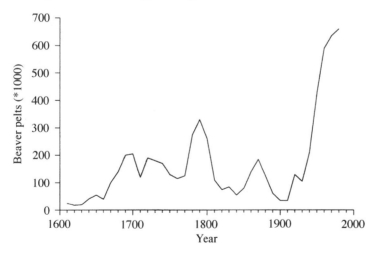

Figure 3.4 Estimated total commercial harvest of beaver pelts from North America since 1610. Figures are averaged by decade. Simplified from Obbard *et al.* (1987).

and Douglas, 1987). Currently numbers have increased and reintroductions have restored both species to most of their former range where the
habitat is suitable (Kohn *et al.*, 1993). This has allowed larger numbers to
be harvested (Figure 3.5).

3.4.4 Sea otter

The sea otter (*Enhydra lutris*) is an extreme example of over-exploitation.
Discovered by Steller in 1741 when the world population could have
been 300 000, numbers were reduced to fewer than 2000 in 1911

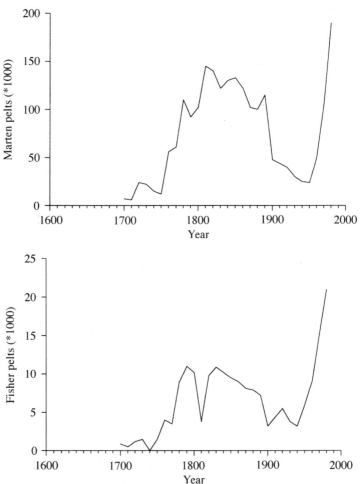

Figure 3.5 Estimated total commercial harvest of marten (upper) and fisher
(lower) pelts from North America since 1700. Figures are averaged by decade.
Simplified from Obbard *et al.* (1987).

(Reidman and Estes, 1990). Figure 3.6 shows their former and current distributions around the north Pacific rim.

With a fur density almost twice that of any other mammal (Reidman and Estes, 1990) it is not surprising that their pelts were highly prized. The initial exploitation was Russian, but after the sale of Alaska in 1867 it became American and, until protection under the Fur Seal Treaty 1911, over one million animals were harvested. Since then, thanks to reintroductions and total protection, numbers have recovered to some 150 000 and much of their former range has been regained (Figure 3.6; Reidman and Estes, 1990). Their continuing recovery, with population growth rates up to 20% per year in favoured sites, has demonstrated what a large impact a single species can have on a whole ecosystem (Estes, 1990; Reidman and Estes, 1990). It appears that their predation on urchins and clams is essential to the establishment and maintenance of the coastal kelp forest (Reidman and Estes, 1990), but their presence is incompatible with commercial coastal clam and shellfish fisheries (Garshelis and Garshelis, 1984). Indeed it is now argued that a fur harvest of sea otters in parts of Alaska would be desirable, partly to compensate for the fishery damage but also to prevent further fishery collapses (Garshelis and Garshelis, 1984; Garshelis, 1987). However, a strong protectionist lobby, particularly in California, means that federal protection is unlikely to be relinquished to state control, and renewed harvesting is extremely unlikely (Garshelis, 1987).

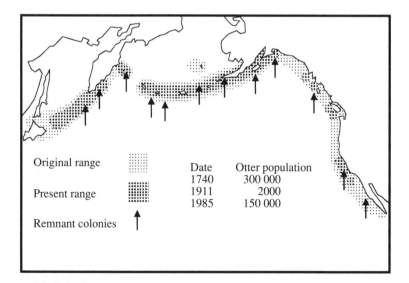

Figure 3.6 Estimated changes in the distribution and abundance of sea otters since their discovery in 1740. Re-drawn from Reidman and Estes (1990).

3.5 CONCLUSION

Where a natural resource is shared amongst competing parties the incentive is always to over-exploit and take too large a share. Historically, when imperial powers or fur companies were competing, over-exploitation was often deliberate and extreme. Only during periods of monopoly and political stability did conservation of resources become an issue.

In North America this central problem has been addressed by international treaty, government regulation, quotas and, above all, by partitioning the resource into exclusive management units – notably the trapline system. The result has been an increase in stock for most species, and their re-establishment in areas where they were once lost. In turn this has now allowed fur harvests to exceed the over-exploitation levels of the past. However, outside North America where such control is not in place, an illegal fur trade still continues to threaten rare species such as the neotropical cats.

Apart from the perceived cruelty involved in their capture, the case against the use of furs is primarily a conservation one.

- History shows that most furbearers were over-exploited in the past and some species were nearly exterminated before they were managed correctly.
- Today such management is not possible in poorly regulated societies, and trade (legal and illegal) from these areas continues to threaten species survival.
- The trade in legal furs from well managed populations makes the trade in illegal furs easier.
- Harvesting animals even on an optimum sustained yield basis necessarily involves keeping populations at lower densities than would otherwise be the case. Such populations may be some 50% lower than unexploited ones.
- A modern society based on the consumption of petrochemicals no longer needs furs. Many fur products are better insulators, but arguably synthetics are more appropriate to a world population which is five to six times what it was in 1700. The world leader in polyester fleece manufacture (Malden Mills Inc.) estimates to have clothed 65 million people with 150 million m^2 of fabric to date (K. Readman, personal communication).

The case for the continued sustainable use of furs rests on several arguments.

- In non-wilderness areas fur trapping gives a value to many damaging furbearers like muskrat and beaver. These would otherwise be considered pests and their numbers would therefore be kept much lower.

- In wilderness areas trapping provides revenue without severely changing the ecological structure, whereas alternative commercial uses such as logging and mining can be more damaging. Even tourism (in the unlikely event that huge tracts of boreal forest could be made attractive to tourists) requires an infrastructure of roads, hotels, camp sites, petrol stations and rubbish dumps.
- Revenue from fur trapping can be used to manage and conserve the species.
- From a human perspective, fur trapping is particularly suited to the way of life of many native peoples. It helps them retain their cultural heritage better than any other form of employment.

REFERENCES

Allen, A.W. (1987) The relationship between habitat and furbearers, in *Wild Furbearer Management and Conservation in North America*, (eds M. Novak, J.A. Baker, M.E. Obbard and B. Malloch), Ontario Trappers Association, North Bay, Ontario, pp. 164–179.

Bakeyev, N.N. and Sinitsyn, A.A. (1994) Status and conservation in the Commonwealth of Independent States, in *Martens, Sables and Fishers. Biology and Conservation*, (eds S.W. Buskirk, A.S. Harestad, R.A. Raphael and R.A. Powell), Cornell University Press, Ithaca, New York, pp. 246–254.

Boutin, S. and Birkenholz, D.E. (1987) Muskrat and round-tailed muskrat, in *Wild Furbearer Management and Conservation in North America*, (eds M. Novak, J.A. Baker, M.E. Obbard and B. Malloch), Ontario Trappers Association, North Bay, Ontario, pp. 313–325.

Dahl, E. (1986) Game and habitat. *Naturopa*, **52**, 13–18.

Douglas, C.W. and Strickland, M.A. (1987) Fisher, in *Wild Furbearer Management and Conservation in North America*, (eds M. Novak, J.A. Baker, M.E. Obbard and B. Malloch), Ontario Trappers Association, North Bay, Ontario, pp. 511–529.

Estes, J.A. (1990) Growth and equilibrium in sea otter populations. *J. Anim. Ecol.*, **59**, 385–401.

Garshelis, D.L. (1987) Sea otter, in *Wild Furbearer Management and Conservation in North America*, (eds M. Novak, J.A. Baker, M.E. Obbard and B. Malloch), Ontario Trappers Association, North Bay, Ontario, pp. 643–655.

Garshelis, D.L. and Garshelis, J.A. (1984) Movements and management of sea otters in Alaska. *J. Wildl. Manage.*, **48**, 665–678.

Hall, E.R. (1981) *The Mammals of North America*, John Wiley and Sons, New York.

Innis, H.A. (1956) *The Fur Trade in Canada*, University of Toronto Press, Toronto.

Kohn, B.E., Payne, N.F., Ashbrenner, J.E. and Creed, W.A. (1993) *The Fisher in Wisconsin*, Department of Natural Resources, Madison, Wisconsin.

Martin, J. (1986) *Treasure of the Land of Darkness*, Cambridge University Press, Cambridge.

Newman, P.C. (1985) *The Company of Adventurers*, Vol. 1, Penguin Books, Markham, Ontario.

Newman, P.C. (1988) *Caesars of the Wilderness: Company of Adventures*, Vol. 2, Penguin Books, Markham, Ontario.

Nockert, M. (1991) *The Högorn Find and Other Migration Period Textiles and Costumes in Scandinavia*, (Högorn Part 2), University of Umeå, Sweden.

Novak, M. (1987a) Beaver, in *Wild Furbearer Management and Conservation in*

North America, (eds M. Novak, J.A. Baker, M.E. Obbard and B. Malloch), Ontario Trappers Association, North Bay, Ontario, pp. 283–312.

Novak, M. (1987b) Wild furbearer management in Ontario, in *Wild Furbearer Management and Conservation in North America*, (eds M. Novak, J.A. Baker, M.E. Obbard and B. Malloch), Ontario Trappers Association, North Bay, Ontario, pp. 1049–1061.

Obbard, M.E., Jones, J.G., Newman, R., Booth, A., Satterthwaite, A.J. and Linscombe, G. (1987) Furbearer harvests in North America, in *Wild Furbearer Management and Conservation in North America*, (eds M. Novak, J.A. Baker, M.E. Obbard and B. Malloch), Ontario Trappers Association, North Bay, Ontario, pp. 1007–1034.

Ponting, C. (1991) *A Green History of the World*, 1992 edn, Penguin Books, London.

Quick, H.F. (1956) Effects of exploitation on a marten population. *J. Wildl. Manage.*, **20**, 267–274.

Ray, A.J. (1987) The fur trade in North America: An overview from historical geographical perspective, in *Wild Furbearer Management and Conservation in North America*, (eds M. Novak, J.A. Baker, M.E. Obbard and B. Malloch), Ontario Trappers Association, North Bay, Ontario, pp. 21–30.

Reidman, M.L. and Estes, J.A. (1990) *The Sea Otter* (Enhydra lutris*): Behaviour, Ecology, and Natural History*, Biol. Rep. 90(14), US Department of the Interior, Fish and Wildlife Service, Washington DC.

Ritchie, J. (1930) *The Influence of Man on Animal Life in Scotland. A Study in Faunal Evolution*, Cambridge University Press, Cambridge.

Seton, E.T. (1910) *Life-histories of Northern Mammals*, Vol. 1, Constable, London.

Sheiff, A. and Baker, J.A. (1987) Marketing and international fur markets, in *Wild Furbearer Management and Conservation in North America*, (eds M. Novak, J.A. Baker, M.E. Obbard and B. Malloch), Ontario Trappers Association, North Bay, Ontario, pp. 862–877.

Strickland, M.A. and Douglas, C.W. (1987) The relationship between habitat and furbearers, in *Wild Furbearer Management and Conservation in North America*, (eds M. Novak, J.A. Baker, M.E. Obbard and B. Malloch), Ontario Trappers Association, North Bay, Ontario, pp. 531–547.

Todd, A.W. and Boggless, E.K. (1987) Characteristics, activities, lifestyles, and attitudes of trappers in North America, in *Wild Furbearer Management and Conservation in North America*, (eds M. Novak, J.A. Baker, M.E. Obbard and B. Malloch), Ontario Trappers Association, North Bay, Ontario, pp. 59–76.

Veale, E.M. (1966) *The English Fur Trade in the Later Middle Ages*, Oxford University Press, London.

Wheelersburg, R.P. (1991) Uma Saami native harvest data derived from Royal Swedish taxation records 1557–1614. *Arctic*, **44**, 337–345.

Wishart, D.J. (1979) *The Fur Trade of the American West 1807–1840*, Croom Helm, London.

Wright, J.V. (1987) Archaeological evidence for the use of furbearers in North America, in *Wild Furbearer Management and Conservation in North America*, (eds M. Novak, J.A. Baker, M.E. Obbard and B. Malloch), Ontario Trappers Association, North Bay, Ontario, pp. 3–12.

4

Assessing the impacts of uses of mammals: the good, the bad and the neutral

Robert and Christine Prescott-Allen

SYNOPSIS

The sustainability of uses of wild mammals should be judged by assessing the impacts of the uses on both the ecosystem and the human system. This chapter proposes a framework for assessment, categorizing uses in terms of their combined impacts on the ecosystem and human system as good, bad, neutral or unknown.

Application of this framework for assessment is illustrated with five cases: vizcacha hunting in Argentina; kangaroo harvesting in Australia; furbearer trapping in Canada; squirrel monkey viewing in Costa Rica; and Communal Areas Management Programme for Indigenous Resources (CAMPFIRE) in Zimbabwe.

Three of these uses are considered to be good (probably sustainable: to be encouraged), one bad (probably unsustainable: to be reformed) and one neutral (low priority for attention). Mutually supportive changes in values, incentives and laws are required to turn bad uses into good uses.

In all five cases, the limited and often fragmentary information available – from both formal and informal sources – is nonetheless sufficient to assess impacts of the uses on the ecosystem. Information about impacts of the uses on the human system needs improvement, although interim conclusions can still be drawn.

4.1 APPROACH TO ASSESSMENT

A sound model of the system in which the use occurs (and a goal that is consistent with that model) clarifies what information is most important for assessment and frees the assessor from collecting unnecessary data. Given such a model and goal, all societies are capable of obtaining the information required.

The system in question is the ecosystem and its human subsystem. People are an integral part of the ecosystem and the well-being of one is bound up in the well-being of the other. Hence we need to consider both together, giving them equal weight. A society is sustainable only if **both** the human condition and the condition of the ecosystem are satisfactory or improving. If **either** is unsatisfactory or worsening, the society is unsustainable (Figure 4.1).

Assessments are done against a specified standard or objective. The impacts of a use are positive or negative (acceptable or unacceptable) depending on what one wants to achieve. For example, one might assess the impacts of uses of mammals against the objective of maintaining robust populations of every mammal species. In this chapter, consistent with our model of the system, we assess uses of mammals against a broader goal – to improve and maintain the well-being of people **and** the ecosystem:

- **Ecosystem well-being** is a condition in which ecosystems maintain their quality and diversity and thus their potential to adapt to change and provide a wide range of choices and opportunities for the future.
- **Human well-being** is a condition in which all members of society are able to define and meet their needs and have a large range of choices and opportunities to fulfil their potential.

In assessing the use and the system, we need a clear sense of direction. Are the conditions of people and the ecosystem getting better or worse? What difference does the use make? A sense of how well the ecosystem is doing may be obtained by looking at four issues:

- **Naturalness/conversion**: how much of the ecosystem is natural, modified, cultivated, or built (see Table 4.1 for definitions). This

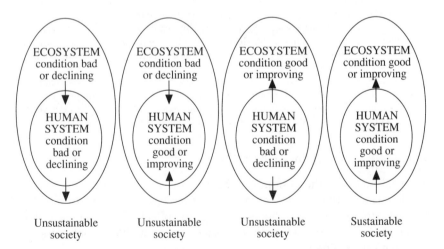

Figure 4.1 The egg of sustainability. The yolk is the human system, and the surrounding white is the ecosystem. If either the white or the yolk is bad, the egg is bad. For the egg to be good, both have to be good.

Table 4.1 Classification of ecosystem conversion levels

Conversion level	Degree of human influence	Definition	Indicator species
Natural	Negligible to light	The scale and rate of human impact on the ecosystem are of the same order as the impact of other organisms	Introduced species absent Domesticated species absent
Modified	Moderate to heavy	Not cultivated but human impact on the ecosystem is greater than that of other species	Introduced species present Domesticated species absent
Cultivated	Human-dominated	More than 50% cultivated	Introduced species present Domesticated species present
Built	Human-centred	More than 50% covered by roads, buildings or other human structures	Introduced species present Domesticated species present

provides a bird's-eye view of the scale and rate of human impact on the ecosystem, and a context for assessments of degradation, diversity and resource depletion.

- **Quality/degradation**: the extent and severity of degradation of land, water and air. Degradation includes pollution.
- **Diversity maintenance/loss**: whether the diversity of ecological communities, wild species and domesticated varieties and breeds is being maintained or is declining.
- **Resource conservation/depletion**: whether wood, forage, wildlife and other resources supplied by the ecosystem are being maintained or depleted.

A sense of how well people are doing may be obtained by looking at four more issues:

- **Health**. Longevity, good health, access to healthful living conditions (clean water, sanitation). A long and healthy life increases the opportunity for a person to pursue goals and develop abilities.
- **Resources**. Income, employment, and access to resources (including technology). Money and other resources expand opportunities and provide means to exploit them.
- **Knowledge**. The knowledge system includes education, training,

research, and monitoring and assessment capacities. It equips individuals, organizations and society to fulfil their potential, to improve their understanding of the ecosystem and human system, to learn from experience and to adapt to changing conditions.

- **Participation and empowerment.** Each society has an institutional system of values, customs, laws, incentives and organizations to manage human relationships with each other and with the ecosystem. The focus here is on the distribution of decision making, the extent to which people have control over their lives, and the balance of laws and incentives.

4.2   FRAMEWORK FOR ASSESSMENT

An assessment should begin with a sense of the ecosystem and human system involved in the use. Are conditions improving or declining, in what ways, and why? The impacts of the use on the ecosystem and on people can then be assessed in the light of conditions and trends. They will be either positive, neutral/negligible, negative, or unknown.

Depending on the combination of impacts on the ecosystem and human system, the assessment would conclude that the use is good (probably sustainable), bad (probably unsustainable) or neutral (makes little or no difference), or that its combined impact is unknown (inadequate information). Table 4.2 sets out the conclusions that may be drawn from different combinations of impacts.

4.2.1 Cases

The following cases are taken from papers contributed to a meeting of the Species Survival Commission (SSC) of the IUCN Specialist Group on Sustainable Use of Wild Species (Prescott-Allen and Prescott-Allen, 1996, includes these case studies and a more detailed discussion of our approach). The data are from the authors of the papers. The cases are used here to illustrate how to assess uses of mammals within this framework. We alone are responsible for our interpretations of the cases and the conclusions we draw from them.

(a) Vizcacha hunting, Argentina (Jackson, 1996)

The vizcacha (*Lagostomus maximus*) is a common rodent throughout much of the pampas – the grassland and scrubland of central and northern Argentina. The wetter eastern portion of its range is cultivated. The rest is modified ranchland, much of it degraded by overgrazing. Little

Table 4.2 Assessing the combined impacts of a use of wildlife on the ecosystem and the human system: possible conclusions (conclusions are in UPPER and lower case)

Impact on the human system	Impact on the ecosystem			
	Positive	*Neutral/negligible*	*Negative*	*Unknown*
Positive	GOOD probably sustainable	GOOD probably sustainable	BAD probably unsustainable	UNKNOWN inadequate information
Neutral/ negligible	GOOD probably sustainable	NEUTRAL makes little or no difference	BAD probably unsustainable	UNKNOWN inadequate information
Negative	BAD probably unsustainable	BAD probably unsustainable	BAD probably unsustainable	BAD probably unsustainable
Unknown	UNKNOWN inadequate information	UNKNOWN inadequate information	BAD probably unsustainable	UNKNOWN inadequate information

natural pampas remains. Eradication campaigns using fumigants have eliminated vizcachas from parts of the cultivated portion of their range, but they are spreading or becoming more numerous on the northern and western margins. Overgrazing and habitat degradation in the semi-arid zone apparently facilitate colonization by vizcachas. Once established, their burrows and feeding further modify and degrade the ecosystem.

Vizcachas are hunted for their meat and pelt and for recreation. Officially classed as a pest, they are also the target of eradication campaigns. Little is known about the distribution, abundance, population dynamics, harvest statistics or socio-economic value of vizcachas in most provinces of Argentina, but their abundance in most parts of their range (despite efforts to eliminate them) indicates that populations are large, widespread and resilient. Many colonies that have been hunted heavily for many years still survive.

Public and private landowners in Argentina do not regard uses of wild species as an alternative or additional source of income to ranching. Consequently, they do not bother to manage wildlife. Entrepreneurs make the most money out of wild species, but only a few have seen the need to invest some of the proceeds in management. The many poor rural dwellers who depend on wildlife for food and income have no tenure, so no incentive to regulate their uses. Such a weak incentive system might not matter if laws and organizations were strong, but they are just

as weak. Laws are poorly enforced, provincial coordination is lacking
and management organizations are understaffed and underfunded.

(b) Kangaroo harvesting, Australia (Grigg, 1996)

Five wallaby and kangaroo species are harvested commercially in
Australia: whiptail wallaby (*Macropus parryi*), euro or wallaroo (*M. robustus*), western grey kangaroo (*M. fuliginosus*), eastern grey kangaroo (*M. giganteus*) and red kangaroo (*M. rufus*). All are regarded as pests by rural
communities and the main motive for killing them has been to reduce
competition with wheat or sheep. The majority of kangaroos live in the
semi-arid chenopod shrubland that covers about 40% of the country.
Overgrazing by sheep during the last 150 years has degraded much of
this area. Most landholders see a significant reduction in kangaroos as
the most valuable contribution to reducing grazing pressure – their aim
being to manage the rangelands as a monoculture of sheep.

However, kangaroos are also Australia's national symbol. As such, all
species are protected by legislation and kangaroo kill quotas are set to
maintain, not reduce, populations. Some three to four million animals
are killed each year (less than the annual quota). Because this is not
enough to reduce kangaroo populations, there is a large illegal kill, much
of it inhumane. Almost no commercial use is made of animals killed illegally. Until recently, kangaroo meat could not be sold for human consumption in Australia (except South Australia). Large quantities were
(and still are) fed to dogs, cats and crocodiles.

Rainfall and its effects on pasture condition are the main influence on
kangaroo population densities. Drought reduces the populations, often
severely, but afterwards they recover rapidly. In average rainfall seasons
or better, populations frequently increase in the face of 15–20% harvests.
People have killed small numbers of kangaroos for thousands of years
and large numbers for more than 100 years. Controls were introduced in
1970 and 1971, and annual population monitoring began in 1981. Aerial
surveys of populations are conducted in much of the harvested area.
Impacts of harvesting on sex ratios, age structure and genetics are being
studied. The commercial industry is controlled by the issue of tags
(which must be attached to the carcass) and licensing of shooters, processors and tanneries. Money from the sale of tags goes into a fund for monitoring, research and management. States meet annually to set quotas,
based on population surveys and information on rainfall. States have the
capacity to close the season, if drought makes it likely that the quota is
too high.

The aim of allowing commercial use and raising the value of kangaroo
products (especially meat) is to change the status of kangaroos from pest
to resource, and so remove any incentive to kill kangaroos illegally. By

providing a supplementary income to landowners, kangaroo harvesting would enable them to reduce the number of livestock on their properties and so take pressure off the range. In addition, legal shooting is done humanely and is considered to be as humane, if not more so, than the treatment of domestic livestock.

(c) Furbearer trapping, Canada (Slough and Jessup, 1996)

The Yukon Territory of Canada consists of arctic and alpine tundra and boreal (subarctic) forest. Most of the ecosystem is natural, although logging and mining have modified some areas. Tiny fractions of the territory are cultivated and built. Some 500–800 residents (2–3% of all Yukon residents, 6–9% of rural residents), 70% of whom are aboriginal, are eligible to trap for furs. The furs are sold on the national or international fur markets or are used locally for personal clothing and the cottage garment industry. Fourteen species are trapped: the 11 carnivores are arctic fox (*Alopex lagopus*), coyote (*Canis latrans*), grey wolf (*Canis lupus*), red fox (*Vulpes vulpes*), lynx (*Felis lynx*), wolverine (*Gulo gulo*), river otter (*Lutra canadensis*), marten (*Martes americana*), fisher (*Martes pennanti*), short-tailed weasel (*Mustela erminea*) and mink (*Mustela vison*); the three rodents are beaver (*Castor canadensis*), muskrat (*Ondatra zibethicus*) and red squirrel (*Tamiasciurus hudsonicus*). Three of these species (lynx, beaver, muskrat) are taken for food as well as furs.

Trapping in the Yukon is managed through a system of Registered Trapping Concessions (RTCs), granted to individuals or groups of trappers. RTCs have renewable five-year terms, and nobody may trap without one. The number of RTCs is fixed at 372, plus 15 group areas administered by First Nations (aboriginal peoples) for aboriginal communities. Covering 93% of the Yukon's 482 515 km², RTCs and group areas are large enough to disperse trapping pressure and provide refugia even for species with low population densities, such as wolverine with a density of 6 per 1000 km².

Furbearer harvests are monitored through a mandatory system of export permits and fur dealer returns, and also seals for lynx, wolf and wolverine pelts. This accounts for all furs that leave the territory (90% of the total kill). Furbearer populations are monitored through annual reports by trappers on furbearer and prey abundance and population trends, supplemented by separate monitoring of lynx recruitment (via pelt measurements) and regional population surveys and indexing. The latter include winter track counts, surveys of beaver food caches and muskrat 'push-ups'(winter feeding structures), and aerial wolf surveys. Biological studies are conducted on vulnerable populations (muskrats in northern Yukon, wolverine and arctic fox) and to determine species' needs for refugia, especially wolverine, because of its large home range (75–270 km²), and lynx and marten, because they are easy to trap.

Under the terms of an agreement by the Yukon and federal governments and the 14 First Nations, formal responsibility for managing wildlife will be shared by the Yukon government and the First Nations. The trappers themselves already participate in day-to-day management, including conducting population surveys, adjusting trapping levels to population size and establishing refugia.

The Yukon Trappers' Association delivers a government-sanctioned education course on harvest management and humane trapping. The course, which is conducted in the communities, focuses on efficient and humane trapping techniques using the best available technology, biologically sound trapline management and proper pelt preparation. First-time trappers must take the course and experienced trappers are encouraged to take regular refresher courses to upgrade their knowledge and skills. About 50% of all trappers have taken the course.

Monitoring of catch per unit effort, changes in relative abundance, and sex and age structure, is sensitive enough to ensure that harvests are at or below sustainable levels.

The boreal forest is subject to fluctuations, such as the 3–5-year vole (Cricetidae) and 8–11-year snowshoe hare (*Lepus americanus*) population cycles, mass migrations of caribou (*Rangifer tarandus*) and catastrophic fires. (Some species, such as lynx and moose, *Alces alces*, are fire-dependent.) Harvests of herbivores (the three rodent species) are kept below levels at which they might influence vegetation, and harvests of the carnivores are kept below levels at which they might influence herbivore populations.

(d) Squirrel monkey viewing, Costa Rica (Wong and Carrillo, 1996)

The northern subspecies of the Central American squirrel monkey (*Saimiri oerstedi citrinellus*) occurs only in the Manuel Antonio forest on the Pacific coast of Costa Rica. The population consists of 681 individuals, 300 of which are in Manuel Antonio National Park (683 ha). The remainder occupy 1100 ha outside the park. The forest is part natural, part modified, and completely hemmed in by ocean on one side and oil palm plantations and cattle ranches on the others.

The park – especially its monkeys – is a popular tourist attraction. The number of visitors has grown rapidly, from 25 000 per year in 1982 to 192 000 per year in 1992. The direct impact of viewing by tourists consists of some disturbance of the monkeys, including cases of taunting. Indirect impacts are a revival of the capture of monkeys for pets and loss of habitat due to expansion of tourism infrastructure. This last impact is the most serious.

The growth of tourism infrastructure has kept pace with the numbers of tourists: 192 houses in 1988, 663 in 1992, with corresponding increases in restaurants and other facilities. In effect, tourism has fostered expansion of the built environment at the expense of the natural forest, which is quickly being fragmented and reduced. There is an urgent need to plan hotel and other developments to ensure maintenance of suitable habitat patches and corridors. The most favourable habitat is a mixture of primary and secondary forest (least favourable is young degraded secondary forest), which needs to be linked by corridors of trees, preferably food trees.

Existing regulatory and incentive systems are unable to conserve habitat. Penalties exist for direct damage to an endangered species but not for indirect damage through habitat destruction. In particular, current laws cannot cope with the cumulative effects of individual property developments. It might be possible to appeal to the self-interest of the tourism industry, since the squirrel monkeys are a tourist attraction. However, the value of the monkeys relative to other attractions, such as the beach, needs to be determined. More important, most of the capital for tourism comes from outsiders, who receive most of the industry's economic rewards: 60% of the infrastructure has been developed with foreign capital, and only 2% of the income from tourism goes back to the local community. Hence there is little local interest in, or commitment to, conservation.

(e) CAMPFIRE, Zimbabwe (Child, 1996)

Zimbabwe has developed a two-pronged wildlife conservation strategy: one prong consists of protection and applies to the 13% of the country that is within protected areas; the other prong consists of sustainable use and applies to the remaining 87%. In 1955 legislation was adopted that allowed commercial (white) farmers to make use of the wildlife (notably elephant, antelope, zebra, giraffe) on their land. Today the livestock monocultures of the past are in the minority: some 75% of Zimbabwe's commercial ranches have a wildlife enterprise, usually alongside cattle production, but increasingly instead of it. Financially, wildlife enterprises consistently outperform livestock enterprises, partly because cattle can be sold only once (for meat) but a wild mammal can be sold three times (for viewing, as a hunting trophy, and for meat). Wild mammals have steadily regained ground formerly lost to livestock.

Meanwhile on communal lands, where black rural people live, wildlife continued to decline. The communal lands are almost entirely modified and cultivated, with little natural ecosystem left, and cultivated and degraded areas are displacing viable modified areas. An attempt to stop the decline failed because it did not provide communities with sufficient benefits or management responsibility. Consequently, CAMPFIRE (Communal Areas Management Programme for Indigenous Resources) was introduced in 1989. CAMPFIRE's rationale is that communities will sustain wild species and their habitats if they get the benefits from them; if they do not, they will destroy them – directly by killing pests and predators, and indirectly by replacing habitats with farms and livestock. In effect, CAMPFIRE gives district councils the same rights to use wildlife as private landowners. Income from wildlife goes directly to the councils, which are also responsible for management. Government's role is to promote sustainability by monitoring harvests and ensuring that benefits reach the communities. The government reserves the right to control quotas until accountable institutions and effective incentive structures develop and councils and communities learn the necessary skills. Communities are being trained to set quotas.

Currently, 24 districts (almost half the rural districts in the country) participate in CAMPFIRE. Gross income has grown from Z$648 620 in 1989 to more than Z$10 million in 1993. About 65% of the income reaches communities. In the case of safari hunting, each district develops a hunting quota in collaboration with the government. It then offers the quota for tender by safari outfitters. This responsibility has allowed the districts to develop marketing and business management skills. Their income from quotas almost tripled (in constant US dollars) between 1990 and 1993, the ratio of income to quota unit doubled, and they have raised their share of net profits from safari operations to 75% (the operators get the remaining 25%). Districts and communities lacked the skills and experience necessary for wildlife management, but are developing them by gradually assuming management responsibilities.

Information requirements for CAMPFIRE are kept as simple as possible. Four components of the programme are monitored:

- **Wildlife populations**. Three sources are used: formal methods (e.g. aerial surveys) and informal methods (e.g. safari operators and community estimates). Communities hold workshops in which they draw maps of wildlife numbers and locations. This not only provides useful estimates; it also directly involves community members in management and quota-setting.
- **Offtake quantity and quality**. Records are submitted of all animals killed (except by poaching). Trophy sizes are recorded, which allows trends in the age of animals killed to be detected.
- **Earnings**. This information is used to assist communities in bargaining with operators and helps to raise prices and income to the communities.
- **How the money is used**. This information is used to ensure that benefits go to the communities who live with the wildlife.

It is worth noting that two of the components are biological and two socio-economic. The latter are regarded as crucial, since they help to keep track of the incentive system, without which the biological information would be redundant.

CAMPFIRE has had to confront some challenging issues, exemplified by the case of Chikwarakwara (Box 4.1). The quota of three bull elephants (6%) out of an itinerant elephant population of about 50 exceeded the sustainable rate of 0.75% of bulls and 5% of the total population, but the hunting quota was **less unsustainable** than the status quo, in which wild mammals were declining sharply. The risk proved worth taking, since it contributed to the social changes necessary to arrest the decline and restore wildlife numbers. Short-term unsustainability is sometimes necessary for long-term sustainability.

BOX 4.1 CAMPFIRE AT WORK IN CHIKWARAKWARA

Chikwarakwara is a remote village on the Limpopo river opposite South Africa's Kruger National Park. Its people are extremely poor and largely uneducated. Most children leave the village as soon as they leave school. Magnificent riverine forest is cut down to grow dry-land crops that succeed only once or twice in a decade. Livestock herds are small, but the village's grazing land is being destroyed by herds from other communities. Forest destruction and overgrazing have reduced the carrying capacity for wildlife. By 1989 wildlife was disappearing due to habitat loss and persecution, many animals being regarded as dangerous nuisances and competitors. That year, three elephants, four buffalo and some other animals were sold to a safari outfitter for Z$60 000 and killed on the community's land. The local district council decided that the money should go to the village and that the community should say how the money should be used. The villagers held a four-day meeting under their baobab tree.

- They listed the animals shot and their values, thereby linking wild-life and economic benefits.
- They determined village membership and hence who had a share in the benefits.
- They decided that the money should be divided equally among all households and paid to them in cash. This meant that each household would get $400, a huge sum in their cash-starved economy.
- They discussed various community projects, agreed that each household would return $200 of its $400 to fund two priority projects ($170 for a grinding mill, $30 for the school), and elected committees to oversee the projects.
- They held a revenue distribution ceremony. $60 000 in $20 bills

---

Box 4.1 *continued*

> were piled on a table. Each household received its $400 and then
> gave $200 to the community. This further reinforced the links
> between wildlife and economic benefits and the shared respon-
> sibilities of households and the community.

Within three months Chikwarakwara had built its grinding mill. At
the ceremony to open it, children acted out the history of CAMPFIRE
in the community, wearing animal costumes to which banknotes
were stapled. The mill is profitable, making $10 000 a year. The com-
munity has unified and gained confidence in its abilities. It now prevents poaching (formerly rife). It is improving irrigation and phasing
out dryland farming. It has decided to limit cattle numbers and
exclude outsiders' cattle. It has reserved the prime Limpopo flood-
plain for wildlife. Although it lacks technical expertise, the village has
begun to manage its wildlife, making sketch maps showing animal
sightings, counting animals, and setting sustainable harvest rates.
Attitudes toward wildlife have been transformed and wildlife is
recovering.

(Source: Child, 1996)

4.3   DISCUSSION

4.3.1 Impacts on the ecosystem

Of the five cases considered here, we conclude that two have a positive
impact on the ecosystem, two have a neutral or negligible impact, and
one has a negative impact.

The two cases with a positive impact are CAMPFIRE in Zimbabwe
and kangaroo harvesting in Australia. In both cases, agriculture and livestock grazing are reducing the extent of natural and viable modified ecosystems and increasing degradation. In Zimbabwe, these pressures are
also destroying wildlife habitat. In both cases, wild mammals are also
under attack because, officially or unofficially, they are classified as pests.
CAMPFIRE's beneficial impact is being demonstrated. Communities that
can benefit economically from uses of wild mammals, and can participate in their management, have incentives to restore and maintain both
the mammal populations and the ecosystems. Both are showing signs of
recovery.

In Australia, the opening of kangaroos to commercial harvesting for
meat is too recent to see the benefits, but the logic appears to be sound.
Although kangaroos are good for the ecosystem, the landowners make

no money from them. On the contrary, they regard kangaroos as pests, taking money away from them by competing with sheep for pasture. Official protection of the kangaroos has not prevented landowners from killing them: it has simply driven the killing underground, increasing the animals' risk of pain and suffering. Sheep are bad for the ecosystem, but landowners prefer them because they can make money out of them. If landowners can make enough money from kangaroos to allow them to replace or coexist with sheep, the condition of the ecosystem will improve.

The two cases with a neutral or negligible impact are vizcacha hunting in Argentina and furbearer trapping in the Yukon (Canada). Vizcacha populations are so robust that they have survived intensive eradication campaigns, except in cultivated areas where their habitat has been eliminated. Theoretically, hunting vizcachas might benefit the ecosystem by allowing degraded areas to recover, but current levels of hunting do not seem to suppress their numbers enough.

The Yukon still has large tracts of natural ecosystem. Kill levels of furbearers are within the natural fluctuations of the ecosystem and the furbearer populations. Any negative impacts are negligible. So, too, are any positive impacts on the ecosystem. Socially and economically, trapping is the most viable use of much of the ecosystem. Ecologically, it is the least harmful consumptive use. Recent legal cases have recognized the rights of holders of Registered Trapping Concessions (RTCs) over other resource users on government-owned land (most of the territory), but the economic clout of trapping is not necessarily sufficient to save an area from being logged or mined.

The difference in classification between hunting in Australia and Zimbabwe on the one hand and trapping in the Yukon on the other is due to the different contexts. Ecosystem conditions are deteriorating in the case areas in Australia and Zimbabwe. CAMPFIRE's uses of mammals are helping to stop the decline in Zimbabwe; kangaroo harvesting will probably do likewise in Australia. Thus they are promoting an improved ecosystem condition. Ecosystem conditions are much better in the Yukon and the threats of change are few and local. Furbearer trapping is compatible with an already good ecosystem condition, but does not necessarily promote it. A useful guide is to ask: what would happen to the ecosystem if the use did not exist? In Australia and Zimbabwe, the ecosystem would probably get worse. In the Yukon, it would probably stay the same.

The one case with a negative impact is squirrel monkey viewing in Costa Rica. Although the direct impact of viewing on the ecosystem is probably negligible, its indirect impact is high. Tourism is driving the expansion of the built environment at the expense of a tiny island of natural and modified forest, without which the monkeys will become extinct. Squirrel monkey viewing is not the only tourism activity, of course, but it is a contributor.

It is an accident that the only non-consumptive use is 'bad', whereas all of the consumptive uses are 'good' or 'neutral'. Without question, there are 'good' non-consumptive uses and 'bad' consumptive uses, but it does suggest that the distinction between consumptive and non-consumptive is not very important. The reason is that these terms relate to the minimum potential impact the use might have on the ecosystem. In real life, the impact of a use is usually greater. Thus a use that necessarily kills animals (e.g. hunting) may end up killing fewer creatures than a use that does not require the killing of animals (such as viewing).

4.3.2 Impacts on the human system

We conclude that two of the cases have a positive impact on the human system, two have a neutral or negligible impact, and the impact of one is unknown. None has a negative impact.

The two cases with a beneficial impact are CAMPFIRE in Zimbabwe and furbearer trapping in the Yukon. Directly and indirectly, CAMPFIRE has helped to improve the living conditions, education, income, security and self-reliance of participating communities. In the Yukon, furbearer trapping is a mainstay of the culture and economy of aboriginal communities, which otherwise have little means of support. Both uses have improved the quality of life of the people living with wild species. They also empower communities, by giving them opportunities to develop new skills, by restoring management responsibilities, or both.

The two cases with a neutral or negligible impact are vizcacha hunting in Argentina and kangaroo harvesting in Australia. Although vizcacha hunting contributes to the incomes of the rural poor, the people have no formal rights to the resource and are not involved in its management. Systems for obtaining and acting on information about the ecological and human impacts of the uses are extremely weak. In Australia, commercial hunting is expected to increase landowners' incomes enough to change their attitudes to kangaroos, but its other social and economic impacts are probably marginal.

We classify the impact of monkey viewing and tourism in Costa Rica as unknown, because how much tourism benefits from the monkeys and other forest species – and how those benefits are distributed – are crucial bits of missing information. If the main draw is sun, sea and sand, there is little point in appealing to the self-interest of the tourism industry to protect squirrel monkeys and their habitat. If monkey viewing is a major attraction, and local people benefit significantly, the prospects of allying conservation and development are better.

4.3.3 Assessment of the combined impacts on the ecosystem and human system

Table 4.3 shows our assessment of the combined impacts of each of the five case uses, applying the framework set out in Table 4.2.

Good uses – CAMPFIRE in Zimbabwe, furbearer trapping in the Yukon, kangaroo harvesting in Australia – should be encouraged. Neutral uses – vizcacha hunting in Argentina – need be neither encouraged nor discouraged (they are not a priority for attention). Bad uses should be made into good (or at least neutral) uses. If that is not possible, they should be stopped, but this may not be possible either.

In the Costa Rican case, of course, it is not monkey viewing *per se* that is bad: the infrastructure of tourism is doing the damage. How can tourism's impacts on ecosystems and wild mammals be changed for the better? What lessons can we learn from the good uses that would help us to reform bad uses?

A look at the good uses shows that the keys to sustainable use are supportive values, incentives and laws. In the Yukon, values, incentives and laws reinforce each other in favour of conservation. Progress is being made in this direction in Australia and Zimbabwe, with Zimbabwe further ahead than Australia. In Argentina the incentives are weak and regulations weaker still: the only reason why vizcachas are not being

Table 4.3 Assessments of five uses of mammals, based on their combined impacts on the ecosystem and the human system (our assessment is shown in CAPITALS)

Impact on the human system	Impact on the ecosystem			
	Positive	*Neutral/negligible*	*Negative*	*Unknown*
Positive	GOOD CAMPFIRE, Zimbabwe	GOOD Furbearer trapping, Canada	BAD	UNKNOWN
Neutral/ negligible	GOOD Kangaroo harvesting, Australia	NEUTRAL Vizcacha hunting, Argentina	BAD	UNKNOWN
Negative	BAD	BAD	BAD	BAD
Unknown	UNKNOWN	UNKNOWN	BAD Monkey viewing, Costa Rica	UNKNOWN

overexploited is because they are abundant and resilient. In Costa Rica the law is not equipped to control habitat destruction on private land, and incentives are completely askew: only 2% of income from tourism goes back to the local community.

In the Yukon, furbearers are respected as a resource because communities receive a fair share of benefits from them. In both Zimbabwe and Australia a crucial step has been to change values by changing the incentive structure (the flow of benefits), so that landowners and communities regard the wild mammals not as pests but as resources. This step has not been taken in Argentina where landowners and the state consider vizcachas to be pests. Some people think that it devalues wild mammals to treat them as a resource. On the contrary, valuing them as a resource is often essential for their conservation, especially when they are negatively valued as pests, and in any case does not exclude valuing them in other, non-resource, ways.

The next step is to make beneficiaries responsible for sustaining the resources by involving them closely in management. This is already done in Zimbabwe and the Yukon but not in the other countries. Participation in management helps resource users to understand interactions between the ecosystem and the human system and to take (or endorse) the actions needed to support both systems.

4.3.4 Information required for assessment

Information requirements for assessment are not onerous. Formal sources (e.g. monitoring indicators of population trends) and informal sources (e.g. maps and reports by wildlife users) can be combined to provide a reasonably reliable picture of conditions and trends. Although information about the impacts of the five uses is far from complete, and more information would be useful, in no case do we consider more information to be necessary to come to a conclusion about whether a use is good, bad or neutral. In general, there is better information about the impacts of the uses on ecosystems than on human systems. Although interim conclusions have been drawn, information about impacts on the human condition needs to be improved. In particular, a study of the economic impact of wildlife viewing at Manuel Antonio (Costa Rica) is essential. If the income from wildlife viewing is substantial but little of it goes to the residents of Manuel Antonio, ways need to be explored to increase their share.

REFERENCES

Prescott-Allen, R. and Prescott-Allen C. (eds) (1996) *Assessing the Sustainability of Uses of Wild Species. Case Studies and Initial Assessment Procedure*, IUCN, Gland, Switzerland. Chapters include:

Child, B. (1996) CAMPFIRE in Zimbabwe.

Grigg, G. (1996) Harvesting kangaroos in Australia.

Jackson, J.E. (1996) Vizcacha hunting. Part of: Jackson, J.E., Bucher, E.H., and Chani, J.M. Capture of blue-fronted amazons and hunting of vizcachas and tegu lizards in Argentina.

Slough, B.G. and Jessup, R.H. (1996) Furbearer trapping in the Yukon, Canada.

Wong, G. and Carrillo, E. (1996) Squirrel monkey viewing and tourism in Costa Rica.

Part Two

Harvesting Wild Mammal Populations

5

Game ranching

John D. Skinner

SYNOPSIS

Game ranching originated in Africa. Parameters for measuring productivity in nine popular ranching ungulate species are examined, as are management and economic considerations. Many wild ungulates have evolved in Africa, yet attempts at domestication have been spasmodic and seemingly ineffectual. Although these species are apparently well suited to game ranching, only limited research has so far been carried out. Suggestions are made regarding future research.

5.1 INTRODUCTION

It is generally considered that the modern concept of game ranching – that is, extensive containment systems for wild ungulates – originated in Africa (Dasmann, 1964). Although game animals have always been important as a source of food they were, until recently, never farmed. This was probably because the overall profitability of such an enterprise in the past never reached a level comparable with conventional animal husbandry and ownership of wild ungulates was vested in the state. Moreover, although proposals to domesticate some indigenous wild ungulates were made in the middle of last century (Methuen, 1848) the only species of African wild animals to be domesticated have been the guinea fowl (*Numida meleagris*), cat (*Felis sylvestris lybica*), most probably the donkey (*Equus africanus*) and, to an extent, the ostrich (*Struthio camelus*). Domestication is the bringing of an animal into subjection to or dependence on man, a process which usually spans a great length of time. However, some species of antelope are not strictly feral because the farmer can enclose them in paddocks, thereby exercising ownership. There has also been an increasing tendency to improve productivity per hectare by mixing browsing antelopes with grazing domestic livestock such as cattle and sheep.

This chapter examines developments over the last three decades, particularly with regard to productivity in nine species of ungulate used in game ranching in southern Africa.

5.2 PRODUCTIVITY

Traits that are important in assessing the productivity of species include reproduction, growth, carcass yield and meat quality, behaviour and feasibility of domestication.

5.2.1 Reproduction

The life history features of nine ungulate species used in game ranching in southern Africa are illustrated in Table 5.1. These can be separated into aseasonal and seasonal breeders. Aseasonal breeders inherently have a major advantage in that reproduction can theoretically be manipulated. However, to manipulate reproduction, we need to establish which environmental cues trigger sexual activity in these species. Seasonally breeding species of antelope have all evolved in mesic areas where there is a fixed summer rainfall pattern, in contrast to arid areas with sporadic low rainfall.

Research into aseasonal breeding patterns has thus far been meagre. For example, eight years of researching springbok reproduction have shown that the male rut plays a crucial role in synchronizing oestrus in the female. The rut may occur at any time between February (summer) and July (mid-winter) but the cues responsible for triggering the rut are still not established (Skinner *et al.*, in press). Nevertheless, the springbok is a prolific breeder capable of conceiving twice and rearing two lambs in 12 months if climatic conditions are favourable. Preliminary evidence for the oryx indicates a similar reproductive pattern to the springbok, although the gestation period is some three months longer. Eland also breed throughout the year but it has been shown that environmental factors, such as nutrition, influence birth peaks and calving percentage. Under optimal nutritional conditions, calving percentage may be 5% greater and cows calve three months earlier (Skinner and van Zyl, 1969).

In seasonally breeding ungulates, the cue determining the onset of breeding is shortening day length. As the mechanisms involved are the same as for northern hemisphere ungulates, considerable research has been carried out to establish the timing and duration of the breeding season. The adaptive significance of this phenomenon is to ensure that the young are born after the onset of the summer rains when there will be sufficient food available for the lactating females (Skinner and Van Jaarsveld, 1987). A number of more subtle cues have also been established for fine-tuning the breeding pattern. For example, in impala the lunar cycle is also important, inducing roaring in the male which in turn

Table 5.1 Life history features of nine species of ungulates used for game farming in southern Africa (adapted from Skinner, 1989)

Life history feature	Aseasonal breeders					Seasonal breeders			
	Springbok	Eland	Oryx	Mountain reedbuck	Giraffe	Blesbok	Impala	Kudu	Warthog
Mature weight (kg)									
Female	37	450	210	29	828	60	40	160	56
Male	42	650	240	30	1 200	70	55	230	80
Age when first breeding (months)									
Female	7	28	24	8	56	18	18	17	18
Male	12								
Breeding season	all year	all year	all year	all year	all year	autumn	autumn	winter	autumn
Gestation (days)	168	271	264	240	457	225	196	212	171
Parturition	all year	all year	all year	all year	all year	summer	mid summer	late summer	early summer
Birth weight (kg)									
Female	3.4	25.2		3.1	89	7	5	16	0.8
Male	3.4	30.0			95				
Annual calving % (mature females)	46–100	83		80	48	85	90	100	96
Calving interval (days)		355			645	365	365	365	365
Theoretical maximum productivity (offspring)	24	16	15	14	10	13	13	14	36
Age of last breeding (yr)	13	18	17	15	23	15	15	16	15
Adult sex ratio M:F	1:30	1:10	1:30		1:20	1:25	1:30	1:30	1:30

induces females to come into oestrus (Murray, 1982). Wildebeest also respond to the lunar cycle, resulting in synchronized oestrus (Sinclair, 1977), while impala, blesbok and springbok ewes respond to the male in a typical ram-effect (Skinner *et al.*, 1990; Marais and Skinner, 1993) so that lambs are born over a restricted period. Estes (1966) has suggested that the resultant synchronization of births is an adaptation to resist predation because the predators are 'flooded' with lambs in a vulnerable stage. The lambs grow rapidly and, because they are so numerous, more survive to a stage where they can fend for themselves.

When calving percentages of wild ungulates and domestic livestock in southern Africa are compared, the latter only exceed the former in intensive farming systems.

5.2.2 Growth, carcass yield and meat quality

In smaller species (springbok, reedbuck), the point at which the growth curve begins to slow is reached before one year of age; in the blesbok, impala and warthog this is at 18 months, in the eland and gemsbok at about 36 months and in the giraffe at about five years. This will take place earlier in years of good rainfall and be delayed when climatic conditions are adverse. To gain maximum advantage, cropping should be carried out at the peak in exponential growth and should also take breeding seasons into account in seasonal breeders.

Dressing percentages for most species under consideration exceed 55% (Table 5.2). This is somewhat higher than percentages attained by domestic livestock under ranching conditions. African ungulates are typically lean (Table 5.2) and in springbok, for example, carcass lean increases continuously, peaking at 84% and 82% in adult males and females, respectively (von La Chevallerie, 1970). The range of lean meat for 17 species of East African ungulate is somewhat lower at 75–80% (Ledger, 1963).

Meat quality has only been compared in four antelope species (Table 5.3) of which the springbok came out best; this was probably due to the thin muscle fibre diameter. Muscle fibre thickness is an important aspect of meat quality as it determines the coarseness of grain and texture of the meat. In male warthogs, muscle fibre diameter increased from 62.9 μm in yearlings to 79.1 μm in adults (Mason, 1982). In contrast to springbok, mature male giraffes are known for the tough texture of their meat, with a fibre diameter of 65.5 μm (Hall-Martin *et al.*, 1977). Moreover, in giraffes, muscle fibre diameter increases relatively slowly above one year of age, whereas in springbok the major increase takes place early in the animal's life (von La Chevallerie and van Zyl, 1971). Female giraffes have a greater fat content and both they and young males have thin muscle fibres (50–53 μm) so this must favour the use of younger animals

Table 5.2 Live mass, dressing percentage and percentage buttock fat in male ungulates (adapted from Huntley, 1971; Hall-Martin *et al.*, 1977; Skinner, 1989)

Species	Live mass (kg)	Dressing (%)	Fat (%)		
			May	July	Sept
Springbok	33	56	1.5	1.5	1.1
Eland	408	51	3.9	2.2	2.1
Mountain reedbuck	30	55			
Giraffe	1174	62	0.5	0.5	
Blesbok	73	53	2.4	1.6	1.3
Impala	50	59	2.4	1.6	1.3
Kudu	236	57		1.4	
Warthog	80	55			

Table 5.3 Meat quality of males of four antelope species (adapted from von La Chevallerie, 1972)

Factor	Springbok	Eland	Impala	Blesbok
Number	72	6	18	23
Moisture content (%)	74.7	74.8	75.7	75.5
Fat content (%)	1.7	2.4	1.4	1.7
Colour (colorimeter unit)	7.3	5.9	7.4	7.9
Fibre diameter (μm)	45.5	66.3	56.7	53.8
Toughness (g/cm)	1181	3366	2751	2323
Taste panel scores for flavour (*n*)	36	3	9	11
Intensity (out of 10 points)	4.2	4.4	4.0	4.9
Acceptability (out of 10 points)	6.1	5.3	5.2	5.8

in any programme which seeks to market giraffe meat. Lean meat is a desirable commodity in the diet conscious First World consumer market.

In viewing the carcass composition as a whole, the buttock (which is considered a high quality cut) comprised some 30% of the carcass in the nine species examined. On the other hand, in the giraffe the foreleg (containing flat muscles, plenty of tendon and connective tissue and a high proportion of bone) contributes a much higher proportion to the carcass than any of the other ungulates investigated in South Africa.

Much confusion has arisen between trophy hunting and the cropping of ungulates for meat production. In the former, meat produced is a by-product of sport-hunting; the fee charged is related to horn length or another parameter, frequently inversely related to meat quality. On the other hand, when producing meat from wild ungulates, the same criteria should apply as those for domestic livestock, namely tenderness, succulence and tastiness – all of which are characteristics associated with young animals.

5.2.3 Behaviour

Domestic animals are generally not territorial but harem-holding in their breeding system. Aspects such as territoriality and home range are therefore irrelevant to herd management. Wild ungulates are constrained by such traits, and must use resource-defence polygyny. In springbok, for example, only territorial rams mate with ewes and territory size is about 0.22 km^2. Blesbok and impala are also territorial with territories of about 0.023 and 0.6 km^2, respectively, depending on the type of habitat. Oryx bulls are also territorial but, like eland, they can be herded, though too many confined oryx bulls disrupt the activities of territorial bulls, resulting in a drastic lowering of the calf crop.

Puberty in aseasonal breeding ungulates may be reached at any time, after which females will mate if conditions are suitable. In males this is not possible in the wild due to social constraints – the male first has to develop to the stage where he can capture and defend a territory. The latter is also true for seasonally breeding males but such females will only breed during the rut. Impala, for example, only experience two oestrous cycles at this time but conception is usually 90%. All these characteristics must be considered when designing suitable management systems.

5.2.4 Temperament and feasibility for domestication

With the advent of domestic livestock such as cattle and sheep, Africans had little reason to domesticate indigenous species. This was a great pity because cattle and sheep under human management have caused major degradation in a habitat in which they did not evolve.

Temperament is probably crucial in decisions governing domestication. Ideal candidates such as buffalo (*Syncerus caffer*) are too aggressive, springbok are too nervous and most species are not ideal in this respect. Because of its large size and docile disposition the eland was at one stage regarded as a prime candidate (Posselt, 1963; Skinner, 1967), but several recent attempts to domesticate herds of eland met with only variable success. This is partly because thorough studies of the behaviour of eland were lacking. For example, when eland are confined or herded, an optimum herd size is about 20. With larger numbers the animals do not flourish and numbers decline. Eland are also strongly nomadic – small enclosures do not suit them – and very susceptible to tick infestation. At present the reasons for the success of some ventures with eland and failure of others have not been established.

Attempts to domesticate oryx have been more limited and have also met with variable success – which is not surprising as we know even less about the breeding system of this species. Springbok and blesbok can be managed under conventional farm fencing as they are not prone to jumping like the impala, for which 2 m fencing is a prerequisite.

Despite its poor meat quality, a prime candidate for domestication is surely the giraffe, which has an exceptionally docile temperament and does not compete for nutritional resources with domestic livestock. No attempts have been made to investigate domestication of this species yet, probably because of the difficulty involved in its capture and transport.

5.3 MANAGEMENT

Managing a game ranching enterprise requires a high level of expertise – probably greater than for ranching with domestic livestock. This is because the latter can be more easily herded and moved between paddocks to facilitate veldt management. Ideally, management should facilitate ecological complementarity. Stocking rates using wild ungulates have to take into account the considerable expense of capturing them, and allowance should be made for a greater margin of error to avoid the inevitable consequences of drought. Assessing stocking rates in a dynamic ranching system is not easy and therefore sustained productivity is best maintained with a mix of domestic and wild ungulates, by supplementing domestic grazers with wild browsers. In addition, the rancher should err on the side of understocking to hedge against seasonal drought, which reduces primary productivity.

Earlier methods of managing wild ungulates often led to severe stress and death as a result of injudicious handling of the animals. Since the work of Gericke *et al.* (1978), far superior methods of herding, capturing and translocation have been developed to prevent the onset of the serious stress disease known as capture myopathy. These methods, together with humane killing, have contributed significantly to the improved welfare of harvested game animals.

5.4 ECONOMICS

Recent studies (summarized by Skinner, 1989) have shown that a mix of domestic and wild ungulates, with appropriate and correct management, can maximize profits. For example, comparative grazing studies on Merino sheep and springbok have clearly shown that springbok take a significantly greater variety of dicotyledonous plant species than sheep, which consume far more grass and even lignified grass during drought conditions. A combination of springbok and sheep leads to a more complete use of the vegetation (Davies and Skinner, 1986). Wild ungulates in a monoculture would be the least profitable farming system because the restrictions imposed by their behaviour will limit the flexibility of management options. A monoculture of domestic livestock also limits productivity, because the animals are unable to utilize all the vegetational strata. To realize greater profits, one would ideally combine (for example) giraffe, kudu, springbok or impala and dikdik with domestic cattle or sheep.

5.5 CONCLUSIONS

Despite three and a half decades of research, game ranching has only received 'official' recognition as a viable and alternative form of land use in the last decade. Prejudice still exists amongst veterinarians and pasture managers because of the impression that game carry endemic diseases to which livestock are susceptible and because of supposed degradation of the habitat by game. These accusations are unfounded or refuted but they still persist and have slowed research progress.

There has been considerable progress in increasing the profitability of the wildlife industry as a whole. This has been achieved by integrating different forms of land use such as safari foot trails, wildlife photography, bird watching and trophy hunting, to name but a few, with conventional game ranching.

Perhaps a major problem still confronting the game rancher is how to crop and market the animals efficiently. Over a period of 25 years at the Lombard Nature Reserve, losses due to bullet wounds from cropping springbok amounted to 8% of those hunted (Skinner, 1975). The time has arrived to consider other methods, such as rounding up with helicopters and transporting buck to mobile abattoirs as is done when selling and marketing live game. The cost of refrigerated, mobile abattoirs can, however, be exorbitant and must be offset by some high value product, such as zebra skins (Swank *et al.*, 1974).

Despite setbacks in the domestication of game this line should be pursued with renewed vigour. We have yet to examine the causes of some failures in this regard, notably with eland, and progress with this species is currently being assessed. The oryx, too, deserves closer examination. Far too little attention has been given to selecting wild ungulates for specific traits such as temperament and growth and game animals must still rank as the most unexploited resource, a failure which Africa, with its inherent shortage of protein, can ill afford not to address.

ACKNOWLEDGEMENTS

Research on which this article was based has been sponsored by the Agricultural Research Council and the Foundation for Research Development.

REFERENCES

Dasmann, R.F. (1964) *African Game Ranching*, Pergamon Press, Oxford.

Davies, R.A.G. and Skinner, J.D. (1986) Spatial utilization of an enclosed area of the Karoo by springbok *Antidorcas marsupialis* and Merino sheep *Ovis aries* during drought. *Trans. Roy. Soc. S. Afr.*, **46**, 115–132.

Estes, R.R.D. (1966) Behaviour and life history of the wildebeest (*Connochaetes taurinus* Burchell). *Nature*, **212**, 999–1000.

Gericke, M.D., Hofmeyr, J.M. and Louw, G.N. (1978) The effect of capture stress and haloperidol therapy on the physiology and blood chemistry of springbok, *Antidorcas marsupialis*. *Madoqua*, **11**, 5–18.

Hall-Martin, A.J., von La Chevallerie, M. and Skinner, J.D. (1977) Carcass composition of the giraffe *Giraffa camelopardalis giraffa*. *S. Afr. J. Anim. Sci.*, **7**, 47–50.

Huntley, B.J. (1971) Carcass composition of mature male blesbok and kudu. *S. Afr. J. Anim. Sci.*, **1**, 125–128.

Ledger, H.P. (1963) A note on the relative body composition of wild and domestic ruminants. *Bull. Epiz. Dis. Afr.*, **11**, 163–165.

Marais, A.L. and Skinner, J.D. (1993) The effect of the ram on synchronisation of oestrus in blesbok ewes. *J. Afr. Ecol.*, **31**, 253–260.

Mason, D.R. (1982) Studies on the biology and ecology of the warthog *Phacochoerus aethiopicus sundevalli*. Lonnberg 1908, PhD thesis, University of Pretoria.

Methuen, H.H. (1848) *Life in the Wilderness; or Wanderings in South Africa*, Richard Bentley, London.

Murray, M.G. (1982) The rut of impala: aspects of seasonal mating under tropical conditions. *Z. Tierpsychol.*, **59**, 319–337.

Posselt, J. (1963) The domestication of the eland. *Rhod. J. Agric. Res.*, **1**, 81–87.

Sinclair, A.R.E. (1977) Lunar cycle and timing of mating season in Serengeti wildebeest. *Nature*, **267**, 832–833.

Skinner, J.D. (1967) An appraisal of the eland as a farm animal in Africa. *Anim. Breed. Abstr.*, **35**, 177–186.

Skinner, J.D. (1975) Game farming in South Africa. *J. S. Afr. Biol. Soc.*, **16**, 8–15.

Skinner, J.D. (1989) Game ranching in southern Africa, in *Wildlife Production Systems* (eds R.J. Hudson, K.R. Drew and L.M. Baskin), Cambridge University Press, Cambridge.

Skinner, J.D. and Van Jaarsveld, A.S. (1987) Adaptive significance of restricted breeding in southern African ruminants. *S. Afr. J. Sci.*, **83**, 657–663.

Skinner, J.D. and van Zyl, J.H.M. (1969) Reproductive performance of the common eland *Taurotragus oryx* in two environments. *J. Reprod. Fert.*, Suppl. 6, 319–322.

Skinner, J.D., Jackson T. and Marais, A.L. (1990) The 'ram effect' in three species of African ungulates, in *Proc. Int. Symp. 'Ongules/Ungulates 1991'*, (eds F. Spitz, G. Janeau, G. Gonzalez and S. Aulagnier), SFEPMIRGM, Toulouse, pp. 565–568.

Skinner, J.D., van Aarde, R.J., Knight, M.H. and Dott, H.M. (in press) Morphometrics and reproduction in a population of springbok *Antidorcas marsupialis* in the semi-arid southern Kalahari. *Afr. J. Ecol.*

Swank, W.G., Casebeer, R.L., Thresher, P.B. and Woodford, M.H. (1974) *Cropping, Processing and Marketing of Wildlife in Kajiado District, Kenya*, Government of Kenya and FAO, Nairobi.

von La Chevallerie, M. (1970) Meat production from wild ungulates. *Proc. S. Afr. Soc. Anim. Prod.*, **9**, 73–87.

von La Chevallerie, M. (1972) Meat quality in seven wild ungulate species. *S. Afr. J. Anim. Sci.*, **2**, 101–104

von La Chevallerie, M. and van Zyl, J.H.M. (1971) Growth and carcass development of the springbok *Antidorcas marsupialis*. *Agroanimalia*, **3**, 115–121.

6

Status and exploitation of the saiga antelope in Kalmykia

James G. Teer, Valeri M. Neronov, Lir V. Zhirnov and Anatolij I. Blizniuk

SYNOPSIS

From a population of over 750 000 in 1958, the number of saiga antelope in the north-western Caspian region of the Autonomous Republic of Kalmykia, Russian Federation, is now about 160 000. The range of the species has been reduced from over 120 000 km² to about 20 000 km². Intensive grazing, principally by sheep and also by cattle, and attempts to farm chernozemic (black) soils are major causes of habitat loss. Overgrazing has resulted in severe degradation of carrying capacity of the habitats for both livestock and saiga antelope. Over 20% of the semi-arid steppe habitat is desert and of little value to herbivores. In recent months poaching for horns used in oriental medicines has become a serious factor in population loss and distortion of the sex ratio of adults. This distortion in Kalmykia has put reproduction of the species at an unknown risk. Irrigation canals emanating from the Volga River, roads, fences, power lines and other contrivances are also important in disruption of migration routes and mortality. Although hunting and other uses of this species in Kalmykia are now prohibited, the antelope is at serious risk and management must be implemented to restore it to former numbers and use. Local people must receive some benefits from saiga to ensure its future.

6.1 INTRODUCTION

The saiga antelope (*Saiga tatarica* L.) is an important economic resource in Russia (Silant'ev, 1898; Bannikov *et al.*, 1961; Maksimuk *et al.*, 1987). A herd antelope about the size of an American pronghorn antelope (*Antilocapra americana*) or African impala (*Aepyceros melampus*), it is adapted to arid and steppe environments with interesting morphological and behavioural attributes. Enlarged nasal vestibules, protruding orbits for large range and scope of vision, and an ambling gait in which legs of the same side are moved simultaneously are adaptations to open grass-land habitats. Saiga are extremely shy with large flight distances, probably made so by harassment by humans. Moving herds, always in a dust cloud, give the appearance of a blanket being pulled across the flat terrain of the steppes.

A monotypic genus with two subspecies, the saiga presently ranges in the arid and steppe habitats of the pre-Caspian and Kazakhstan regions. Pleistocene remains have been found throughout Europe and Asia, from England to Alaska. Its range has been greatly reduced and now *Saiga tatarica tatarica* is found primarily in two populations, the most numerous one being in the vast arid zones of Kazakhstan. The second, much reduced in distribution and numbers, inhabits semi-desert and steppe habitats south of the Volga River and west and north of the Caspian Sea in the Republic of Kalmykia. Distribution of the two populations was formerly continuous.

The second subspecies, *S. t. mongolica*, occurs in Outer Mongolia. Its numbers are uncertain but are said by Russian scientists to be no more than 2000 animals and possibly fewer than 700. The Mongolian race has been over-exploited for its meat, hides and horns, and is in danger of extinction.

Adult males of *S. t. tatarica* weigh 32.5–51.0 kg ($n = 29$, $\bar{x} = 43.1$ kg); adult females weigh 21.4–40.9 kg ($n = 25$ $\bar{x} = 30.8$ kg) (Bannikov *et al.*, 1961); *S. t. mongolica* is considerably smaller in size and weight. Only the males have horns, which are ribbed, slightly curved outward, between 28 and 38 cm ($\bar{x} = 32$ cm) in length, and with almost transparent, ivory-coloured tips for the final 5–8 cm (Bannikov *et al.*, 1961).

One of the most important characteristics of the species is its fecundity. Its resilience to environmental pressures and its rapid recovery from low numbers have been demonstrated on several occasions. Saiga are polygamous with one male tending from five to 20 females in a harem system. Harems last through a single breeding season. Females breed at 8 months of age, and males are sexually mature at 1 year 7 months (Zaplyuk, 1968). Breeding occurs from the end of November through most of December. The gestation period is from 135 to 141 days ($\bar{x} = 138$ days). Twinning is the rule for adults. Zaplyuk (1968) reported that young females on Bazsa-Kelmes Island averaged 95–102 embryos/100 young females. Females older than 2 years had 160–181 embryos/100 adults.

The saiga antelope is migratory and somewhat nomadic. Because its ranges have been reduced, migratory distances and routes between breeding and wintering habitats have changed and shortened in the past few years.

The species has been cropped for its meat, hides and horns, the latter being prized for medicinal concoctions by oriental peoples for centuries. About three sets of adult male horns weigh 1 kg and this weight has fetched as much as US$600 in trade to markets in China, Korea and other oriental nations. In 1992, the value of horns in the oriental markets was about US$150/kg. The peasant poacher was paid perhaps US$10/kg. Hides were worth about 1.5 roubles and meat about 92 kopeks/kg. These values are now inflated due to devaluation of the rouble in recent months.

Over 7.5 million saiga were harvested in the Soviet Union between 1951 and 1990 (B.V. Vinogradov, personal communication). These produced 120 000 tonnes of meat and upwards of 100 million roubles for skins and horns. About as many saiga have been poached in recent years as have been legally taken (L.V. Zhirnov, personal communication).

We report here on the Kalmykian population whose habitat has been reduced from over 100 000 km^2 to about 23 000 km^2, and whose numbers have been reduced from more than 750 000 to about 160 000. Our objective is to review its status and vulnerability, in the hope of mustering support for management of the Kalmykian population so that it can regain its economic and cultural importance.

6.2 HISTORY OF THE KALMYKIA SAIGA POPULATION

The early history of saiga numbers and their exploitation and management throughout the species' range was described by Bannikov *et al.* (1961). Their numbers have fluctuated widely, primarily as a result of over-exploitation at times, or weather patterns and disease at other times. The Customs office in Kjakhta recorded exports of 3.95 million pairs of horns to China in the 1800s (Silant'ev, 1898). Saiga were almost completely extirpated from their entire range by the 1920s, after which they increased to peak numbers in 1958.

The fluctuations in population since the 1950s are shown for Kazakhstan and Kalmykia in Tables 6.1 and 6.2. The Kalmykian population increased to around 800 000 in the 1950s and early 1960s. As suggested by harvest data from 1951 to 1980 (Table 6.1), the population in Kalmykia peaked in 1958, fell sharply in the next decade (Zhirnov, 1982) and peaked again, reaching 715 000 in the late 1970s. Since then the population has declined. Data on saiga numbers and other population parameters were obtained in a more systematic manner in later years and were better organized for understanding the current status of the species.

6.2.1 The census effort

Censuses have been conducted by automobile or by aircraft, or both, since 1969. In the 1970s, censuses were conducted in the early spring during lambing and in autumn prior to the harvest season. Estimates of sex and age ratios and other population parameters were made in June and July.

Table 6.1 Harvest of saiga in the north-western Caspian region of Kalmykia, 1951–1990[1]

Years	Numbers (000's)	Amount of meat (000's tons)
1951–55	103	1.4
1956–60	612	8.5
1961–65	347	4.8
1966–70	52	0.7
1971–75	195.7	2.7
1976–80	653.1	9.1
1981–85[2]	235.0	2.7
1986-90	26.3	0.3

[1] Adapted from Zhirnov (1982).
[2] Data for 1981–90 provided by Anatole Maksimuk.

Automobiles were used in censuses from 1969 to 1973, after which both aircraft and automobiles were employed. From 13 to 20 automobiles were used on prescribed transects. Fixed-wing aircraft were used after 1980. Concentrations were located by reconnaissance flights and observers counted all saiga in transects 500 m wide.

Numbers of adult saiga were determined in the spring by estimating the areas of and average densities in the concentrations. These areas were determined by vehicles moving around the concentrations and plotting them on maps. Aircraft were also used to map concentrations. Average densities were obtained from photographs taken from fixed-wing aircraft flying at 100 m. With these data, projections were made of total numbers of animals in each concentration area.

Census accuracy is open to question. It has probably decreased over the years due to changes in methodology and in behaviour and distribution of saiga in their migratory patterns. Differences in census vehicles, observers, sampling intensity and saiga distribution patterns are but a few of the possible sampling vagaries. For example, transects counted from vehicles were more accurate than transects counted from aircraft because automobiles could be stopped when it was necessary to obtain counts of animals in the concentrations (A. Bliznyuk, personal communication). Further, except for the years 1989–1993, sampling design precluded the calculation of variance terms, which makes comparisons of census results tenuous.

The saiga's ranges were changed by desertification and greater movement occurred in or near lambing concentrations in the 1980s. Thus it became much more difficult to conduct spring censuses in lambing concentrations. In recent years only autumn censuses were made.

Despite these problems, we judge that the censuses give reasonable

Table 6.2 Estimated numbers, percentage of adult males and ratio of lambs to adult females in the Kalmykian population of saiga antelope, 1969–1993[1]

Year	Number in population (000's) ± SE	Adult males in population (%)	Ratio of lambs/adult female
1969	213.0	12.7	1.09
1970	190.0		
1971	246.0		
1972	213.0	19.4	1.15
1973	324.0	17.2	1.13
1974	425.0	24.0	1.27
1975	500.0	16.4	0.90
1976	650.0	17.9	1.17
1977	660.0	16.7	1.14
1978	715.0	17.8	1.20
1979	430.0	30.4	0.60
1980	380.0		
1981	430.0		
1982	385.0		
1983	280.0		
1984	265.0		
1985	222.0		
1986	200.0	14.9	0.89
1987	143.0		
1988	157.0	16.6	0.85
1989	150.0 ± 13.5	20.6	0.79
1990	160.0 ± 11.7	16.7	0.75
1991	168.3 ± 22.1	24.3	0.68
1992	152.0 ± 13.0	6.4	0.70
1993	148.3 ± 34.3		0.98

[1] According to Anatole Maksimuk, Deputy Director, Central Laboratory of Game and Hunting Economy, the standard error of the mean for population estimates obtained after 1978 varied between 11 and 14%. Confidence limits for 1989–1993 were provided by Anatolij Blizniuk.

estimates of trends and provide usable estimates of numbers. They were certainly valuable for setting harvest quotas and regulations for management of the species.

6.3 CURRENT STATUS

6.3.1 Numbers

The population of saiga in Kalmykia began to increase in 1969 and reached a peak in 1978 (Table 6.2). Numbers steadily decreased, reaching

143 000 in 1987. They have since fluctuated rather tightly around 150 000 (Table 6.2 and Figure 6.1) in response to natural environmental and man-made factors.

The Central Government of Russia (CIS) controls the harvest of saiga for export of its horns. Harvest quotas are set, markets for sale of the products are controlled, and organized cropping is conducted by government personnel.

The legal harvest of saiga in Kalmykia reached a peak of 201 000 in 1978, but has declined since with hunting being closed in six of the last 10 years (Table 6.3). Harvests in most years, however, have been far lower than those in peak years. With the decline in numbers, cropping of saiga for export of horns in Kalmykia was stopped in 1987, but 11 000 permits were issued for sport-hunting in 1990 and 200 were issued in 1993. Most of the permits were used by residents of the region.

Senior scientists of the Kalmykia Department of Environmental Protection (KDEP) report that the Kalmykian population is presently poised for growth, citing an increase in the ratio of lambs per adult female in 1993 (Table 6.2). Further, they feel that the 1993 census estimate was low due to limited coverage of the range during the census.

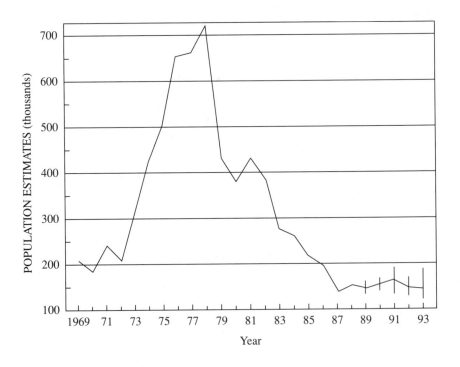

Figure 6.1 Estimates of saiga numbers in the Autonomous Republic of Kalmykia.

Table 6.3 Harvest of saiga in Kalmykia and Kazakhstan for export of horns and for sport hunting

Year	Kazakhstan (000's)	Kalmykia (000's)
1955	2.3	no data
1956	16.5	no data
1957	2.7	100.0
1958	16.9	173.0
1959	70.9	180.0
1960	138.0	39.0
1961	202.9	121.0
1962	175.6	131.0
1963	164.8	43.0
1964	110.4	40.0
1965	72.8	Closed
1966	75.0	22.0
1967	61.3	30.0
1968	85.3	Closed
1969	33.4	Closed
1970	161.6	Closed
1971	194.6	15.5
1972	264.9	Closed
1973	296.8	33.0
1974	344.0	66.0
1975	500.6	86.8
1976	321.4	137.6
1977	351.0	162.6
1978	100.0	201.0
1979	36.0	80.9
1980	149.9	71.5
1981	193.1	97.4
1982	223.1	87.6
1983	205.1	47.5
1984	123.0	Closed
1985	142.7	20.5
1986	86.8	15.3
1987	100.0	Closed
1988	64.7	Closed
1989	110.4	Closed
1990	94.8	11.0
1991	112.0	Closed
1992	108.0	Closed
1993	62.0	0.2

6.3.2 Sex and age ratios

Females predominate in the population although at birth the sex ratio is even. Natural mortality is higher in males and the disparity in sexes occurs rapidly as the animals age and as selective hunting removes the males (Table 6.2). Zhirnov (1982) calculated mortality of Kalmykian saiga for a series of years; 25% of males were lost as neonates, and another 10% of males and females were lost by the time they reached 1.5 years. In some years, the sex ratio of adults has been as wide as one male to nine females. The distortion is largely due to selective harvest of males through sport hunting and cropping for horns.

Productivity of the Kalmykian population of saiga has apparently declined in the past 25 years. During the period 1972–1978, age ratios were in the order of 1.2 lambs per adult female (Table 6.2). In recent years (1986–1993), the ratio was less than one to one. Causes of the decline in productivity are not specifically known for each year; however, a number of events including weather, predation and poor nutrition have been involved.

6.4 CAUSES OF THE DECLINE IN SAIGA NUMBERS

6.4.1 Land-use, overgrazing and desertification

Grazing is the primary industry of the Kalmykian people. Formerly nomadic graziers from Mongolia, they have grazed sheep in herding systems for centuries. The western Caspian and eastern Cis-Caucasia were highly productive winter ranges (Vinogradov *et al.*, 1985). The Black Soils are fragile and soil erosion from the high winds readily occurs when cover is removed. Changes in land tenure systems and farming of Black Soils have resulted in overgrazed and desertified areas that are increasing very rapidly. Remote sensing studies of the region show the extent of desertification has increased from 80 to 90% in the past two decades (B.V. Vinogradov, personal communication).

The Blacklands (Chernozems) of Kalmykia represent one of the largest foci of desertification in Europe. In fact, the Chernozems are the only extensively desertified regions in Europe and form one of 13 regions in Russia with very grave ecological conditions (Zonn, 1994).

Overgrazing of rangeland by livestock (primarily sheep) is the underlying major cause of the long-term decline in the Kalmykian saiga population. About 20% of saiga habitat is now degraded to blowing sand and dunes. The carrying capacity of the areas not severely desertified is reduced for livestock and wildlife.

Numbers of livestock have steadily increased since earliest times (Table 6.4). Records of livestock include only those held by the state: they do not include more than a million sheep believed to be in private ownership. If true, the total number of sheep in Kalmykia in the mid 1980s exceeded 4 million animals.

Zonn (1994) succinctly stated the problem in a report to the Central Government in Moscow. His statement is translated, somewhat paraphrased, as follows:

The number of sheep has greatly increased in the Blacklands, especially in the winter months, without regard to carrying capacity. This has resulted in excessive grazing and degradation of pasturage. By 1959, overgrazed pastures made up 32% of the region's pastures. By 1972, 23 years later, 59% were overgrazed. By 1986, overgrazing had occurred on 76% of the region (Reznikov, 1993). Of the 3 000 000 ha of pastures, 1 195 000 ha are overgrazed, 534 000 ha were very greatly overgrazed, and 665 000 ha were transformed to blowing sands and barren solonchaks (Kazakov, 1989). The desert claims over 50 000 ha of grazing lands each year. If existing methods of farming of the Blacklands are continued, this unique place of Russia will vanish and 3 000 000 ha will be transformed into a desert.

The number of livestock, especially sheep, has apparently decreased since 1987, which should ease competition between saiga and livestock. In the area north of the Chernye Zemli Biosphere Reserve, sheep numbers are now about half of their numbers five to six years earlier. Dr N.I. Reznikov (personal communication) states that pastures are recovering from heavy grazing and range recovery is being aided by increased precipitation. His agency is developing guidelines for conservative stocking of livestock. The Central Government has plans for tax incentives to livestock owners to follow the guidelines for conservative stocking rates. Prices for livestock are very low at present, but stocking rates fluctuate with market prices and some conservationists fear that an improved market will cause increased stocking of sheep and cattle. It remains to be seen if Dr Reznikov's predictions will be borne out by future numbers.

Restoration of the degraded soils in Kalmykia is underway by the Black Soils Restoration Commission. Some progress is being made, but apparently restoration is not keeping pace with losses. Over 80 000 ha have been restored to produce 680 kg forage/ha. Between 300 000 and 500 000 ha are candidates for restoration. More than 155 000 ha are now drifting sand and dunes (N.I. Reznikov, personal communication).

Table 6.4 Numbers of livestock in Kalmykia, 1883–1986

Year	Cattle (000's)	Sheep (000's)	Horses (000's)	Camels (000's)
1883	143.9	802.3	114.7	16.9
1916	291.8	972.5	104.4	15.6
1928	327.0	1 018.4		
1940	258.8	1 031.5		
1957	137.2	1 497.9		
1967	330.8	2 422.2		
1986	361.6	3 316.8	19.0	0.4

6.4.2 Poaching

Poaching is alleged to be a major factor in the recent decline in saiga numbers and in the distortion of the adult sex ratio. Undoubtedly poaching occurs but its impact on the total population is not known. It increased after the dissolution of the Soviet Union and, because of conditions in rural areas, it continues to this time.

Anti-poaching is a major activity of the KDEP. Resident wardens are stationed throughout the saiga's range in Kalmykia. Five 'swat' teams patrol areas where saiga concentrate to calve or to overwinter.

According to Dr Georgi Pavlov, Director of KDEP, confiscation of horns from illegal traffic decreased from 186 kg in 1991 to 70 kg in 1992 and 90 kg in 1993. Although confiscation of contraband horns has decreased, current enforcement cannot control illegal harvest or trade. Funds for salaries and equipment (jeeps, petrol, spare parts, subsistence of personnel, etc.) are abysmally low. The current economic health of the Russian Federation does not provide much hope for increased budgets.

Little or no effort is exercised to control traffic or curb transfer of horns across frontiers, especially between Kalmykia and the new independent state of Kazakhstan. It is likely that Kalmykia will soon become a member of CITES and opt for placing the saiga on Appendix II. Although Kalmykia's population may not withstand even carefully controlled harvests and permits are not likely to be issued for export of horns, illegal horns will easily pass to Kazakhstan where export permits can and probably will be issued. Russia (CIS) proposed the saiga for Appendix II in CITES, and the proposal was adopted at the meeting of the parties of CITES in Florida in November 1994.

6.5 OTHER FACTORS

6.5.1 Disease

Anthrax, foot-and-mouth disease, brucellosis, plague and intestinal pasteurellosis are known diseases of saiga. While documentation of these diseases is well established, only a few epizootics have been described. One such, intestinal pasteurellosis, killed over 500 000 saiga in Kazakhstan in the 1980s (Bannikov *et al.*, 1961).

6.5.2 Weather

Lack of food in the winter is the most important natural factor causing mass mortality of saigas (Bannikov *et al.*, 1961). Snowstorms and frozen snow crusts prevent saiga from reaching forage, and many thousands perish. For example, during the winter of 1949/1950 blizzards covered the landscape with over 80 cm of snow. As a result, more than 40% of the saiga in the Caspian Lowland perished from hunger (Bannikov *et al.*, 1961). Well-nourished animals can withstand severe cold, but protracted snow periods and droughts debilitate them. Direct mortality and losses of productivity ensue.

6.5.3 Human intervention

Migration patterns of saiga have been altered by irrigation canals constructed to bring water to the arid zones of Kalmykia. About 750 km of canals form a network throughout the saiga's range. Water is used to irrigate forage crops, primarily sorghum and lucerne. Fodder produced in irrigated lands is used to feed domestic livestock, and many collective farms have pastures for this purpose.

The canals act as barriers to movements between summer calving ranges and winter ranges. Direct losses of saiga occur from drowning in the canals: more than 14 000 saiga drowned in 1990 when they attempted to cross a steep-sided canal. They swim well, but they piled upon each other and could not negotiate the muddy bank on either side. However, the major problem is disruption of traditional migratory routes. Roads, fences, power and telephone lines as well as canals have created disturbance and disrupted the normal movement of saiga. The saiga's habitats are open rangelands without fences, but fencing of cropland is increasing and saiga are particularly vulnerable to such barriers.

6.5.4 Predation

Wolves are present but not now numerous in saiga habitat. They and foxes are said to take saiga calves, and wolf control has been practised in some years by shooting from aircraft. It is apparent that wolf and fox control is done to protect sheep as well as saiga. The impact of predation is not well known.

6.6   THE FUTURE OF SAIGA IN KALMYKIA

If the species is to be more than an artefact in these great steppes and semi-arid lands, a management plan for the saiga must address the fundamental causes of its decline. Desertification and over-use of range-land resources are clearly the most important factors in the future of the saiga. Coping with them will be a long-term effort and, at present, requires a partnership arrangement between Russian and outside funds and technical skills. Hunting of the Kalmykian population of saiga has been closed for five years (1992–1996) as a first step in the species' recovery. It is, however, only the first step.

With a management plan that protects and restores range resources and controls poaching, the saiga can recover. While direct control of poaching is needed, market systems for trade in saiga horns must be centralized to control illegal offtakes. Further, if the species is to co-exist on ranges where livestock, especially sheep, are the primary land use, the local people must benefit from its presence and utilization.

Production of horns on game farms may be a viable strategy to relieve pressures on wild stock. Little is known of the efficiency of farming saiga although some have been bred in captivity on a small scale. Farming of saiga on state farms and reserves could develop if the species shows biological and economic promise in production systems.

Efforts are being made by Kalmykian authorities to obtain funds for the development and implementation of a management plan. The plan must permit rational use of the species with provisions for local people to benefit economically from such use. Protection from illegal hunting and wise use of the land are inextricably wedded to the needs of rural peoples who earn their livelihoods from the land. In our view, this strategy can ensure the species' future.

ACKNOWLEDGEMENTS

The Species Survival Commission of the International Union for the Conservation of Nature (IUCN) and the Office of International Affairs, US Fish and Wildlife Service, provided travel and field support for the study. The Institute of Evolutionary Morphology and Animal Ecology of the Russian Academy of Sciences were helpful in providing data and in arranging for field support in Kalmykia. The Welder Wildlife Foundation gave the senior author leave and salary support.

Officials of the Government of Kalmykia were especially helpful: K.N.

Ilymzhinov, President of the Autonomous Republic of Kalmykia; Dr Emma B. Gabushina, Minister and Chairperson, The President's Committee for Nature and Environment Protection; Georgij Pavlov, Director, the Department of Protection and Rational Use of Hunting Resources; Victor Bademaev, Director, Chernye Zemli Biosphere Reserve; and Dr Nikolaj Reznikov, Deputy Chairman, Main Agency of Chernye Zemli and Kizliar's Pastures. Dr Anatole V. Maksimuk, Deputy Director, Central Laboratory of Game and Hunting Economy, Moscow, accompanied the authors on several field excursions in steppe habitats of Kalmykia and provided important data on populations and harvesting of saiga antelope.

REFERENCES

Bannikov, A.G., Zhirnov, L.V., Lebedeva, L.S. and Fandeev, A.A. (1961) *Biology of the Saiga*, Main Administration of Hunting and Reserves at the Council of Ministers of the RSFSR, Astrakhan Reserve Laboratory for the Biology of the Saiga. Translated from Russian, Israel Program for Scientific Studies, Jerusalem, 1967.

Kazakov, B.G. (1989) Chernye Zemli, in *The Caspian Sea Area for Us and Our Children*, Moscow, pp. 358–383.

Maksimuk, A.B., Khakhin, G.V. and Poznjak, V.G. (1987) *Animals of Aquatic and Wetland Biogeocenoses of Semidesert*, (translated), Researchers and Practical Workers of Low Volga Region and North Caucasus Region, Kalmyk State University, Elista.

Reznikov, N.I. (1993) *First results of the realization of the general scheme of desertification control of Blacklands and Kizlyar Pastures*, Black Soils Restoration Commission, Elista, Kalmykia.

Silant'ev, A.A. (1898) *A Survey of Commercial Hunts in Russia*, Saint Petersburg.

Vinogradov, B.V., Lebedev, V.V., Kulik, K.N. and Kaptsov, A.N. (1985) Measurement of ecological desertification from repeat aerospace photographs. Doklady Biological Sciences, *Proceedings of the Academy of Sciences of the USSR*, **285**, 1–6. Translated from Russian, Consultants Bureau, New York.

Zaplyuk, O. (1968) *Dynamics of sexual activity of saiga*. Zoology dissertation, Academy of Sciences of Kazakhstan.

Zhirnov, L.V. (1982) Modelling of dynamics of saiga populations. *News of Timiryazev's Agricultural Academy*, **5**, pp. 157–166.

Zonn, I. (1994) *Desertification and its Control: Republic of Kalmykia – Halmg Tangch*. A report to the United Nations Environmental Programme, Moscow.

7

Capybara use and conservation in South America

José Roberto Moreira and David W. Macdonald

SYNOPSIS

The capybara (*Hydrochaeris hydrochaeris*), a 50 kg grazing semi-aquatic rodent, displays important ecological and behavioural prerequisites for agricultural domestication, namely fast growth rate, high reproductive output, sociality and a cheap diet. Today, despite being protected in most countries, capybara are hunted all over their range for meat (and, in some cases, hides) or to achieve perceived pest control. Although there may have been local extinctions throughout their range, the species is not endangered. The advancing conversion of neotropical forest into grassland is potentially creating capybara habitat, but the high hunting pressure in some areas has led to much lower population densities than the environment can support. The only large-scale commercial harvesting of capybara is being conducted in Venezuela. Subsistence hunting is allowed in Peru only in two forest districts. There is a capybara hunting season in two of the Argentinian provinces. In Colombia there are already a few pilot farms exploiting capybaras. Although there are some ranches in Brazil producing capybaras in captivity, their economic feasibility is doubtful. Seasonally flooded savannahs, like those present in Brazil in Pantanal Matogrossense and along the Amazon valley, are suitable for capybara hunting in a similar manner to that used successfully today in Venezuela. A deterministic age-structured seasonal model of the Leslie matrix type was used to research the way in which harvesting could affect capybara populations. The maximum sustainable yield was found at a rate of 0.17 for capybaras from Marajó Island. Selective harvesting by sex may reduce fertility and lead the population to collapse and should not be recommended. The productive potential of capybaras in the neotropics calculated by the model was 841 kg/km^2 per annum.

7.1 INTRODUCTION

The majority of research done in South America on the use of game by Indians and rural communities demonstrates that small animals constitute 50% or more of the consumption of meat in their diets (Rios *et al.*, 1974; Smith, 1976; Ayres and Ayres, 1979; Patton *et al.*, 1982). Among the mammals of the neotropical region, small herbivores are the most abundant and constitute the greatest biomass. Notable among these small herbivorous mammals are those which browse, particularly rodents. Despite their small individual biomass, rodent productivity is relatively high in comparison with that of large herbivorous mammals (Eisenberg *et al.*, 1979). The meat of neotropical rodents has always been consumed by indigenous populations, and some species are relished as delicacies (Dourojeanni, 1985).

The New World hystricognath rodents (= caviomorphs) today represent one of the most diverse and widespread suborders among the over 800 mammal species of South America (Woods, 1982). Having been isolated from other rodents from the Oligocene to Miocene (some 50 million years), neotropical rodents evolved separately from those in other continents. They occupied a diversity of niches, especially those vacated following the extinction of notoungulates (Webb and Marshall, 1982). As a result, many contemporary species of hystricognath are comparable in ecological and morphological terms to other terrestrial herbivores such as ungulates, macropods and lagomorphs (Dubost, 1968; Mares and Ojeda, 1982). In particular, New World hystricognaths are generally more cursorial and larger than most rodents, and use burrows less (e.g. maras: Taber and Macdonald, 1992). Consequently, the New World hystricognaths contain the majority of large extant rodent species (Dubost, 1988) and constitute an important component of the neotropical mammal fauna.

The hystricognath rodents share several reproductive characteristics which are unusual among rodents. They are characterized by long life spans, long and variable oestrous cycles, protracted gestations, precocious young and moderate to small litter size when compared with myomorph rodents (Weir, 1974; Kleiman *et al.*, 1979). These characteristics result in the majority of the hystricognaths having low reproductive potential. This may explain the low taxonomic diversity of most hystricognath families relative to that of myomorph families, and their evolutionary stability (Woods, 1982).

The capybara (*Hydrochaeris hydrochaeris*), a semi-aquatic herbivorous rodent, is an exception among the hystricognaths in relation to its productivity. As an outlier to the negative correlation found among the hystricognaths between litter size and body-weight (Kleiman *et al.*, 1979), the capybara has one of the largest litter sizes (4.2 young/litter) within this suborder and is the largest living rodent (50 kg). This savannah-grazing mammal is considered one of the species with the most potential for sustainable harvest in South America. The species displays important ecological and behavioural prerequisites for agricultural domestication, namely fast growth rate, high reproductive output, sociality and a cheap diet. Today, despite having been protected in most countries, the capybara is hunted all over its range for meat (and, in some cases, hides) or to achieve perceived pest control.

Despite the wide distribution of this species, studies have been carried out throughout its geographical range (Venezuelan Llanos: Ojasti, 1973; Macdonald, 1981; Colombian Llanos: Jorgenson, 1986; Brazilian Pantanal Matogrossense: Schaller and Crawshaw, 1981; Alho *et al.*, 1987; Brazilian Amazonia: Moreira, 1995). This chapter reviews and updates existing information on capybara reproductive characteristics and productive potential throughout its range. It concentrates on aspects related to the sustainable harvesting of the capybara in South America, and discusses the present status of its conservation.

7.2   REPRODUCTION AND PRODUCTIVITY

Unpredictability of resource availability, which is characteristic of capybara habitat, may have exerted selective pressure which has led to a life history strategy involving greater fecundity and productivity than expected from its body size (Kleiman *et al.*, 1979; Moreira, 1995). An average of 4.2 young are born per litter, weighing approximately 1.5 kg each (Ojasti, 1973; Moreira, 1995). The gestation period averaging 150.6 days (López, 1987) is slightly longer than predicted from the litter size (Kleiman *et al.*, 1979).

Capybaras are fertile throughout the year, but a peak in reproductive effort coincides with the periodicity of annual flooding in the savannahs. On Marajó Island (in Brazilian Amazonia) the capybara breeding season coincides with the onset of the rainy season, in December–January (Moreira, 1995), whereas in the Venezuelan Llanos and in the Brazilian Pantanal Matogrossense breeding peaks at the end of the rainy season (Ojasti, 1973; Schaller and Crawshaw, 1981). Female capybaras have a post-partum oestrus 15 days after giving birth; the oestrous cycle lasts 7.5 days (López, 1982). Females are sexually mature at 12 months of age (Ojasti, 1973).

Ojasti (1973) recorded 1.6 births per annum for capybaras in the Venezuelan Llanos, with a fertility rate of 6 young/female per annum calculated from a hunted sample. In those calculations Ojasti assumed a gestation length of 120 days, which has subsequently emerged as an underestimate (López, 1987). The recalculated frequency of births in Venezuela, using the correct gestation length of 150 days, is 1.2 births per annum, giving a fertility rate of 5 young/female per annum. In Brazilian Amazonia we also recorded 1.2 births per annum in a sample of hunted females. However, the frequency of pregnant females in this sample differed significantly from that recorded from our observational surveys: the survey data indicated a much more marked seasonality in the incidence of pregnancies than was apparent in the hunted sample (Moreira, 1995). The hunting techniques used by the cowboys in Marajó Island significantly affected the frequency of pregnant females caught ($\chi^2_1 = 7.2$,

$P < 0.01$) when we separated them into two groups in relation to the exhaustion inflicted on the animals. A similar contrast is evident between Ojasti's (1973) hunting and census data. Ojasti (1973) himself considered that his hunted sample was not a random one. Therefore we recalculated the fertility rate of the Venezuelan capybaras using Ojasti's (1973) census data, which gave a fertility rate of only 2.3 young/female per annum and 0.7 births per annum. Although this calculation is devalued by several sources of error (e.g. not all females considered as adults in the sample were fertile and the number of newborn may have been underestimated), these factors do not alter the conclusion that both calculations of capybara fertility rate from hunted samples led to a substantial over-estimate.

What is the correct frequency of births and the fertility rate for capybaras in the seasonally flooded savannahs of the Amazon and Orinoco basins? Unfortunately, neither our census data nor Ojasti's (1973) are sufficiently detailed to provide a definitive answer. Nevertheless, our best estimate is one litter of 4 young/female per annum (see also Eisenberg *et al.*, 1979; Robinson and Redford, 1986). This estimate is far below the value that has been used in calculations concerning capybara management over the past 20 years, but it is nevertheless higher than predicted on the basis of body mass (Robinson and Redford, 1986).

The capybara is a social animal, forming closed family groups of five to 14 adults (Herrera and Macdonald, 1987). Each group is generally composed of a dominant male, several related females and their litters, and subordinate males at the periphery of the group (Macdonald, 1981). Although the dominant male is responsible for most of the copulations in the group, the subordinate males collectively account for half of the copulations (Herrera and Macdonald, 1993). In the Venezuelan Llanos, capybaras are sedentary, occupying territories of 5–16 ha, varying according to the dispersion of resources. A typical capybara home range encompasses an extensive grazing area, a permanent water body (which is used for drinking, as a mating site, to regulate body temperature and as a refuge from predators) and an elevated dry area in which to shelter when water levels rise during the wet season (Herrera and Macdonald, 1989). In Venezuela the peak of their population numbers occurs during the rainy season with an ecological density (number of individuals per unit of area actually used by the capybaras) of approximately 1.3 individuals/ha (Herrera, 1986). The ecological density of capybaras in the Brazilian Pantanal Matogrossense is only 0.15 individuals/ha, possibly because of the scarcity of forage there (Alho *et al.*, 1987).

The age structure of the capybara population on Marajó Island indicated a rapid population turnover, with a high mortality rate (0.68) within the first year. The mean rate of mortality was slightly higher for males (0.63 per annum) than for females (0.56 per annum). There was no significant difference between survival curves of females and males (Log-Rank: $\chi^2_5 = 0.40$, $P = 0.53$; Wilcoxon: $\chi^2_5 = 2.61$, $P = 0.11$), but significantly more females survived in the second year than did males ($\chi^2_5 = 5.02$, $P = 0.03$). This probably arises because of greater female than male recruitment into social groups. The life span of capybara on Marajó Island was 6–7 years, being slightly longer for males. Due to the high mortality rate in the first year, life expectancy at birth for capybaras on Marajó Island was very short (1.1 years for males and 1.3 years for females). The generation length was 2.6 years (Moreira, 1995). Table 7.1 presents a résumé of capybara life history traits.

The intrinsic rate of natural increase (r_{max}) for capybaras living in natural savannahs was calculated using Cole's (1954) equation:

$$1 = e^{-r_{max}} + be^{-r_{max}(a)} - be^{-r_{max}w+1}$$

where a is the age at first reproduction (2 years), b is the fertility rate given as the annual birth rate of female offspring ($2♀/♀$ per annum), and w is the age at last reproduction (7 years). The r_{max} found for capybaras living in natural savannahs was 0.69, being one of the highest found among neotropical mammals (Robinson and Redford, 1986). This rate was high for capybaras despite the fact that the intrinsic rate of natural increase of mammal populations is inversely related to the body size of the species (Hennemann, 1983).

From this result the maximum production for capybaras in neotropical savannahs was calculated using the equation developed by Robinson and Redford (1991) for neotropical forest mammals:

$$P_{max} = (0.6D \times \lambda_{max}) - 0.6D$$

Table 7.1 Life history traits of free-living capybaras

Trait	Average
Gestation period[1]	150 days
Incidence of pregnancy	1 birth/year
Litter size	4.2
Age at first reproduction[2]	2 years
Weight of newborn	1.5 kg
Adult weight	50 kg
Life span	7 years
Generation length	2.6 years
Density[3]	1.3 individuals/ha
Intrinsic rate of increase	0.69

[1] López (1987).
[2] Ojasti (1973).
[3] Herrera (1986).

where D is density (100 individuals/km^2) and λ_{max} is the exponent of the intrinsic rate of natural increase ($e^{0.69}$). The maximum production for capybaras was 59.6 individuals/km^2 per annum. Taking into account that, according to Robinson and Redford (1991), as much as 40% of the annual production in a short-lived species can be harvested sustainably, the potential harvest is 23.8 individuals/km^2 per annum, or 834 kg/km^2 per annum (assuming a body mass of 35 kg for harvested animals aged 1.5 years). Kleiman *et al.* (1979) estimated a harvestable biomass of 244 kg/km^2 per annum for capybaras in neotropical savannahs using Ojasti's (1973) data and a cropping rate of 0.20. Assuming an optimum density of 100 individuals/km^2 and a harvest rate of 0.30 (but disregarding productivity), Ojasti (1991) predicts a potential harvest of 1200 kg/km^2 per annum.

Substituting the data given by Robinson and Redford (1986) for forest-dwelling capybaras ($a = 2$ years; $b = 2♀/♀$ per annum; $w = 9$ years; $r_{max} = 0.69$; $D = 9.88$ individuals/km^2; body mass of 31.5 kg), the maximum production is 5.9 individuals/km^2 per annum and the potential harvest would be 74 kg/km^2 per annum. Although much lower than the potential harvest of savannah-living capybaras, this is one of the highest productivities calculated for any neotropical forest mammal (Robinson and Redford, 1991). It is evident that these figures for savannah- and forest-living capybaras seek to describe the potential harvest of the species in only the best of all possible environmental conditions. Despite the differences in the results, they all suggest that the productive potential of the capybara in the Neotropics is high (Table 7.2).

Of South American mammals, the capybara arguably has the greatest potential for sustainable harvest and indeed has been exploited since pre-Colombian times. Its meat is commonly found in the markets of Amazonian settlements, where is sometimes called *vaquinha* (little cow). It is largely consumed during Lent in Venezuela, but in most of its range it is essentially a food of the poorer inhabitants and is not considered especially tasty (Wetterberg *et al.*, 1976). Nevertheless, in the southern states of Brazil capybara meat is regarded as a delicacy and commands a correspondingly high price. The hides yield chamois-quality leather, known internationally as *carpincho* leather. In the southern part of the species' range (Argentina, Uruguay, Paraguay and the south of Brazil) this leather is the main product of hunting. Capybara fat is extensively used, especially for medicinal purposes: it is believed to cure asthma, rheumatism and allergies. In Brazil a proprietary medicine called 'Capivarol' is made from capybara fat extract.

7.3 HUNTING TECHNIQUES

Capybara are hunted throughout their range, despite being protected in most countries. The only large-scale commercial harvest of capybara is in

Table 7.2 Calculations of potential capybara harvest

Group of capybaras	Productivity kg/km²/year	Source
Savannah-living capybaras	834	Robinson and Redford (1991) equation
	244	Kleiman *et al.* (1979)
	1200	Ojasti (1991)
	841	Moreira and Milner-Gulland model
Forest-living capybaras	74	Robinson and Redford (1991) equation

Venezuela. Elsewhere most of the hunting is for subsistence. The hunting techniques used vary from place to place, the phase of the moon and the season. In open savannah (for example, in Venezuela, Colombia or in Brazilian Amazonia), capybaras are herded during the dry season, either by horsemen or on foot, and tire after a chase of approximately 200 m. The herd then bunches together, and a slaughterman clubs selected victims. In Venezuela this task is reserved for the oldest and most experienced cowboy. In water, in contrast, the herds disperse and individuals may be speared or shot from canoes or from the bank.

By day, skilled hunters may benefit from the capybara's anti-predator tactic of standing motionless when hidden in the midst of the vegetation. By night, capybaras are dazzled using a spotlight and then speared or shot, either from a canoe or from horseback.

In parts of Brazil, capybaras are hunted using dogs (Ojasti, 1991), whereas in much of Amazonia a young capybara is used as a lure. One of the hunters carries the young decoy while another walks ahead with a torch and a club, to intercept any capybaras that approach. In the Middle Amazon, whistles (made out of bottle tops) are used to imitate the young's alarm call. Wire traps, pitfalls and firearm traps are also commonly used.

7.4   IMPACT OF HUNTING PRESSURE

The impact of hunting pressure is difficult to estimate in the Neotropics (Ojasti, 1991). Data on capybara hunting and the effects of hunting pressure are either dubious or non-existent. Nevertheless, a preliminary assessment of factors relevant to capybara hunting can be made.

The use of torches, firearms, ingenious traps and other gadgets, together with horses and dogs have made it relatively easy to over-harvest capybaras. In response, capybaras become more nocturnal and increase their vigilance and flight distance (Azcárate, 1981). Furthermore, where hunting pressure is intense, the capybara, a savannah-living animal, is found more frequently in forest (Cordero and

Ojasti, 1981), although savannah may support a higher density of capybaras (Ojasti and Sosa Burgos, 1985).

Capybara hunting is generally selective. The most accessible groups and larger individuals which return a better profit are at greatest risk. Such selection affects the social and age structure of the population. After a commercial harvest in Venezuela, groups almost devoid of adults are often found in areas favourable to hunting, whereas groups in inaccessible refuges retain a larger representation of adults (Herrera and Macdonald, 1987). This tendency for hunting to skew the age structure towards younger capybaras can affect productivity, because younger female capybaras have a relatively low incidence of pregnancy and small litter sizes (Ojasti, 1973; Moreira and Macdonald, 1993). Cordero and Ojasti (1981) found a significant association between the types of habitat and the age distribution of adults, which again may reflect the consequences of selective hunting. The disruption of social organization and reduction of group size caused by the harvest seems likely to affect survivorship; Herrera and Macdonald (1987) found an increase in the survival of young capybaras per female with the increase in group size. Furthermore, hunting techniques based on active pursuit are likely to select pregnant females which tire easily (Moreira, 1995).

Herrera (1992) compared two savannah-living populations of capybaras in Venezuela: one hunted and the other protected. Although Herrera's data do not allow any further conclusions about the hunting effect on the age structure of a capybara population, he found a significant increase in adult body-weight in the protected population following a moratorium on hunting. This is probably because of the selective slaughter of larger individuals. The long-term consequence of such a cropping regime is a decrease in the body size of the hunted population.

Such alterations might be expected to affect capybaras' social organization and breeding biology (Ginsberg, 1991). The results are population densities depressed below carrying capacity over most of their range. Surprisingly, however, our data from Marajó Island (Moreira, 1995) revealed no significant difference between survival curves for hunted and unhunted sub-populations (Log-Rank: $\chi^2_3 < 0.01$, $P = 0.95$; Wilcoxon: $\chi^2_3 = 0.57$, $P = 0.45$; -2Log(LR): $\chi^2_3 = 0.04$, $P = 0.84$). Life expectancy at birth for hunted capybaras was also similar to that for unhunted capybaras (both 1.2 years).

7.5   PRESENT STATUS OF CAPYBARA CONSERVATION

The recent growth of the human population in South America has forced an expansion of the agricultural frontier into previously uninhabited areas. Poorly planned colonization leads to depletion of habitat. Wildlife is often the main source of protein for these settlers. Larger species rapidly become scarce close to new settlements. However, with their potential for high productivity, rodents can play an important and sustainable role in the subsistence diet.

It is difficult to evaluate the general status of capybara populations; although there may have been local extinctions the capybara is not an endangered species. Indeed the advancing conversion of neotropical forest into grassland is creating potential capybara habitat. Since capybaras are less affected by deforestation, have less of an economical value and have greater productivity than several other neotropical mammal species of similar size, they have probably fared comparatively well. However, capybaras have been pushed well below carrying capacity by indiscriminate hunting, be it for their meat or pelts, or due to a perceived role in competing with cattle for food, or in transmitting diseases to cattle, or as pests of sugarcane and rice plantations.

Despite being able to forage on short grasses and despite living in habitats sometimes inaccessible to cattle, capybaras do compete with domestic species for grass (Escobar and González-Jiménez, 1976). This competition is especially strong during the late dry season when the availability of forage is low. Furthermore, they can act as asymptomatic carriers of some pathogens of cattle, horses and pigs (Bello *et al.*, 1984). The presence of brucellosis has been demonstrated in wild individuals. Capybaras that are apparently healthy may be carriers of *Trypanosoma evansi*, possibly participating in the transmission of *mal das cadeiras* to horses. These diseases of capybaras may not constitute a problem to vaccinated domestic stock. Mange and coccidiosis can limit the efficiency of captive production of capybaras.

Today, in spite of official protection in most countries capybaras are hunted all over their range. In French Guyana the capybara is not a protected species. In Guyana, there is not even a wildlife protection law, although capybara hunting is limited to a quota of 10 individuals/exporter/annum. In Suriname capybaras can be hunted throughout the year, free of quotas. Peru allows subsistence hunting only in the Forest Districts of Selva and Ceja de Selva. Capybara hunting is banned in the majority of the Argentinian provinces. There is, though, a hunting season for capybaras in the Formosa Province, from April to July, with a limit of five animals per hunter. In Corrientes Province capybara are hunted for purposes of pest control. Wildlife hunting is banned in Uruguay and Colombia, but the rearing of capybaras in captivity is legal. A few ranches in the Colombian Llanos farm capybara. The 1992 Colombian harvest was 8608 animals, which were largely destined for the Venezuelan market. Recently in Uruguay a programme of research on capybaras was started with the aim of developing a management system for the species. The export of capybara meat to Israel as a substitute for pork is one of their goals.

From 1960 to 1967, when hunting was not banned, Brazil produced officially 1 546 696 capybara hides, but the real figures are certainly much higher (Caça, 1963–1970). From 1967 onwards, the Law Decree 5197 prohibited the utilization, persecution, destruction, hunting and capture of wildlife in Brazil; nevertheless, the commercial captive exploitation of this resource by farmers remained legal. This law was never enforced due to the lack of financial and human resources and the size of the country. Today, even though hunting is prohibited, capybara meat is commonly found in rural markets in small villages throughout Brazilian Amazonia and their numbers are illegally controlled in other areas. The only method of wildlife exploitation permitted in Brazil involves captive rearing. Recently some farms in the southern states of Brazil were finally authorized to rear capybaras in a semi-intensive system in large enclosures, and their slaughter was allowed. Capybara productivity in captivity is much higher than that of free-living capybaras, but the economic feasibility of these farms is doubtful (Ojasti, 1991).

In the Venezuelan Llanos the capybara has been hunted for its meat since the last century. The harvest occurs shortly before Easter, because being a semi-aquatic animal its consumption is allowed during Lent by papal decree. The present harvesting procedure was established in 1988, from minor alterations to the previous system in use since 1968. Permits are given according to the number of capybaras on the property, as counted by a wildlife officer. The limit is 30% of counted individuals and only animals over 35 kg may be harvested. Such a cropping regime is allowed exclusively in the states of Cojedes, Portuguesa, Barinas and Apure. In 1992 only four permits were issued, three in the state of Apure and one in Barinas. The harvest of only 2272 animals was authorized, in stark contrast to the 92 734 of the 1981 harvest. The reduction of cropping quotas imposed by the end of the 1980s, and still in force today, is a consequence of the decrease in population size, probably due to an overestimation of capybara abundance in the previous years (Ojasti, 1991).

We explored the boundaries of the Venezuelan hunting scheme using an age-specific model (developed by J.R. Moreira and E.J. Milner-Gulland) structured to simulate capybara life history. Our findings suggest that the present quota limit may lead the population to extinction within 27 years. The maximum sustainable yield was found at the annual harvest of 17% of the population when hunting was not selective by sex (Moreira, 1995). If hunting is selective towards either sex the population is likely to be over-hunted even at lower hunting quotas (Figure 7.1). Using a 17% hunting quota and the yield estimated by the model, the productivity of capybaras in the seasonally flooded savannahs is estimated at 841 kg/km^2 per annum (Table 7.2). It should be borne in mind that the model was fitted with data from the Brazilian Amazonia and the predictions are limited to that area.

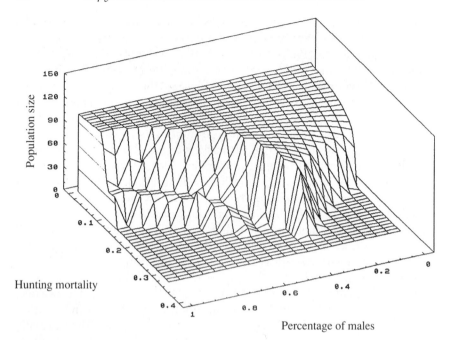

Figure 7.1 Effect of hunting mortality and sex-biased harvesting of only adult capybaras on the population size.

Capybara exploitation in the Venezuelan Llanos rewards the farmers with a net profit greater than that obtained from cattle, as well as with product diversification (Ojasti, 1991). In Brazil, although there are some farms rearing capybaras in captivity in the south and south-east regions, the economic viability of such farming in a semi-intensive system is questionable in the face of stiff competition from pork. The lack of an established game market is perhaps the main hindrance to developing capybara as a farm product. However, seasonally flooded savannahs like those of the Pantanal Matogrossense and along the Amazon valley are suitable for capybara ranching in a similar manner to that practised today in Venezuela.

The harvesting scheme in use in Venezuela over the last 20 years may have overestimated the sustainable crop, but it nonetheless serves to demonstrate the capybara's productive potential and resistance to hunting pressure. The recent reduction in capybara numbers in Venezuela indicates the need for a re-evaluation of the cropping limit, together with finer tuning of the hunting quotas and great care in accurate population estimates (Ojasti, 1991). As more data on capybara reproductive biology and ecology become available, harvest models (such as the one reported here) offer an opportunity for exploring the optimal rewards of capybara harvest. The conservation of this intriguing rodent, the maintenance of its role in a natural ecosystem and the sustenance of local people in impoverished communities can, with careful management, go hand-in-hand.

ACKNOWLEDGEMENTS

This work was supported by a grant to J.R.M. from CNPq (Brazil) and partially funded by International Foundation for Science (Sweden), Wildlife Conservation International (USA) and Fundação de Apoio à Pesquisa Família Alencar (Brazil). The authors wish to thank P. Johnson and E.J. Milner-Gulland for helpful criticism, F. Shickle for editorial suggestions, and all the South American environmental institutions who answered our requests for information. We are grateful for the assistance and hospitality of the Lobato family and especially the help of their staff at Fazenda Eco-búfalos.

REFERENCES

Alho, C.J.R., Campos, M.S.C. and Gonçalves, H.C. (1987) Ecologia de capivara (*Hydrochaeris hydrochaeris*, Rodentia) do Pantanal: I. Habitats, densidades e tamanho de grupo. *Revista Brasileira de Biologia*, **47**, 87–97.
Ayres, J.M. and Ayres, C. (1979) Aspectos da caça no alto rio Aripuanã. *Acta Amazonica*, **9**, 287–298.
Azcárate y B., T. (1981) Sociobiologia y manejo del capibara. *Doñana Acta Vertebrata*, **6–7**, 1–228.
Bello, N.A., Lord, V. and Laserna, R. (1984) Enfermedades infecciosas que afectan el chiguire (*Hidrochaeris hidrochaeris*). *Revista Veterinaria Venezolana*, **278**, 32–46.
Caça; Produção de peles e couros de alguns animais silvestres (1963–1970) *Anuário Estatístico do IBGE* **24**, 48–51; **25**, 58–61; **26**, 80–83; **27**, 90–93; **28**, 76–78; **29**, 118–120; **30**, 128–130; **31**, 102–104.
Cole, L.C. (1954) The population consequences of life history phenomena. *Quarterly Review of Biology*, **29**, 103–137.
Cordero R., G.A. and Ojasti, J. (1981) Comparison of the capybara populations of open and forested habitats. *Journal of Wildlife Management*, **45**, 267–271.
Dourojeanni, M.J. (1985) Over-exploited and under-used animals in the Amazon Region, in *Key Environments; Amazonia*, (eds G.T. Prance and T.E. Lovejoy), Pergamon Press, Oxford, pp. 419–433.
Dubost, G. (1968) Les niches écologiques des forêts tropicales sud-américaines et africaines, sources de convergences remarquables entre Rongeurs et Artiodactyles. *La Terre et la Vie*, **22**, 3–28.
Dubost, G. (1988) Ecology and social life of the red acouchy, *Myoprocta exilis*: comparison with the orange-rumped agouti, *Dasyprocta leporina*. *Journal of Zoology*, **214**, 107–123.
Eisenberg, J.F., O'Connell, M.A. and August, P.V. (1979) Density, productivity, and distribution of mammals in two Venezuelan habitats, in *Vertebrate Ecology in the Northern Neotropics*, (ed. J.F. Eisenberg), Smithsonian Institution Press, Washington, pp. 187–207.
Escobar, A. and González-Jiménez, E. (1976) Estudio de la competencia alimenticia de los herbivoros mayores del llano inundable con especial al chigüire (*Hydrochoerus hydrochaeris*). *Agronomia Tropical*, **26**, 215–227.

Ginsberg, J.R. (1991) The consumptive use of wildlife: harvesting, sociobiology, and implications for population dynamics, in *Wildlife Research for Sustainable Development*, (eds J.G. Grootenhuis, S.G. Njunguna and P.W. Kat), Kenya Agricultural Research Institute, Nairobi, pp. 74–75.

Hennemann, W.W., III (1983) Relationship among body mass, metabolic rate and the intrinsic rate of natural increase in mammals. *Oecologia*, **56**, 104–108.

Herrera, E.A. (1986) The behavioural ecology of the capybara, *Hydrochoerus hydrochaeris*. D.Phil. thesis, University of Oxford.

Herrera, E.A. (1992) The effect of harvesting on the age structure and body size of a capybara population. *Ecotropicos*, **5**, 20–25.

Herrera, E.A. and Macdonald, D.W. (1987) Group stability and the structure of a capybara population. *Symposium of the Zoological Society of London*, **58**, 115–130.

Herrera, E.A. and Macdonald, D.W. (1989) Resource utilization and territoriality in group living capybaras. *Journal of Animal Ecology*, **58**, 667–669.

Herrera, E.A. and Macdonald, D.W. (1993) Aggression, dominance, and mating success among capybara males (*Hydrochaeris hydrochaeris*). *Behavioral Ecology*, **4**, 114–119.

Jorgenson, J.P. (1986) Notes on the ecology and behaviour of capybaras in Northeastern Colombia. *Vida Silvestre Neotropical*, **1**, 31–40.

Kleiman, D.G., Eisenberg, J.F. and Maliniak, E. (1979) Reproductive parameters and productivity of caviomorph rodents, in *Vertebrate Ecology in the Northern Neotropics*, (ed. J.F. Eisenberg), Smithsonian Institution Press, Washington, pp. 173–183.

López B., S. (1982) Determinacion del ciclo estral en chigüires (*Hydrochoerus hydrochaeris*). *Acta Científica Venezolana*, **33**, 497–501.

López B., S. (1987) Consideraciones generales sobre la gestacion del chigüire (*Hydrochoerus hydrochaeris*). *Acta Científica Venezolana*, **38**, 84–89.

Macdonald, D.W. (1981) Dwindling resources and the social behaviour of capybaras (*Hydrochoerus hydrochaeris*) (Mammalia). *Journal of Zoology*, **194**, 371–391.

Mares, M.A. and Ojeda, R.A. (1982) Patterns of diversity and adaptation in South American hystricognath rodents, in *Mammalian Biology in South America*, (eds M.A. Mares and H.H. Genoways), University of Pittsburgh, Linesville, (Univ. Pittsburgh, Pymatuning Laboratory of Ecology, Special Publication Series, 6), pp. 393–432.

Moreira, J.R. (1995) The reproduction, demography and management of capybaras (*Hydrochaeris hydrochaeris*) on Marajó Island – Brazil. DPhil. thesis, University of Oxford.

Moreira, J.R. and Macdonald, D.W. (1993) The population ecology of capybaras (*Hydrochaeris hydrochaeris*) and their management for conservation in Brazilian Amazonia, in *Biodiversity and Environment: Brazilian themes for the future*, (eds S.J. Mayo and D.C. Zappi), Linnean Society of London/Royal Botanic Gardens, Kew, London, pp. 26–27.

Ojasti, J. (1973) *Estudio Biologico del Chigüire o Capibara*, FONAIAP, Caracas.

Ojasti, J. (1991) Human exploitation of capybara, in *Neotropical Wildlife Use and Conservation*, (eds J.G. Robinson and K.H. Redford), University of Chicago Press, Chicago, pp. 236–252.

Ojasti, J. and Sosa Burgos, L.M. (1985) Density regulation in populations of capybaras. *Acta Zoologica Fennica*, **173**, 81–83.

Patton, J.L., Berlin, B. and Berlin, E.A. (1982) Aboriginal perspectives of a mammal community in Amazonian Perú: knowledge and utilization patterns among the Aguaruna Jívaro, in *Mammalian Biology in South America*, (eds M.A. Mares and H.H. Genoways), University of Pittsburgh, Linesville, (Univ.

Pittsburgh, Pymatuning Laboratory of Ecology, Special Publication Series, 6), pp. 111–128.

Rios R., M., Dourojeanni, M.J. and Tovar S., A. (1974) La fauna y su aprovechamiento en Jenaro Herrera (Requena, Perú). *Revista Forestal Peruana*, **5**, 73–92.

Robinson, J.G. and Redford, K.H. (1986) Intrinsic rate of natural increase in neotropical forest mammals: relationship to phylogeny and diet. *Oecologia*, **68**, 516–520.

Robinson, J.G. and Redford, K.H. (1991) Sustainable harvest of neotropical forest mammals, in *Neotropical Wildlife Use and Conservation*, (eds J.G. Robinson and K.H. Redford), University of Chicago Press, Chicago, pp. 415–429.

Schaller, G.B. and Crawshaw, P.G., Jr (1981) Social organization in a capybara population. *Säugetierkundlichen Mitteilungen*, **29**, 3–16.

Smith, N.J.H. (1976) Utilization of game along Brazil's transamazon highway. *Acta Amazonica*, **6**, 455–466.

Taber, A.B. and Macdonald, D.W. (1992) Spatial organization and monogamy in the mara *Dolichotis patagonum*. *Journal of Zoology*, **227**, 417–438.

Webb, S.D. and Marshall, L.G. (1982) Historical biogeography of recent South American land mammals, in *Mammalian Biology in South America*, (eds M.A. Mares and H.H. Genoways), University of Pittsburgh, Linesville, (Univ. Pittsburgh, Pymatuning Laboratory of Ecology, Special Publication Series, 6), pp. 39–52.

Weir, B.J. (1974) Reproductive characteristics of hystricomorph rodents. *Symposium of the Zoological Society of London*, **34**, 265–301.

Wetterberg, G.B., Ferreira, M., Brito, W.L.S. and Araujo, V.C. (1976) Espécies da fauna amazônica potencialmente preferidas para consumo nos restaurantes de Manaus. *Brasil Florestal*, **7**, 59–68.

Woods, C.A. (1982) The history and classification of South American Histricognath rodents: reflections on the far away and long ago, in *Mammalian Biology in South America*, (eds M.A. Mares and H.H. Genoways), University of Pittsburgh, Linesville. (Univ. Pittsburgh, Pymatuning Laboratory of Ecology, Special Publication Series, 6), pp. 337–392.

8

Sustainable use of whales: whaling or whale watching?

Vassili Papastavrou

SYNOPSIS

An examination of the history of the International Whaling Commission (IWC) can offer valuable insights to those managing the exploitation of other wild mammals. To understand the many issues one must look further than the IWC's past inability to manage whaling which resulted in the declaration of an indefinite moratorium on commercial whaling.

More international effort has been put into determining procedures for the sustainable exploitation of whales than for any other group of wild mammals, yet the history of whaling shows that these substantial efforts have largely failed. Previous attempts to set catch quotas have relied on approaches that were later discovered to be flawed. In addition, science apart, it has never been possible to effectively control whaling. Even when the IWC was setting catch limits that were far too high, those rules were being systematically broken. One example, revealed in 1994 by the Russian Federation, concerns thousands of blue, right and humpback whales that were caught (but not reported to the IWC at the time) by the whaling expeditions of the former Soviet Union years after these species had supposedly received IWC protection.

The various policy options before the IWC include the continuation of the moratorium, a resumption of whaling, the declaration of sanctuaries and the management of whale watching. In recognition of previous inadequacies in attempts to exploit whales in a sustainable fashion, the Scientific Committee of the IWC has spent several years putting together a sophisticated and precautionary procedure (known as the Revised Management Procedure, or RMP) for calculating catch quotas. The RMP is a component of the Revised Management Scheme (RMS) which will encompass other matters to ensure that any whaling is effectively regulated and that catch limits are enforced. In contrast to the effort expended on the development of the RMP, little time has been spent on discussion of regulatory matters. The remote Southern Ocean (the feeding grounds for about 80% of the world's remaining whales) is one area where regulation of whaling operations remains an insuperable problem and in May 1994 the IWC declared this area a sanctuary for an indefinite period. Finally, utilization should not imply that the use must be lethal. Whale watching is a rapidly growing form of non-lethal sustainable utilization which occurs in around 40 countries; it generates substantial tourist revenue and requires regulation to limit disturbance.

8.1 THE INTERNATIONAL WHALING COMMISSION: OVER-EXPLOITATION OF WHALES, SUSTAINABILITY AND THE EVENTS LEADING UP TO THE MORATORIUM

The exploitation of whales is an often quoted example of the gross misuse of a natural resource. The whaling industry, largely driven by economic forces, moved from species to species and population to population in an approach that has been compared to mining rather than to sustainable utilization. Whaling was, and remains, a hugely profitable industry, which is one of the reasons why it has so far proved impossible to regulate. In addition, whaling operations are often conducted in remote locations, usually the high seas, where it is extremely difficult to ensure that rules are observed.

In the Antarctic fishery (summarized by Gaskin, 1982, and described in detail by Tonnessen and Johnsen, 1982), the first species to be caught were humpback whales (*Megaptera novaeangliae*). As a result of the development of fast steam-powered whaling vessels and explosive harpoons, the whalers were then able to turn their attention to the blue whale (*Balaenoptera musculus*), by far the largest species. As catches of blue whales declined, there was a progression through the other rorquals in order of decreasing size, ending with the minke whale (*Balaenoptera acutorostrata*) which, at a length of around 9 m, had previously been considered to be too small to be worth catching (Chapter 15). The over-exploitation was substantial: there were once a quarter of a million blue whales in the Antarctic. Despite 30 years of protection there are now very few blue whales in this area: only a few hundred remain in the Southern Ocean (IWC, 1995a).

Ludwig *et al.* (1993) discuss in general the problems of achieving sustainability in practice. They argue convincingly that in most situations resources are over-exploited, often to the point of collapse or even extinction, and list a number of reasons why sustainability is rarely achieved. The exploitation of the resource is usually driven by economic forces (and the more valuable the resource, the more likely the over-exploitation); the scientific understanding of the system is usually incomplete; and large levels of variability (in both the distribution of the resource and the pattern of its exploitation) mask the effects of over-exploitation. The authors then state that although a lack of information often allows continued over-exploitation, many practices continue even when there is abundant hard or circumstantial evidence that they are indeed destructive. The authors further stress that the scientific aspects of any situation should not be discussed in isolation but together with political, management, ethical and other components. These various elements will be examined with respect to the history of whaling and the present situation within the IWC.

Whaling is managed by the International Whaling Commission (IWC) which consists of representatives from approximately 30 active nations. The IWC first met in 1949 when a sufficient number of nations had ratified the 1946 International Convention for the Regulation of Whaling (ICRW). The Convention was far-sighted in that it recognized in its preamble 'the interest of the nations of the world in safeguarding for **future generations** the great natural resources represented by the whale stocks' (emphasis added). The preamble also contained two goals, the first being 'the proper conservation of whale stocks' and the second being 'the orderly development of the whaling industry'. Whaling countries have argued that the two goals should be considered with equal weight. However, Lyster (1985) argues that, logically, the second goal cannot be considered until the first has been achieved.

The IWC is advised by a Scientific Committee which consists of scientists appointed by the governments of countries who are members of the Commission, and invited participants with specific expertise whose expenses are usually covered by the IWC. The Scientific Committee should not be regarded as an independent body of scientists and Butterworth (1992) takes the view that some scientists are under political instructions to argue a particular position.

In 1982 an indefinite moratorium on commercial whaling was adopted by the IWC. The call for a moratorium already had the support of some members of the Scientific Committee in 1979, whose report to the Commission stated that 'some members believed that the degree of scientific uncertainty is so widespread and the problems of the New Management Procedure so completely unresolved that the only way to assure that stocks are not over-exploited is through a moratorium' (IWC, 1980). The following year, some members of the Scientific Committee reiterated the concern that the current management procedure did not take into account the uncertainties inherent in all the then stock assessments and called for measures (including a cessation of whaling if necessary) which would decrease the risk of over-exploitation in the face of such uncertainties (IWC, 1981a).

The moratorium, which came into effect in 1986, has attracted a variety of comments. Those countries that wish to continue whaling as before view it as a protectionist measure. However, to others it is seen as a sound application of the precautionary principle, firmly laying the burden of proof on those who wish to exploit the resource before permitting exploitation.

The present situation is that an indefinite pause in exploitation remains in place because the component parts of the scheme intended to guarantee a sustainable exploitation of this resource are still incomplete. Nevertheless, Norway and Japan continue to catch minke whales. Norway's commercial whaling is technically legal in that Norway filed formal objections to the moratorium decision and to the protection stock status of the north-east Atlantic minke whales. Japan continues whaling under the guise of scientific research, despite the fact that information obtained is not required for management purposes. Butterworth (1992) has suggested that the 'strongest' defence of scientific whaling is that, because the meat is sold commercially, it underwrites the substantial cost of sightings surveys.

More international effort has been put into the management of whaling than that of the exploitation of any other wild mammal species. An examination of the history of whaling, together with an understanding of the present situation, can provide useful lessons for those evaluating the exploitation of other species, but for those who wish to make such comparisons there are two caveats. The first is that whales hold a special place in the hearts of many people, perhaps because of their size, or the complexity of the social relationships of some species, or perhaps because whales have become a focus for the conservation movement. In the words of the late conservationist Sir Peter Scott, 'If we cannot save the largest animals in the world we have little chance of saving the biosphere itself and therefore of saving our own species.'

In addition, on a practical level, there is as yet no method to guarantee humane killing of whales (Kestin, 1995). The working definition of humane killing accepted by the IWC is that 'humane killing of an animal means causing its death without pain, stress or distress perceptible to the animal'. It was further noted that the above definition is the ideal and, in practice, 'any humane killing technique aims first to render an animal insensitive to pain as swiftly as is technically possible' (IWC, 1981b).

The killing procedure requires that the whale is first harpooned and an explosive charge in the harpoon then detonates within the animal. Although instantaneous death occurs in a proportion of cases, secondary killing methods usually have to be employed. These include the use of 9 mm rifle bullets (Norway) or inadequate electric shocks (Japan). Film footage obtained on board a Japanese whaling ship in the 1992/1993 season shows some individual minke whales struggling for a minimum of 8 minutes 21 seconds following electrocution (McLachlan, 1995). Due to breaks in the film footage, the true times to death may be very much longer and other studies, summarized by Kestin (1995), indicate that wounded animals may survive for up to an hour. Many would argue on animal welfare grounds that whaling should be prohibited at least until a way can be found to kill whales humanely.

It is worth noting that in 1980 the IWC agreed to ban the use of the cold (i.e. non-exploding) harpoon on humane grounds. This ban initially applied to all species except minke whales but the following year the ban was extended to include minkes with effect from 1983. Japan and the

Russian Federation maintain formal objections to this ban and are therefore not bound by it.

The IWC has since established a Humane Killing Working Group and has therefore recognized that animal welfare should be considered at some level. However, at present few governments are prepared to continue to push the substance of this argument and prefer to focus their attention on other scientific and regulatory matters.

The second caveat is that whales have a doubly special status in international law not only as highly migratory species but also specifically as cetaceans. Cetaceans are listed in an Annexe to the United Nations Convention on the Law of the Sea (UNCLoS) which came into force in 1994. Article 65 of UNCLoS reads: 'States are required to work through the appropriate international organizations for their [i.e. cetaceans] conservation, management and study.' The relevant sections of the UNCLoS text were repeated in Agenda 21, the consensus document which resulted from the 1992 UN Conference on the Environment and Development (UNCED), held in Rio de Janeiro. Both UNCED and UNCLoS allow the coastal states and appropriate international organizations (specifically, the IWC) to 'prohibit, limit or regulate the exploitation of marine mammals more strictly' than required for their optimum (i.e. sustainable) utilization (Anon, 1992; Birnie, 1993).

Another consequence of the special status of marine mammals is that it is recognized (in UNCLoS, UNCED and the ICRW itself) that all nations are entitled to contribute to management decisions – even countries such as Switzerland that do not have a coastline.

8.2   DEVELOPMENT OF A PROCEDURE FOR CALCULATING CATCH QUOTAS

Over the years, two of the key topics to have been addressed by the IWC Scientific Committee are the determination of a sustainable catch and, linked to this, the determination of current abundance estimates (together with their coefficients of variation) of particular stocks of whales. Prior to the moratorium, methods to ensure that catches were sustainable failed for a variety of reasons, similar to those summarized by Ludwig *et al.* (1993). In 1986, the Scientific Committee of the IWC began work on a sophisticated procedure for calculating catch quotas, known as the Revised Management Procedure (RMP), which was intended to overcome the inadequacies of the previous approach. In 1994, the IWC recognized, through a resolution on the Revised Management Scheme (RMS, discussed in section 8.3), that the work is now complete (IWC, 1995b).

However, before the Scientific Committee could begin the development of an RMP, it was necessary for the Commission to make a political decision on the management objectives. It was clear that the Commission intended that any future exploitation of whales should be sustainable. Although sustainability is a term that is frequently used and promoted as a conservation ideal, there are few clear practical definitions of the term. *Caring for the Earth: A Strategy for Sustainable Living*, the joint strategy of IUCN, WWF and UNEP (IUCN/UNEP/WWF, 1991), contains the following definition of sustainability: 'If an activity is sustainable, for all practical purposes it can continue for ever.' Perhaps this statement is not intended to apply to the utilization of wild species, yet there is no clearer direction elsewhere in the document. Certainly, it is not helpful in clarifying the goals of sustainability. For example, it would be theoretically possible to calculate a sustainable yield from the few hundred blue whales that remain in the Southern Ocean. Such a yield may of course only be one or two whales each year. A more precautionary alternative would be to prohibit any exploitation until this population recovers to a substantial proportion of its pristine numbers (around a quarter of a million).

Early in the RMP development process the Commission defined three not entirely compatible management objectives to guide future work:

1. A stability of catch limits from one year to the next (to allow the orderly development of the whaling industry).
2. An acceptable level of risk to ensure that a stock would not be depleted, at a certain level of probability, below some chosen level (e.g. a fraction of its carrying capacity), so that the risk of extinction of the stock is not seriously increased by exploitation. (After 100 years of application of the RMP, populations should be no less than 72% of their pristine numbers.)
3. To obtain the highest continuing yield from the stock (later agreed to be the largest cumulative catch over 100 years of future whaling).

Following much discussion of the trade-offs between the three objectives, it was generally agreed by the Commission that the second objective was of the highest priority (IWC, 1990).

An element of precaution is directly built into the second aim. In addition, the third aim is indirectly precautionary, because to obtain the highest continuing yield in the future it is necessary to allow depleted populations to recover substantially under complete protection before any exploitation begins. (Specifically, the RMP is designed to ensure zero catches from any population that has been reduced to approximately half its pristine numbers. In 1986, for this reason, the IWC declared the north-east Atlantic stock of minke whales to be a protection stock.) The yield is specified as continuing because otherwise the highest yield over any period would be achieved by exterminating the population. Thus, no catches of Southern hemisphere blue whales would be allowed until their numbers were once more in hundreds of thousands.

Another element of precaution is that the size of the catches is determined by the frequency of surveys and the quality of data that is submitted. Whaling would only be permitted if it were possible to show with sufficient certainty that it would be safe to conduct some level of exploitation, and the fewer data that are available, the lower the catch becomes. The burden of proof is firmly on the shoulders of those who wish to exploit whales and there is a strong incentive to conduct regular surveys. The RMP utilizes data which are practicable to obtain (current abundance estimates, together with their coefficients of variation and historic catch data) and makes few assumptions.

Finally, the RMP has a built-in element of negative feedback. The requirement to conduct abundance estimates on a regular basis allows the RMP to respond to the effect of setting a particular catch limit. This approach is becoming known as adaptive management. For it to work, it is recognized that there needs to be international involvement in the collection and analysis of data that is put into the RMP. If either the abundance estimates or the historic catch data are incorrect, then clearly any catch quotas would also be wrong.

One recent example that is worth considering concerns the abundance estimate for minke whales in the north-east Atlantic. Papastavrou (1993) briefly summarized the various abundance estimates calculated by Norwegian scientists over the years, including the 1992 Norwegian estimate of approximately 87 000 whales in this area. Norway claims that their catch quota of about 300 whales (taken under formal objection to the moratorium) is calculated from this abundance estimate using the RMP. In 1992, the Scientific Committee accepted the Norwegian estimate as the best available estimate but without checking any of the calculations (IWC, 1993), in part because they did not have the data.

However, in 1994 (by which time the data had been made available), the Scientific Committee of the IWC discussed the Norwegian abundance estimate in some detail. In particular, the Scientific Committee considered alternative approaches to calculate $g(0)$, which is the probability of seeing a whale that is actually present on the track-line of the survey vessel. This can vary between 0 and 1 and dramatically affects the corrected estimate for any population of whales. The Norwegian estimate of 87 000 is calculated using a $g(0)$ value of 0.36. If, as seems likely, the value is higher, the population estimate is correspondingly reduced. A value of 0.59, which was the result of alternative calculations presented to the Scientific Committee, would result in an estimate for the northeast Atlantic stock of around 53 000 whales. If the latter figure is used to calculate a catch under the RMP, that catch would be negligible. The differences between the two values of $g(0)$ remain unresolved at the time of writing and as a result there is no longer an agreed estimate for the north-east Atlantic population of minke whales (IWC, 1995c).

In 1995, the Scientific Committee agreed that the 1992 estimate was not valid and that, despite a great deal of progress since the previous year, it was not in a position to present a revised estimate (IWC, 1995d). In addition, there was much discussion of an appropriate mechanism which would ensure that these mistakes would not be repeated in the future.

Technicalities aside, this example illustrates the importance of international involvement both in surveys and in the analysis of data that results from those surveys. This is already the situation with respect to the sightings cruises in the Southern Ocean. These cruises have been conducted since the late 1970s under IWC auspices and have Scientific Committee involvement in their design, execution and analysis of the data. If future estimates are to be relied upon, analysis of such data should be conducted independently, ideally by two separate teams of scientists in different countries, who could then compare results.

8.3   DEVELOPING A SCHEME TO MONITOR AND REGULATE WHALING OPERATIONS

Beyond the scientific issues, it is clear that many other management problems remain unresolved. An appropriate scheme must be found for monitoring and regulating whaling operations. Such a scheme must effectively control the entire process from the moment the whale is caught to the point where the meat is sold to the consumer; in other words, from the harpoon to the chopstick. This scheme is known as the Revised Management Scheme (RMS) and the above matters are listed in the 1994 resolution on the RMS. In addition, it has been agreed that commercial whaling cannot re-commence until all elements of the RMS are complete (IWC, 1995b).

Many examples have come to light of undeclared catches or the taking of protected species. Yablokov (1994) and Zemsky *et al.* (1995) revealed that data submitted by the former Soviet Union to the International Whaling Commission were fictitious. In the 1960s and the 1970s the Soviet whalers violated every rule in the book whilst publishing fabricated figures to disguise the fact. The Soviets took protected species such as blue, humpback and right whales. In one year alone, one Soviet factory ship processed 717 protected right whales, the most endangered of the great whales, from a population that now numbers only 2500. The scale of the Soviet deception was staggering and involved the KGB and Soviet officials at the highest levels. Approximately 90 000 more whales were killed than were reported to the IWC. Soviet whaling ships were built with steam pipes to the deck so that carcasses of prohibited species could be hidden in a cloud of steam if other vessels or planes approached. Yablokov concluded by stating that problems are also known to exist with the whaling records from other countries.

The falsification of data and lack of enforcement continues to the present. In October 1993, an illegal consignment of 3.5 tonnes of minke whale meat (labelled as prawn) was uncovered at Oslo airport by an alert airline handler (Mulvaney, 1993). In 1994, *Dagbladet*, a Norwegian newspaper, reported the case of the sleeping inspector (Anon, 1994). A Norwegian whaling vessel which illegally caught an extra whale on 19 July 1994 was reported to the police by another boat. The whaling vessel had a Government inspector on board who claimed that he had seen nothing and had slept through the chasing, harpooning and flensing of the whale.

Baker and Palumbi (1994) used molecular genetic analyses to determine the species and origin of whale meat on sale in retail markets in Japan. Their results showed that meat from protected species such as humpback whale was on sale in 1993. In addition, fin whale meat and illegally imported minke whale meat from the North Atlantic were also identified. In conclusion, the authors of that paper question the assumption (made by proponents of whaling) that only abundant species would be caught should large-scale commercial whaling be allowed to resume. Any legal trade would provide a cover for illegal trade. Abundant species such as the minke whale could subsidize the final extermination of protected species such as blue, right and humpback whales.

The 1994 resolution on the RMS lists the steps that remain to be completed to develop a management scheme intended to prevent such cheating and falsification of data occurring in the future (IWC, 1995b). In addition, it is essential that there is no provision for the whalers simply to side-step IWC management decisions by filing formal objections to RMP-based catch limits, and then awarding themselves higher catches. Also, a mechanism must be found to ensure that special permit catches (i.e. whales caught for 'scientific' research) are not used simply to top up catches that are calculated under the RMP. It is clear that IWC management decisions must be rigorously applied by the Commission. Prior to the moratorium, political compromises were on occasion used to influence catch quotas: this must be prevented from recurring in the future. Although many of these points have been addressed in the RMS resolution, the scheme remains under discussion and it is essential that these important provisions are retained in the final version.

8.4   SANCTUARIES

There are some oceans which are so remote that it would never be possible to regulate whaling operations. The waters surrounding Antarctica, which are the feeding grounds for approximately 80% of the world's remaining whales, are so remote that enforcement of whaling regulations remains an insuperable problem. However, the IWC is explicitly author-ized to designate sanctuary areas where whaling is prohibited. A vast area of the Southern Ocean, known simply as The Sanctuary, pre-dated the establishment of the IWC and was in existence until 1955. In 1979 the IWC declared the Indian Ocean (south to latitude 55°S) a sanctuary for whales. Initially this declaration was temporary but in 1992 the sanctu-ary was made indefinite with a review in 10 years.

In recognition both of the impossibility of regulating whaling opera-tions in the Southern Ocean and of the special status of the waters sur-rounding Antarctica, the IWC declared this area a sanctuary in May 1994. The northern boundary of the Southern Ocean Sanctuary is set at latitude 40°S except around the tip of South America, where it dips further south. The Southern Ocean has been declared a sanctuary for an indefinite period with a review in 10 years.

8.5 WHALE WATCHING

Article V of the International Convention for the Regulation of Whaling empowers the Commission to 'take into consideration the interests of the consumers of whale products and the whaling industry' in providing for the 'conservation, development and optimum utilization of the whale resources.' A number of IWC member nations have realized that live whales also have a commercial value. Whale watching has been recog-nized by the IWC (in resolutions adopted by consensus in 1993 and 1994) as a form of non-lethal, sustainable utilization. Unlike whaling, whale watching brings many conservation benefits and, if conducted well, it can have a substantial educational value. Quite apart from the interest that whales and dolphins generate in their own right, they can also pro-vide a focus to promote the understanding of more general conservation issues, such as pollution or overfishing.

In many parts of the world, there is a symbiotic relationship between commercial operators and scientific researchers. Whale-watching opera-tions often invite researchers aboard to study whales. The researchers conduct benign behavioural research, in particular using photographs to identify individual animals.

Whale-watching operations are known to occur in around 40 countries (Hoyt, 1992, 1994). Hoyt estimated that the total global revenue from whale-watching activities was around £44 million. Whale watching is growing rapidly with around 4 million people watching whales each year in countries that include Iceland, Norway and Japan. Appropriate whale-watching regulations are being formulated in many countries to ensure that this activity does not unduly disturb the animals whilst still permitting the industry to continue. A scientific workshop was held in April 1995 as a part of the process to guide the formulation of sensible science-based rules for managing whale watching (IFAW *et al.*, 1995) and the report was discussed in detail at the 1995 IWC Scientific Committee and Commission meetings.

8.6 CONCLUSIONS

The IWC has recognized (as a result of the failure of previous attempts) that any scheme to manage whaling must be precautionary and must have incentives to ensure that those who wish to exploit the resource obtain the data necessary for management. There must be adequate safeguards to ensure international participation in the collection and analysis of data and to prevent the submission of falsified data. In addition, there must be foolproof mechanisms to monitor and enforce any catch limits that are set. The IWC has agreed that until this whole package is complete the only prudent approach is a continuation of the moratorium on commercial whaling. It has accepted that there are also situations where it may never be possible to regulate whaling and that in these areas the most appropriate course of action is the creation of sanctuaries. Finally, the IWC has recognized the economic value of non-lethal sustainable utilization, such as whale watching.

There is an overwhelming case for the same principles to be embodied in regimes to regulate the exploitation of other wild species, but elsewhere no advocates of lethal utilization have addressed these very serious problems in such detail.

ACKNOWLEDGEMENTS

The author would like to thank Leslie Busby and Kees Lankester for commenting on an earlier draft of the manuscript. The paper also benefited from the comments of an anonymous reviewer and careful editing by Victoria Taylor.

REFERENCES

Anon (1992) *Agenda 21: The United Nations Conference on Environment and Development: Rio Declaration, Rio de Janeiro.* HMSO, London.
Anon (1994) Article in *Dagbladet*, 1 September 1994.
Baker, C.S. and Palumbi, S.R. (1994) Which whales are hunted? A molecular genetic approach to monitoring whaling. *Science*, **265**, 1538–1539.
Birnie, P. (1993) UNCED and marine mammals. *Marine Policy*, **17**, 501–514.
Butterworth, D. (1992) Science and sentimentality. *Nature*, **357**, 532–534.
Gaskin, D.E. (1982) *The Ecology of Whales and Dolphins*, Heinemann Educational Books Ltd, London.
Hoyt, E. (1992) Whale watching around the world: a report on the value, extent and prospects. *International Whale Bulletin No. 7*, Whale and Dolphin Conservation Society, Bath, UK.

Hoyt, E. (1994) Whale watching worldwide: an overview of the industry and the implications for science and conservation. *Proceedings of the European Cetacean Society*, **8**, 24–29.

IFAW, Tethys Research Institute and Europe Conservation (1995) *Report of the Workshop on the Scientific Aspects of Managing Whale Watching*, Montecastello di Vibio, Italy.

IUCN/UNEP/WWF (1991) *Caring for the Earth: A Strategy for Sustainable Living*, Gland, Switzerland.

IWC (International Whaling Commission) (1980) Report of the Scientific Committee. *Rep. int. Whal. Commn*, **30**, 46.

IWC (1981a) Report of the Scientific Committee. *Rep. int. Whal. Commn*, **31**, 57.

IWC (1981b) Report of the Workshop on Humane Killing Techniques for Whales. Cambridge 10–14 November 1980. Paper IWC/33/15 presented to the Thirty-third Meeting of the Commission (unpublished).

IWC (1990) Chairman's Report of the Forty-first Annual Meeting. *Rep. int. Whal. Commn*, **40**, 18.

IWC (1993) Report of the Scientific Committee. *Rep. int. Whal. Commn*, **43**, 65.

IWC (1995a) Chairman's Report of the Forty-sixth Annual Meeting. *Rep. int. Whal. Commn*, **45**, 64.

IWC (1995b) Chairman's Report of the Forty-sixth Annual Meeting, Appendix 5. IWC Resolution 1994–5. Resolution on the Revised Management Scheme. *Rep. int. Whal. Commn*, **45**, 43–44.

IWC (1995c) Report of the Scientific Committee. *Rep. int. Whal. Commn*, **45**, 23–24.

IWC (1995d) Report of the Scientific Committee. *Rep. int. Whal. Commn*, **47**(4), 18.

Kestin, S.C. (1995) Welfare aspects of the commercial slaughter of whales. *Animal Welfare*, **4**(1), 11–27.

Ludwig, D., Hilborn, R. and Walters, C. (1993) Uncertainty, resource exploitation and conservation: lessons from history. *Science*, **260**, 17 and 36.

Lyster, S. (1985) *International Wildlife Law*, Grotius Publications, Cambridge, UK.

McLachlan, H. (1995) The use of electricity to kill minke whales: humane considerations. *Animal Welfare*, **4**(2), 125–129.

Mulvaney, K. (1993) Norway caught flogging minke – baggage man finds world's largest shrimp. *BBC Wildlife*, **11**(12), 62.

Papastavrou, V. (1993) Harvesting of whales. *Nature*, **361**, 391.

Tonnessen, J.N. and Johnsen, A.O. (1982) *The History of Modern Whaling*, Hurst and Co., London.

Yablokov, A. (1994) Validity of whaling data. *Nature*, **367**, 108.

Zemsky, V.A., Berzin, A.A., Mikhaliev, Y.A. and Tormosov, D.D. (1995) Report of the Scientific Committee, Annex E. Report of the Sub-Committee on Southern Hemisphere Baleen Whale. Appendix 3. Soviet Antarctic pelagic whaling after WWII: Review of actual catch data. *Rep. int. Whal. Commn*, **45**, 131–135.

Part Three

Hunting and Its Impact on Wildlife

9

The impact of game meat hunting on target and non-target species in the Serengeti

Heribert Hofer, Kenneth L.I. Campbell, Marion L. East and Sally A. Huish

SYNOPSIS

In the Serengeti National Park (SNP), illegal game meat hunting is largely carried out using snares in the south-western, western and north-western areas. Game meat hunting provides cash income and protein to communities outside the SNP. The economic benefits of game meat hunting have drawn people to villages close to the park boundary, causing a rise in human population density well above the regional average. Game meat hunting has already drastically reduced populations of Cape buffalo and must in the long term be considered unsustainable for a number of other herbivore species. In this chapter an estimate of the current wildlife offtake from the National Park is made and the impact of unselective hunting methods on carnivore species, the most common non-target species, is considered. The analysis demonstrates that game meat hunting poses a threat to both target and non-target species of the Serengeti wildlife community. Optimality models, commonly used in behavioural ecology and economics, are introduced to assess a hunter's profit in relation to hunting effort (costs) and to ask whether unchecked illegal hunting is likely to be sustainable in the long term. A review of studies on African systems demonstrates that whenever costs are reduced, the impact on wildlife due to illegal hunting is dramatically increased and reaches unsustainable levels. Proposals to limit wildlife offtake to sustainable levels, including limited legalization of game meat hunting in areas adjacent to SNP and the development of alternative sources of income and protein for local communities, are considered. The evaluation of these proposals suggests that the situation in the Serengeti does not meet the pre-conditions and assumptions of programmes developed elsewhere for maximizing economic returns from wildlife utilization as an incentive to preserve wildlife; hence such programmes are unlikely to be successful here. This is because the Serengeti is a wildlife system dominated by migratory herbivores, exacerbating the problem of assigning unambiguous ownership of wildlife outside the protected area to a given local community – a pre-condition for any successful privatization or commercialization scheme. Also, if future community conservation services are focused only on those communities that currently benefit most from illegal exploitation, i.e. communities adjacent to the protected areas, then such programmes are likely to reinforce a vicious cycle. They are likely to attract more people to villages close to the protected area and ultimately put greater demands on the protected area, just as currently people are attracted to these villages because of enhanced opportunities for illegal hunting. The analysis suggests that in ecosystems dominated by migratory herbivores and with low levels of law enforcement a large investment is required in both law enforcement and rural development of local communities, that the success of the latter may be linked to investment in the former, and that without both of these the long-term conservation of Serengeti wildlife populations is unlikely to be ensured.

Backs to the wind, plumes of dust rising from millions of hooves, the herds move on. And already … the newly-made arrows with their vulture feather flights are drying, stuck in the thatch of the village huts. In the north the *Acokanthera* poison is being prepared, the deadly tar being rendered down to arm the arrow-heads; and, most effective of all, rolls of steel wire are being expertly fashioned into wicked snares that will hold fast their victims, biting to the bone around necks and legs. The time of the wildebeest is near and the people of the plains are preparing to receive the harvest, as they have done for centuries.

My Serengeti Years, Myles Turner

9.1 INTRODUCTION

Utilization of wildlife by people has probably had a considerable impact on biodiversity over a much longer time frame than previously thought (Diamond, 1989). Concern over the detrimental impact of people on wildlife has led to the historically recent idea (McNaughton, 1989a) of minimizing or removing human activity from areas designed to conserve wildlife and biodiversity, an idea also enshrined in the UNESCO Biosphere Reserve concept (Batisse, 1986; Ishwaran, 1992). Such protected areas are increasingly important in preserving natural ecosystems and contributing to human welfare (McNeely, 1989).

The creation and management of protected areas is costly and has frequently promoted hostile attitudes among local communities. Local communities are often displaced when protected areas are established and they may suffer loss of life, property or crops when wildlife transgresses the protected area boundary, a problem that becomes more severe as human populations increase (Newmark *et al.*, 1994). Perhaps the most common cause of conflict between local communities and managers of protected areas is the fact that local communities are partially or totally prohibited from exploiting natural resources within protected areas. The traditional approach to conserving wildlife has been questioned in recent years, because the minimization of this conflict is now widely seen as critical to the functioning of many protected areas (Myers,

1979; Anderson and Grove, 1987). A frequently suggested alternative expects that wildlife has a long-term future only if it is able to pay its way through game viewing, sport hunting and exploitation for commodities such as meat and hides (Murphree, 1993; Martin, 1994), a process called commercialization of wildlife (e.g. Rasker *et al.*, 1992). Whichever view is held, minimization of conflict between managers and local people can only be achieved when the types, order of magnitude and mechanisms of direct benefits obtained legally or illegally by communities from wildlife in protected areas is clearly understood (Myers, 1972; Robinson and Redford, 1991a).

In this context, Robinson and Redford (1991b) defined the issues by asking three questions. What is the importance of wildlife to people? What impact does the use of wildlife by people have on populations or the biological community? Is the present pattern of human use sustainable over the long term or could it be made so?

This chapter reviews the extent and impact of illegal game meat hunting on wildlife populations in the Serengeti, an ecosystem dominated by migratory herbivores. Based on the results of studies in the Serengeti and elsewhere in Africa, it asks whether unchecked illegal hunting is likely to be sustainable in the long term. Prior to the establishment of the Serengeti National Park (SNP) and adjacent game reserves, game meat hunting was an important component of the lives of many local communities (Turner, 1987). Levels of utilization are likely to increase when its benefits are sufficient to encourage people to migrate closer to protected areas. This is the case in the Serengeti ecosystem, where a large human population surrounds and partially inhabits protected areas, and human population growth due to high fecundity and migration has increased demands on the natural resources in the ecosystem (Makacha *et al.*, 1982; Campbell and Hofer, 1995). It is concluded that unchecked illegal offtake of wildlife in the Serengeti will become unsustainable in the long run.

The chapter reviews proposals to limit wildlife offtake to sustainable levels, including limited legalization of game meat hunting, and the development of alternative sources of income and protein for local communities. For instance, in an effort to minimize conflict between protected area management and local communities, a scheme to legalize game meat hunting by villages outside the SNP has been initiated (SRCS, 1991). The conclusion is that the situation in the Serengeti does not meet the pre-conditions and assumptions of programmes developed elsewhere for maximizing economic returns from wildlife utilization as an incentive to preserve wildlife; hence such programmes are unlikely to be successful. Instead, it is emphasized that any serious attempt to solve conflicts between management and local communities around the Serengeti needs to take two aspects into account:

1. The Serengeti is a wildlife system dominated by migratory herbivores, exacerbating the problem of assigning unambiguous ownership of wildlife to a given local community, a pre-condition for any successful privatization or commercialization scheme (Rasker *et al.*, 1992).
2. If future community conservation services are focused only on those communities that currently benefit most from illegal exploitation, i.e. communities adjacent to the protected areas, then such programmes are likely to reinforce a vicious cycle: they are likely to attract even more people to villages close to the protected area and ultimately put greater demands on the protected areas, just as currently people are attracted to these villages because of enhanced opportunities for illegal hunting and wider availability of other natural resources.

9.2   STUDY AREA

The Protected Area (PA) referred to below includes the Serengeti National Park (SNP), Maswa Game Reserve (GR), Ikorongo and Grumeti GRs, and the Ngorongoro Conservation Area (NCA) to the east. No settlement, hunting or cultivation is permitted in the SNP; licensed hunting but no settlement or cultivation occur in GRs, while the NCA is inhabited by Masai pastoralists. Cultural differences between the Masai in the east and the agricultural and agro-pastoralist people in the west result in major differences in human demands on wildlife within the PA, with illegal game meat hunting being chiefly conducted by hunters from the west (Turner, 1987).

There are more than 30 species of ungulates in the Serengeti ecosystem but it is dominated by migratory species (wildebeest, *Connochaetes taurinus*; zebra, *Equus burchelli*; Thomson's gazelle, *Gazella thomsonii*; and eland, *Tragelaphus oryx*). During their annual migration, the large herds of wildebeest, zebra and Thomson's gazelle move twice along a rainfall gradient. At the start of the rains in December, the herds leave their dry season woodland refuges in the north and west of the ecosystem and move to the highly nutritious short-grass plains in the south-east where they give birth. The herds leave the plains at the end of the rains in May and move to the dry season areas where they remain until November–December, approaching and often transgressing the PA boundary. More details on the migration can be found in Sinclair and Norton-Griffiths (1979).

The eight species that constitute the majority of resident mammalian herbivore biomass (Campbell and Borner, 1995) are hereafter referred to as total resident wildlife populations: African buffalo (*Syncerus caffer*), giraffe (*Giraffa camelopardalis*), Grant's gazelle (*Gazella granti*), impala (*Aepyceros melampus*), kongoni (*Alcelaphus buselaphus*), topi (*Damaliscus korrigum*), warthog (*Phacochoerus aethiopicus*) and waterbuck (*Kobus ellip-*

siprymnus). Resident mammalian herbivores were concentrated in the hilly regions in the south-west, west, and north of the SNP (Figure 9.1), areas that receive a rainfall of > 800 mm and are covered by savannah or woodland (Campbell and Hofer, 1995). The flat, drier Serengeti Plains in the south-east of SNP have only low densities of resident wildlife. Total mammalian herbivore population size exceeds 2 million animals (Table 9.1).

Both migratory and resident mammalian herbivores are a key force in the ecosystem, influencing vegetation dynamics (McNaughton, 1988,

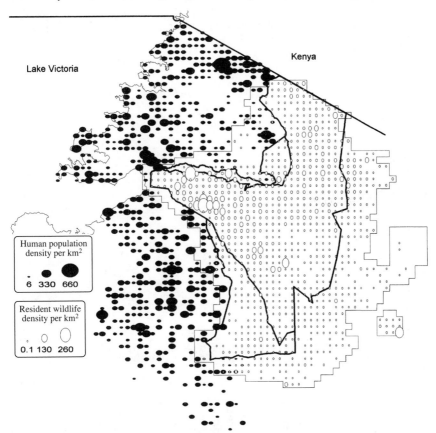

Figure 9.1 Population densities summarized by 5 km grid square. Number of people in villages in 1988, in seven adjacent districts west of the Serengeti, outside Protected Area boundaries. Mean distribution of major resident herbivore populations from 1988, 1989 and 1991 aerial survey data. The thick line in the north separates Kenya from Tanzania. Other thick lines delineate the protected areas: the Serengeti National Park (SNP, centre), Maswa Game Reserve (southwest of SNP), Grumeti Game Reserve (thin strip north-west of SNP), Ikorongo Game Reserve (north-west of SNP). The thin line outlines the censusing area for the aerial surveys of herbivore populations.

Table 9.1 Wildlife hunted inside Serengeti National Park (SNP) in 1992–1993 and total population sizes in SNP plus adjacent game reserves (Protected Area, PA) in the Tanzanian portion of the Serengeti ecosystem in 1991

Species	Status	Population size in PA 1991	Animals in hunting camps 1992–1993	Estimated total offtake by hunters	Offtake as % of estimated population size	Body mass (kg)[i]	% usable meat	Offtake in meat (tonnes)
Wildebeest	migratory	1 278 603[a,n]	932	87 476	6.8	123	60[j]	6 455.73
Zebra	migratory	146 867[b]	202	18 959	12.9	200	55[k]	2 085.49
Eland	migratory	9 416[b]	18	1 689	17.9	340	65[k]	373.27
Thomson's gazelle	migratory	325 769[a]	38	3 567	1.1	15	64[j]	34.24
Grant's gazelle	resident	25 483[b]	11	1 032	4.0	40	64[j]	26.42
Cape buffalo	resident	40 735[b,c]	27	2 534	6.2	450	60[k]	684.18
Giraffe	resident	7 853[b]	16	1 502	19.1	750	55[l]	619.58
Impala	resident	79 098[b]	123	11 545	14.6	40	65[k]	300.17
Kongoni	resident	11 716[b]	11	1 032	8.8	125	60[m]	77.40
Topi	resident	95 037[b]	175	16 425	17.3	100	60[m]	985.50
Warthog	resident	7 151[b]	71	6 664	–	45	65[k]	134.95
Waterbuck	resident	2 466[b]	21	1 971	–	160	55[k]	73.45
Bohor reedbuck[f]	resident	–	24	2 253	–			–
Spotted hyaena	commuter	5 214[d]	4	375	8.0[e]	–	–	–
Lion	resident	–	12	–	–	–	–	–
Cheetah	resident	–	2	–	–	–	–	–
Silver-backed jackal[g]	resident	–	4	–	–	–	–	–
Porcupine	resident	–	1	–	–	–	–	–
Ostrich[h]	resident	4 317[a,n]	11	–	–	–	–	–

[a] Campbell and Borner (1995); [b] Campbell and Hofer (1995); [c] counts from 1992; [d] Hofer and East (1995); [e] Hofer et al. (1993); [f] *Redunca redunca*; [g] *Canis mesomelas*; [h] *Struthio camelus*; [i] Sinclair and Norton-Griffiths (1979); [j] Blumenschine and Caro (1986); [k] Marks (1973); [l] conservative guess based on lowest estimate; [m] assumed to be identical to wildebeest; [n] merged estimates from the 1989 and 1991 censuses.

1989b, 1990) and carnivore populations (Hofer and East, 1995). Migratory and resident mammalian herbivores are the target for game meat hunting (Table 9.1) and the terms 'wildlife' and 'offtake' below usually refer to them.

9.3 LOCAL COMMUNITIES

In seven Districts west of SNP the human population has grown continuously since 1957, reaching a total of 1 777 620 in 1988 (Figure 9.1; data sources in Campbell and Hofer, 1995, updated). Following Campbell and Hofer (1995), a 'source area' for game meat hunters is defined here as a belt 45 km wide around the western edge of the PA. In 1988 this source area contained 454 villages with a total population of 1 161 749 living in households with an average of seven persons. In the 5 km belt adjacent to the PA boundary alone there were 122 023 people. In some areas, settlements have been established on the boundary of the PA.

The average annual rate of population increase in villages close to the PA boundary (< 10 km) between the two national censuses of 1978 and 1988 was substantially higher than the national average of 2.9%, indicating that migration contributed to this increase (Figure 9.2). There was a significantly lower average annual population increase (2.2%) in areas 10–25 km from the boundary, compared with those that were close (< 10 km, 3.5%) or further away (> 25 km, 3.0%; Kruskal-Wallis analysis of variance, $H = 6.24$, $df = 2$, $P = 0.04$), suggesting that the high growth of villages close to the PA was due in part to migration from villages located at intermediate distances of 10–25 km.

9.4 HUNTERS

Although game meat hunting was the most common illegal activity (79%) inside SNP in 1992–1993, local communities extracted other natural resources through firewood collection, livestock grazing, tree cutting for building poles, cultivation and other activities (Table 9.2). Standardized questionnaires (Campbell and Hofer, 1995) completed by SNP rangers between February 1992 and December 1993 indicated that the 452 people arrested for hunting activities belonged to hunting parties consisting of 705 people. More than 75% of arrested hunting personnel (hunters and porters) originated from villages within 15 km of the PA boundary (Hofer *et al.*, submitted). This information was used to derive an estimate of the total number of hunting personnel operating inside SNP. On the basis of the 1988 National Census data this produced an estimate of 17 856 hunting personnel.

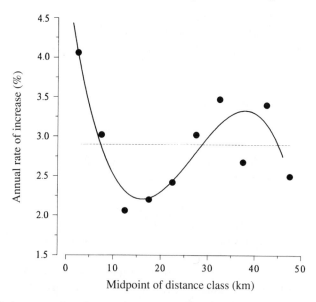

Figure 9.2 Average annual rate of increase (1978–1988) of village populations in 5 km distance classes from the boundary of the Protected Area. --- National and regional average. The solid line is the 3rd order polynomial fit ($y = 4.95 - 0.397x + 0.018x^2 - 0.00022x^3$, $r^2 = 0.80$).

9.5   HUNTING PATTERNS AND WILDLIFE OFFTAKE

Hunting is primarily done by snares; other infrequently used methods and weapons include pit traps, poisoned arrows and firearms (Turner, 1987). The most common type of snare is wire (telephone wire, mining wire or wire extracted from the treads of burnt tyres). Hunters establish camps as a base for their hunting operations. Snares are set along game tracks, around watering holes and along rivers, or snare lines are created by laying fences made from thorn bushes (Turner, 1987). At intervals along these thorn fences gaps are left in which snares are set. Resident and migratory herbivores either wander into snares, or are driven towards the fences, attempt to move through the openings and are caught by the snares, usually around the neck. Game meat is sun dried at the camp; porters are employed to assist in transporting the dried meat out of the PA.

Wildlife located in hunting camps by law enforcement patrols in 1992–1993 are listed in Table 9.1. Other species known to have been killed by hunters since at least 1986 but not found in camps between 1992 and 1993 include elephant (*Loxodonta africana*), black rhinoceros (*Diceros bicornis*), hippopotamus (*Hippopotamus amphibius*), bushbuck (*Tragelaphus scriptus*), dikdik (*Madoqua kirkii*), hare (*Lepus capensis*), aardvark

Table 9.2 Recorded activities, inside the Protected Area, of people arrested by law enforcement patrols of the Serengeti National Park Authority between February 1992 and December 1993

Activity	Frequency	Percentage
Hunting	403	70.6
Carrying meat out (porter)	24	4.2
Hunting and carrying meat out	22	3.9
Buying meat rather than hunting	1	0.2
Hunting and tree cutting	2	0.4
Tree cutting (building poles)	15	2.6
Firewood collecting	37	6.5
Grass cutting (for thatching)	2	0.4
Livestock grazing	33	5.8
Honey gathering	2	0.4
Cultivation inside Park	9	1.6
Stealing livestock	13	2.3
Fishing	1	0.2
Digging for gold	2	0.4
Banditry (attacking/robbing people)	5	0.9
Total	571	

(*Orycteropus afer*), aardwolf (*Proteles cristatus*), honey badger (*Mellivora capensis*), crocodile (*Crocodylus niloticus*) and monitor lizard (*Varanus niloticus*).

The hunting parties from which people were arrested had killed a total of 1703 wild animals (Table 9.1), i.e. 2.42 carcasses per hunting personnel. Prior to their arrest, these hunting parties had spent on average 3.5 ± 0.2 ($n = 417$) days in the Park. Those arrested admitted to 3.7 ± 0.3 (range 1–36, $n = 233$) hunting trips per year.

9.5.1 Wildlife offtake

The number of hunting trips per year multiplied by the number of wildlife killed per trip gave a total of 8.95 kills per hunting personnel per year. Of these, 2.52 were resident and 6.25 migratory mammalian herbivores, and 0.18 represented other species. Total wildlife offtake by all villages in the source area can be estimated as the number of hunting personnel multiplied by the average annual offtake per hunting personnel. This gave an estimated total annual offtake of 159 811 wildlife, including 44 958 resident and 111 691 migratory mammalian herbivores, equivalent to a minimum of 11 950 tons of meat (Table 9.1).

Given our estimates of the number of hunters per village and the total number of kills per hunter, it is possible to calculate how much wildlife is killed by each village. The results highlight that the areas where peaks of hunting pressure originate are close to the PA boundary, but also that the pressure is distributed unevenly (Figure 9.3).

Estimated annual offtake as a percentage of population size varied substantially between species (Table 9.1). The data also produced an off-take estimate for the non-target species with the best population data, the spotted hyaena (*Crocuta crocuta*), of 375 animals per year. We compared this estimate with a second one calculated from the percentage of an individually known study population recorded as killed every year due to hunting (Hofer *et al.*, 1993) and the size of the main segment of the hyaena population (Hofer and East, 1995). This segment has access to their main prey, the migratory wildebeest (Kruuk, 1972; Hofer and East, 1993a), throughout the year and is likely to experience a similar snaring pressure as wildebeest (section 9.7). This second estimate gives as annual offtake a figure of 423 hyaenas.

9.5.2 Distribution of hunting activity

Figure 9.4 shows the spatial distribution of arrests of hunters by law enforcement patrols expressed as arrest densities (number of people arrested/km²; sources of data in Hofer *et al.*, submitted) between 1986 and 1993. Arrests were concentrated in the west and in the north of SNP. Arrests were not restricted to regions adjacent to SNP boundaries but often took place many kilometres inside SNP, although there is a trend for arrest densities to decline towards the interior of SNP. Over the years, places with peaks in arrest densities changed. For instance, the key areas for arrests in the north shifted towards the eastern side of SNP during period II, suggesting that hunters had moved towards the interior and further away from their home villages. In contrast, the areas with peaks in arrest densities in the extreme west of SNP moved closer to the boundary. Changes from period II to period III were local: key areas in period II declined in importance while neighbouring areas saw sharp increases in arrest densities. The mean straight line distance from an 'arrest site' (the centre point of a patrol search area) to the home village was 30.2 ± 1.0 km ($n = 433$ arrests), with a recorded maximum of 170 km.

9.6 LAW ENFORCEMENT

The mean total budget of SNP for the period 1986–1991 was US$321 094 $\pm$ 34 135 ($n = 5$ years) according to figures in the SNP Management Plan (Campbell *et al.*, 1991). This is equivalent to an expenditure of US$25/km².

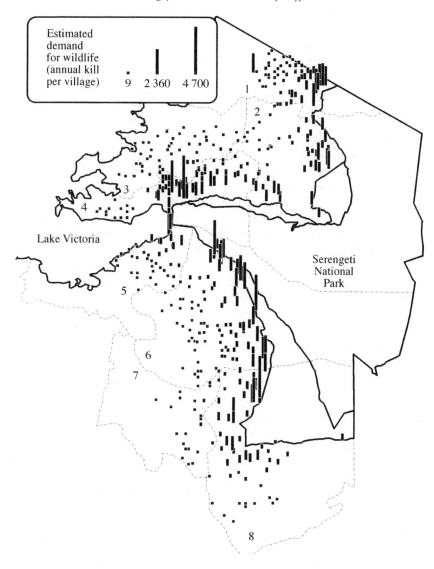

Figure 9.3 The estimated number of wildlife killed annually inside the Serengeti National Park by hunters from each village from the 45 km wide source area around the western edge of the protected area boundary. Numbers were aggregated for each 5 km grid square. Figures were calculated from the estimated number of hunting personnel per village and the records of the average total number of wildlife killed annually per hunting personnel. The thin dashed lines indicate the boundaries of administrative districts (1, Tarime; 2, Serengeti; 3, Musoma Rural; 4, Bunda; 5, Magu; 6, Bariadi; 7, Maswa; 8, Meatu).

9.7　DISCUSSION

Few studies have examined the impact of game meat hunting on mammal-dominated ecosystems. This is probably in part because of the lack of a theoretical framework, the substantial time and financial investment associated with traditional long-term wildlife population studies, and the necessity to generate results quickly. Also, the assessment of the quality of the data in studies of hunting is complex. Nevertheless, our study demonstrates that by using national censuses, information on hunters and their activities collected by law enforcement agencies during patrols, and population surveys of major wildlife species it is possible to assess the impact of game meat hunters on a large ecosystem. The collection of data by law enforcement agencies in combination with wildlife censuses would seem to have considerable potential for wildlife management. The aim in this evaluation of results is to assess the order of magnitude of the importance of hunting and its consequences, and to understand the rules by which hunters operate. By this means it is possible to identify a number of conditions that will have to be met by any programme with the purpose of conserving Serengeti wildlife in the future.

9.7.1　Past and current impact of illegal hunting

Table 9.3 summarizes the results of this and previous studies on the magnitude and impact of hunting on Serengeti wildlife. Effects demonstrated so far include ecosystem-wide and local declines in population size, changes in the distribution and habitat choice of target species, and changes in the demography and population size of non-target species. Unchecked, illegal trophy hunting between 1975 and 1986 drove black rhino to 'factual' extinction and significantly reduced elephant population size (Table 9.3). It is likely that some of these changes have led to a loss of genetic diversity. Hunting may also have had an impact on the demography and population size of local communities, in that feedback effects due to the profitability of illegal hunting and the successful local depletion of wildlife have influenced the pattern of recruitment and decline of villages close to the PA boundary (Figures 9.1 and 9.2; and see below). Hunting is clearly a driving force in the dynamics of the Serengeti ecosystem.

(a)　Impact on game meat species

Intensity of wildlife exploitation by hunters is high close to the PA boundary and decreases towards the interior of the PA (Figure 9.4; Hofer *et al.*, submitted). This is reflected in a number of measures of wildlife abundance. Localities where hunting was considered particularly

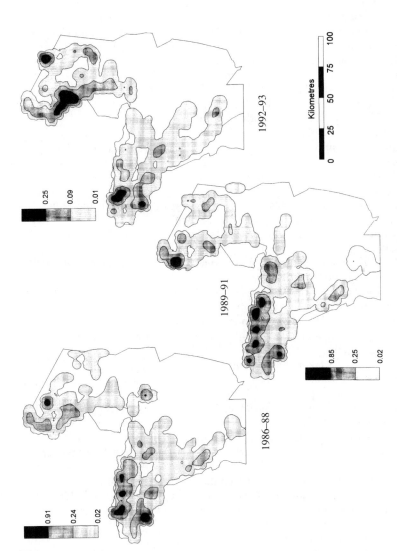

Figure 9.4 Intensity of hunting activity as indicated by the number of hunting personnel arrested per km², during the periods 1986–88 (period I), 1989–91 (period II) and 1992–93 (period III).

Table 9.3 Summary of the impact of illegal activities inside the Serengeti National Park plus adjacent game reserves (Protected Area, PA)

Demonstrated phenomenon	Reference
1. Arrests have increased over the past 30 years at an annual rate of 5%.	Arcese *et al.*, 1995
2. Hunters come from a 'source area' populated by *c.* one million people.	Campbell and Hofer, 1995
3. Mean annual rate of population increase in villages less than 10 km away from the PA boundary substantially exceeds the national and regional average, while rates in areas 10–25 km away are significantly lower and below this average.	This study
4. Hunters appear not to select particular species as targets, but are more likely to catch species whose habitat preferences coincide with hunting areas.	Arcese *et al.*, 1995; Campbell and Hofer, 1995
5. Species whose density in an area is low where suitability for hunting is predicted to be high are under-represented in wildlife killed by hunters.	Campbell and Hofer, 1995
6. The more suitable an area is predicted to be for hunting, the fewer its resident wildlife.	Campbell and Hofer, 1995
7. Density of animal tracks increases with distance to the nearest village after variation in vegetation, rainfall and relief are taken into account.	Campbell and Hofer, 1995
8. Density of kongoni increases with distance to the nearest village after variation in woody canopy cover is taken into account.	Campbell and Hofer, 1995
9. Unregulated trophy hunting significantly reduced elephant populations between 1973 and 1987 and effectively exterminated black rhinoceros populations between 1975 and 1980.	Makacha *et al.*, 1982; Dublin and Douglas-Hamilton, 1987; Arcese *et al.*, 1995
10. Hunting caused a population decline of 50–90% in Cape buffalo over parts of its Serengeti range between 1970 and 1992.	Dublin *et al.*, 1990; Campbell and Borner, 1995
11. Hunting is suspected to have driven the population of roan antelope to near extinction.	Turner, 1987; McNaughton, 1989a
12. Declines in giraffe and waterbuck in parts of their Serengeti range may be due to unsustainable hunting pressure.	Campbell, 1989
13. Zoning the Serengeti on the basis of modelled profitability (suitability for hunting × wildlife density) reliably predicts areas where resident wildlife populations declined between 1989 and 1991.	Campbell and Hofer, 1995

(continued)

Table 9.3 *continued*

Demonstrated phenomenon	Reference
14. Incidental killing by hunters significantly changed population dynamics and age structure of spotted hyaenas and is responsible for an annual population decline of 2.4%.	Hofer *et al.*, 1993; Hofer and East, 1995
15. Hunting is not restricted to the periphery of the PA.	This study
16. Hunter activity declines towards the interior of the PA with distance of a grid cell to the boundary.	This study
17. Hunter activity in an area can be predicted from estimates of the suitability for hunting in an area.	Hofer *et al.*, submitted
18. Hunters preferentially operate in highly profitable areas.	Hofer *et al.*, submitted
19. Hunter activity and predicted offtake in an area are positively correlated even after resident wildlife density and suitability for hunting are taken into account.	Hofer *et al.*, submitted

profitable for hunters – typically easily accessible areas close to the western PA boundary – were the main areas in which resident wildlife populations declined precipitously between 1989 and 1991 (Campbell and Hofer, 1995). Density of animal tracks and the density of kongoni significantly increased towards the interior of the PA even after habitat variation was accounted for (Campbell and Hofer, 1995). Species, such as Grant's and Thomson's gazelle, with a preference for habitats that predominantly occur in the less accessible eastern part of the PA (Campbell and Hofer, 1995), experienced a smaller estimated relative offtake than other species (Table 9.1). Thus, current habitat choice and geographical distribution in the Serengeti may already be influenced by past illegal exploitation and may in turn influence a species' current exposure to hunters (Table 9.3).

Estimated offtake as a percentage of total population size diverged widely between species (Table 9.1). In several resident species, including giraffe, impala and topi, offtake must be considered high. Past exploitation has significantly reduced Cape buffalo by 50–90% in parts of their Serengeti range (Dublin *et al.*, 1990), and local declines in waterbuck and giraffe populations may also be due to overhunting (Campbell, 1989). It is possible that roan antelope (*Hippotragus equinus*), never very common in the Serengeti, has been virtually exterminated by overhunting (Turner, 1987; McNaughton, 1989a). In some species (for example, hippos) offtake may be important yet cannot be assessed as population size is unknown.

Species that are similar in habitat choice and movement patterns may encounter hunters with a similar probability. For instance, wildebeest are the main prey of spotted hyaenas, and hyaenas hunt in areas containing high concentrations of wildebeest (Kruuk, 1972; Hofer and East, 1993b; Hofer *et al.*, 1993). We would then expect hyaena and wildebeest populations to experience a similar offtake. The estimate of 6.8% of wildebeest annually removed by hunting compares well with the estimate of 8% of hyaena offtake by hunting (Hofer *et al.*, 1993).

How do resident and migratory wildlife influence each other's capture probability? The unpredictable movements of migratory herbivores may have complex consequences for the likely offtake of resident herbivores. Hunters probably set a larger number of snares in areas with migratory herds and this might increase mortality among resident wildlife. Alternatively large numbers of migratory herbivores may effectively 'block' snares and provide a temporary reprieve for resident wildlife.

How do preferred hunting techniques influence impact? The preferred method of hunting is wire snaring (see above). In general, hunting is unselective (Campbell and Hofer, 1995), although snaring in specific habitats may result in over-selection of some species (Arcese *et al.*, 1995). Because wire snares are virtually indestructible, they may catch wildlife for a considerable time if left behind after hunters have left an area.

(b) Impact on non-target species

The data in Table 9.1 suggest that snaring may be an important source of mortality for cats (lion, *Panthera leo*; cheetah, *Acinonyx jubatus*) (Plate 4). However, the small numbers may have produced large sampling errors, with possible over-reporting for lions. Under-reporting of other predators may occur because patrols only record carcasses at hunters' camps and snared predators are not necessarily carried to camp by hunters (personal observation). Nevertheless, we now have two independent estimates of the offtake of the most numerous large predator, the spotted hyaena (Hofer and East, 1995). They represent two approaches in estimating hunting offtake: one based on detailed long-term monitoring of an individually known population of several hundred animals (Hofer and East, 1993a; Hofer *et al.*, 1993), the other using the number of hunters and the number of wildlife they take. Both estimates differed by only 11%, lending validity to our offtake estimates for other species.

9.7.2 Non-lethal impact: injuries of target and non-target species

There are few data available on the number of animals that are not killed but manage to escape from snares, and the incidence of permanent injuries sustained by snares. There are recorded cases of lame wildebeest and zebra that had lost a foot or had a foot or leg damaged but it is very rare to see an individual of a target species with a snare around its neck. This might suggest that few target animals manage to escape alive from snares. Amongst non-target species, one elephant was recorded whose lower half of the trunk had been cut off by a snare. Lions that escaped with snare wounds or embedded snares around the head through the mouth, around neck, chest or pelvis are regularly although not frequently recorded (Plate 4). The best data exist for spotted hyaenas from records of individually recognized and/or radio-collared animals that were killed or acquired a snare during the course of the study. Hofer *et al.* (1993) estimated that a spotted hyaena from their study population had a 10% chance of encountering a snare every year, and that a hyaena that was caught in a snare had a chance of escaping of between 25 and 62%, depending on the method of estimation. Assuming that the correct figure is around 50% and that *c.* 400 hyaenas (see above) are killed per year, then there will be at least another 400 hyaenas injured by snares every year. It is possible that long-term effects of an embedded snare or a permanent injury may reduce life expectancy or reproductive success (Plate 6). For instance, one spotted hyaena female that cannot use one front leg due to a snare injury has been unable to raise cubs ever since she sustained the injury. The injury prolongs travel time during foraging trips between the den where the cubs are stationed and the feeding sites, leading to a reduced milk supply rate and the slow starvation of her cubs (unpublished data).

9.7.3 Illegal hunting as sustainable offtake?

Sustainable use has been defined in a variety of ways: maintaining a stable population size; maintaining key demographic parameters, e.g. recruitment, age structure, mortality; maintaining genetic diversity (e.g. Getz and Haight, 1989; Clark, 1990); maintaining total natural capital intact (Costanza and Daly, 1992); maximizing net economic benefits subject to maintaining the services and quality of natural resources over time (Barbier and Markandya, 1993); re-investing economic benefits from offtake to benefit the conservation of wildlife (Martin, 1994). The sustainability of the impact of hunting may therefore be assessed by analysing population trends over time, observing changes in demographic parameters and effective population size, or studying the volume and direction of flow of money from wildlife to exploiters and back to wildlife.

The following sections first evaluate the sustainability of past and current impact, and review factors that determine costs and benefits of hunting and might influence hunter offtake. Then the question of whether unchecked hunting will limit offtake to sustainable levels in the future is considered, and whether increased law enforcement by the management of the PA is likely to restrict offtake to sustainable levels.

(a) Sustainability of current impact

A standard quantitative way of assessing the sustainability of wildlife offtake is to construct a 'harvesting' model, which identifies the maximum sustainable offtake a population can tolerate without experiencing changes in key demographic characteristics (Box 9.1). However, harvesting models may require extensive, long-term demographic data (Getz and Haight, 1989) and, in the case of migratory herbivores, long-term data on environmental cycles (Spinage and Matlhare, 1992). Even if such data are available it is unclear whether predictions of competing harvesting models could be distinguished (Caughley, 1985; Shaw, 1991). A further limitation of harvesting models is that they consider each species in isolation and neglect the impact of utilization on ecosystem dynamics (McNaughton, 1989a). In the case of Serengeti herbivores, data on recruitment and mortality are not available for most Serengeti herbivores; thus studies of basic population processes by ecologists are urgently required for a number of herbivore species. Sufficient data for harvesting models are only available for buffalo and perhaps wildebeest (Sinclair, 1977, 1979; Campbell and Hofer, 1995) and models for these species have not yet been published. While some of the demonstrated impacts are so substantial that they speak for themselves (Table 9.3), the assessment of sustainability of hunting in other cases is more complex, and in the absence of data and models remains a qualitative first approximation.

The ability of resident herbivores to sustain game meat hunting will in part depend on recruitment. Resident species may occur in many places but often can successfully reproduce only in a fraction of that area. Amongst herbivorous mammals, pregnant and lactating females and young animals require higher concentrations of key minerals than other sex-age classes (McDowell, 1985). By beef cattle standards, large areas of the Serengeti grasslands are deficient in calcium, magnesium, phosphorus and sodium (McNaughton, 1988, 1989b, 1990). If these standards also apply to African resident herbivores, then conventional impact estimates may underestimate real impact in terms of a population's ability to recuperate. A comprehensive mapping of mineral concentrations in plant matter, particularly sodium-accumulating grasses, would therefore

BOX 9.1 OPTIMALITY MODELS IN STUDIES OF HUNTING

Optimality models are a powerful tool frequently used in economic, behavioural and evolutionary studies. They are called optimality models because they identify the best possible response in any situation given the constraints of that situation and determine how the optimal response changes if the constraints are modified.

The basic idea is that any response promises a certain reward, called **benefit,** and it carries some risk or expenditure with it, collectively termed **costs.** The difference between the benefits and costs is the net gain or **profit** (profit = benefits minus costs). The **optimal response** is the one that maximizes profit. Profit may be improved by either increasing benefits or reducing costs. Hence, some models specialize in identifying conditions under which benefits are maximized or costs are minimized. In studies of hunting, optimality models have been used in five contexts, although they have rarely been used to solve conservation problems arising from hunting.

The most common models are **harvesting models** of wildlife utilization that predict the maximum sustainable offtake (the **harvest quota**) that a particular population or species can tolerate without population decline (Getz and Haight, 1989; Clark, 1990). Harvesting models analyse the impact and sustainability of different degrees of utilization on wildlife population size and structure and are a well-established tool for wildlife management studies but require extensive, long-term demographic data. Such data can be difficult to obtain for large, long-lived species.

Other models posed the question whether, from a strictly economic point of view, consumptive use of wildlife at maximum sustainable offtake is the most profitable form of wildlife utilization. Clark (1973) showed that this is not necessarily the case, and that instead, 'extermination of the entire population may appear as the most attractive policy, even to an individual resource owner' if resource owners 'prefer present over future revenues'! Discounting potential future income in favour of current income is a common economic practice because information about the future is limited. This favours short-term profits over long-term and therefore uncertain profits (Ehrenfeld, 1988). The conclusion is that an individual's private interest may diverge widely from society's interest in conserving wildlife, and that profit-oriented wildlife exploitation by individuals in itself does not guarantee wildlife utilization that is sustainable and hence conservation-oriented.

Milner-Gulland and Leader-Williams (1992) used an **economic model** to explore the effects of various penalties on the incentives to

(continued)

Box 9.1 *continued*

hunt illegally for trophies (ivory and rhino horns) when law enforcement is imperfect. This model determined sets of conditions when 'poaching pays', i.e. when incentives exceed the costs of hunting expeditions plus penalties due to the chance of detection and arrest. Milner-Gulland *et al.* (1992) considered how costly dehorning of rhinos would be if the horns were removed often enough to protect rhinos from trophy hunters.

Campbell and Hofer (1995) and Hofer *et al.* (submitted) used **spatial models** to predict the pattern (space use) and the extent (total area) of illegal hunting in the protected areas of the Serengeti ecosystem. Their **profitability model** was designed to identify those areas inside protected areas where hunting would be most rewarding if hunters seek to minimize logistical and financial costs (penalties due to detection and arrest). Their **offtake model** predicted the location of hunting areas for hunters from a given village if they minimized logistical costs, competed with hunters from neighbouring villages and had access to target species limited to a percentage of the total herbivore population per unit area. A drawback of the current version of these spatial models is that the costs to hunting parties attempting to utilize the large migratory herds cannot be incorporated. This is because the spatial distribution of the large migratory herds varies tremendously between years as their dry season movements are determined by the spread and amount of rainfall (Maddock, 1979).

The models created by these studies share a number of desirable properties. They make predictions; these predictions provide a yardstick, and they can be tested with empirical data. If model predictions and data match, then the model is useful for identifying the rules by which hunters might operate. In the case of the Serengeti spatial models, areas predicted by the model to be highly suitable for hunting ought to coincide with areas of high hunting activity. As Hofer *et al.* (submitted) show, this is indeed the case. Hence, such models are of practical, strategic use to wildlife managers as they facilitate planning and optimal allocation of the limited resources available to law enforcement agencies.

---

greatly assist in identifying those areas critical for the successful reproduction of resident wildlife (McNaughton, 1990).

Are current levels of offtake of the most common and most frequently hunted herbivore, the migratory wildebeest, sustainable? The sex ratio of animals killed by hunters in the early dry season was biased towards males (Hofer *et al.*, 1993; Campbell and Hofer, 1995), potentially reducing the impact. Intensive hunting of wildebeest early in the dry season also precedes the period when intra-specific competition for scarce high-quality dry season forage would lead to increased mortality (Sinclair, 1979, 1985), thereby reducing the intensity of intra-specific competition and the impact of hunting. Snaring of spotted hyaenas, the most important predator of wildebeest, has led to a decline in the hyaena population (Hofer and East, 1995) and thus may have reduced the contribution of predation to wildebeest mortality. The main predators (hyaenas and lions) were estimated to remove roughly 2–3% of the wildebeest population annually in the late 1960s (Kruuk, 1972; Schaller, 1972). Kruuk (1972) suggested that spotted hyaenas removed 12% annually from a stable wildebeest population in the Ngorongoro Crater. This suggests that the 'scope for stability' in the Serengeti wildebeest (maximum mortality for a stable population minus predator-induced mortality) may be of the order of 9%. However, Peterson and Casebeer (1972) recommended a 'safe' annual harvest of only 6% for the wildebeest population on the Athi-Kapiti plains south of Nairobi, an area comparatively depauperate in predators. This implies that the estimated current offtake of 6.8% approaches the limits of sustainable use.

For non-target species, data on the impact of offtake are only available for the most numerous species, the spotted hyaena. Killing by game meat hunters was the most important mortality factor and was responsible for an average annual population decline of 2.4% during 1987–1991. Hofer and East (1995) calculated that without this mortality the spotted hyaena population had a potential rate of growth of 4.6%, indicating that the offtake is unsustainable. Turner (1987) provides a fascinating account of the history of unchecked large-scale hunting of lions, leopards (*Panthera pardus*) and African wild dogs (*Lycaon pictus*) in the Serengeti until 1960 and concludes that such hunting always proceeded at unsustainable rates, reducing local predator populations by a significant amount or driving them to near extinction.

(b) Does market hunting limit offtake to sustainable levels?

Illegal game meat hunting in the Serengeti is an example of market hunting, as the questionnaires revealed that at least half of the meat was sold or bartered. Market hunting is distinguished from subsistence hunting in that hunting is not restrained to supplying the hunter's family with meat; instead hunters sell products (fresh or dried meat, cured hides, tails) to the open market. Market hunting is therefore influenced by other factors: the financial costs of equipping hunting expeditions, opportunity costs (wages for members of the hunting expedition, financial losses due to the inability to carry out alternative forms of employment during periods of hunting), logistics (accessibility of the hunting area, including travelling time and the problems of transportation of wildlife products to the market), hunting success in relation to hunting effort, penalties in case of detection and arrest by law enforcement patrols, and expected revenue based on current market prices. Is there any evidence that market hunting limits offtake to sustainable levels?

Theoretical models and historical reviews demonstrate that profit-oriented wildlife exploitation by individuals selling to a market is unlikely to comply with the interest in conserving wildlife (Box 9.1; Geist, 1988). We can derive at least two predictions from these models:

1. There should be examples where market hunting has led to unsustainable wildlife offtake.
2. If the costs of hunting expeditions are reduced (while revenue remains unchanged) then there ought to be an increase in offtake and the severity of its impact.

What is the evidence?

The history of legal and illegal trophy hunting (ivory and rhino horns) illustrates that hunting may drive species to extinction (Diamond, 1989), the endpoint of any unsustainable offtake. Well-known examples of unsustainable hunting include the Southern African quagga (*Equus quagga*), extinct since the late nineteenth century, and the near-extinction of the mountain zebra (*Equus hartmanni*). Meat hunting devastated the population of bohor reedbuck in Queen Elizabeth National Park, Uganda, between 1975 and 1982 (Edroma and Kenyi, 1985) and a switch from subsistence to market hunting drastically reduced wildlife densities in the rainforests of northern Congo in the 1970s (Hart, 1978). In the Bangweulu swamps in northern Zambia, a partially protected area dominated by migratory black lechwe (*Kobus leche smithemani*), hunting was thought to have drastically reduced populations in the early part of this century, shifted the focus of lechwe distribution to areas less accessible to hunters and prevented the lechwe population from increasing after the protected area was extended (Howard *et al.*, 1984; Thirgood *et al.*, 1994). The results of our study also suggest that market hunting in the Serengeti must be considered unsustainable for at least some wildlife species.

A number of studies illustrate that accessibility – the financial costs of appropriate equipment and staffing, travelling and transportation of products out of the hunting area – strongly influences offtake and hunting impact. Marks (1973, 1977) demonstrated that buffalo hunting in a local community in northern Zambia was limited by the distance of travelling and transportation of meat. African forest elephant density increased with the distance to a road, village or navigable river

(Michelmore *et al.*, 1994). The opening up of the rainforest in northern Congo for timber extraction by mechanized logging led to a substantial increase in meat hunting and had a significant impact on wildlife (Wilkie *et al.*, 1992). Logistics and accessibility to hunting areas are also important components of hunting in the Serengeti. Campbell and Hofer (1995) and Hofer *et al.* (submitted) constructed spatial models in which they considered the suitability of an area for hunting to be related to logistical costs (and financial costs arising from penalties if arrested, see Box 9.1). The distribution of hunter activity, as indicated by arrest densities, matched the predictions of their models. Additional evidence comes from trends in rural migration: the reduction in size and number of villages to the north-west of the PA might be related to the local depletion of wildlife within the adjacent PA (Campbell and Hofer, 1995), and high growth rates of the human population in areas close to the boundary may be related to the locally greater availability of wildlife and other natural resources (Campbell and Hofer, 1995).

In summary, there is no reason to expect that unchecked illegal market hunting would restrict wildlife offtake to sustainable levels.

(c) Law enforcement spending

Are current law enforcement efforts likely to restrict illegal offtake to sustainable levels? Leader-Williams and Albon (1988) estimated that annual conservation spending in excess of US\$215/km^2 (ivory) and US\$230/km^2 (rhino horns) was required to stop a decline in elephants and rhinos caused by trophy hunting. The calculated mean total budget of the SNP authority of US\$25/km^2 is a factor of 9 smaller than this estimate. Most of the budget is geared towards law enforcement, with the exception of tourism-related activities, so this figure provides an upper estimate for SNP law enforcement expenditure. The differences between trophy hunting and meat hunting are manifold: trophy hunters often use sophisticated equipment, the number of hunters is small and they are highly mobile, the target wildlife populations are difficult to protect and monitor because they range widely and sometimes live in dense habitat. In contrast, meat hunting in the SNP involves large numbers of people using simple equipment to catch animals from large target populations difficult to protect and monitor and that, in the case of migratory animals, range widely. Thus the protection and monitoring of wildlife populations targeted by either trophy or meat hunters requires a large patrol effort, and it is likely that the expenditure required to prevent game meat hunters causing population declines is of the same order of magnitude as calculated for trophy hunting by Leader-Williams and Albon. Even if all donations to SNP matched the official budget and increased total spending to US$50/km^2, this amount is below the estimated threshold of US$215–230/km^2 by a factor of 4 or more. The conclusion is that current spending on law enforcement is unlikely to effectively protect wildlife populations through law enforcement activities alone.

Even a modest increase in spending on law enforcement could reduce the actual impact of snares if the money was used to improve efforts to collect snares in the field. During the 1980s a system was installed whereby rangers were rewarded for each snare collected. The number of snares collected increased dramatically, but has declined again since the reward system was effectively discontinued in the early 1990s (Arcese *et al.*, 1995).

9.8   QUO VADIS?

What useful guidelines for the conservation of ecosystems dominated by migratory mammalian herbivores can be drawn from our analysis of the situation in the Serengeti?

9.8.1 Legalization of hunting

Will legalization of hunting in GRs adjacent to the SNP limit offtake to sustainable levels? Legalized, economic exploitation of wildlife is increasingly heralded as the best way to ensure a future for wildlife (Murphree, 1993; Martin, 1994), but the assumption that commercialization of wildlife will prove sustainable is rarely justified by sufficient data. One successful case of conservation by local participation and commercialization was presented by Lewis *et al.* (1990). The scheme derived its success from a combination of particular circumstances: it was implemented in an area with low human population density; poachers originated primarily from outside the local communities and revenues per carcass from concessions for trophy hunting (of elephants) were high. However, in many cases (including the Serengeti) human population density is high, revenue per carcass is likely to be low and poachers primarily originate from within local communities. Recent critical examinations of such single-product utilization schemes have been sceptical about their sustainability (Holling, 1993; Ludwig *et al.*, 1993). Allocation of legal hunting quotas to local communities in areas outside national parks may therefore reduce the conflict between wildlife managers and local communities, but is unlikely to ensure a sustainable offtake on its own. In the case of the Serengeti, three additional points make us sceptical that legalization of wildlife hunting in areas outside the National Park will limit offtake to sustainable levels:

1. The total population of resident wildlife numbering approximately 19 900 in the Game Reserves adjacent to SNP (Campbell and Borner, 1995) is substantially less than the estimated current total annual offtake of some 45 000 resident herbivores from the ecosystem (Table 9.1). Any sustainable licensed quota will fall far short of the current illegal offtake from the ecosystem and is thus unlikely to satisfy the demands of local hunters. Most of the areas defined as over-exploited by Campbell and Hofer (1995) lie in these GRs, implying that offtake in these GRs has been unsustainable for some time.
2. Villages with hunting licences experience problems in marketing meat to neighbouring communities without hunting quotas. This is because potential customers do not 'waste' limited income on the purchase of legal wildlife meat when they can readily obtain meat at a lower cost and an income for themselves through illegal hunting (M. Loibooki, personal communication).
3. The majority of meat extracted from the Serengeti is derived from migratory herbivores. If migratory herds bypass the allocated hunting block of a particular village for one or several years, hunters from this village would have limited or no legal income from hunting. Under such circumstances hunters are likely to hunt illegally outside their allocated hunting block. Hence, annual village quotas are unlikely to alleviate illegal offtake. Rasker *et al.* (1992) explain why basic economic arguments suggest that whenever assignment of 'ownership' of a resource is ambiguous (as in the case of migratory wildlife), commercialization of wildlife is unlikely to achieve conservation of wildlife.

9.8.2 Law enforcement and the economics of hunting

Current law enforcement efforts in both GRs and the SNP do not restrict illegal game meat hunting to a sustainable level. As the number of game meat hunters is likely to grow in future, wildlife populations in many areas highly suitable for hunting inside SNP are likely to be soon over-exploited. This will cause hunters to switch to currently less attractive sites, where wildlife occurs at currently lower densities or where their chance of arrest is currently higher. To maintain income, hunters will then be forced to increase hunting effort (by walking longer distances from village to hunting area, setting a larger number of snares, increasing the duration of hunting trips) and reduce the chance of detection and arrest (by increasing night activity, seeking cover during the day, operating in larger parties or increasing weaponry). This is likely to make hunting more difficult, less efficient and less profitable (unless revenues per carcass increase concomitantly) and to escalate the conflict between

hunters and law enforcers. If viable alternative forms of income generation became available when the profitability of game meat hunting was
decreasing, hunting activity should decline because potential hunters
should change to other more profitable activities. Effective law enforcement inside SNP at such times should persuade hunters to seek alternative sources of income.

9.9 CONCLUSIONS

We suspect that only a substantial effort aimed at developing the nutritional and economic status of local communities, particularly those at a
distance of 10–25 km away from the PA, will:

- reduce high rates of population increase in the areas close to the PA boundary;
- secure a permanent improvement in relations between local commu-nities and PA management; and thereby
- reduce unchecked large-scale meat hunting.

Viable alternative income-generating schemes urgently need to be
developed away from the park boundary through consultation with
these local communities. More research is required to provide information for targeting social, nutritional and economic development at communities most in need and at those more actively involved in illegal
hunting activities, and to assess the spatial and temporal variation in offtake of both resident and migratory wildlife.

The detrimental effects of hunting originating from local communities
could be ameliorated by:

- a substantial increase in law enforcement in the SNP and surrounding protected areas, the costs of such an increase in part being met by international agencies;
- focusing extension programmes and conservation education on com-munities that supply a large number of hunters;
- providing essential amenities (e.g. maize grinding mills, water sup-plies, dispensaries, schools) in areas removed from the Park boundary using revenue generated by the SNP, whilst ensuring that such ameni-ties are perceived as benefits resulting from the Park;
- providing viable, cost-effective and attractive alternatives to wildlife meat as well as alternative sources of employment;
- recognizing the need to include all stakeholders in efforts to resolve these complex issues.

A key problem that remains for wildlife managers in the Serengeti is how to devise workable rules for the acquisition and distribution of benefits from both migratory and resident species at a sustainable level. From our analysis of game meat hunting in the Serengeti we conclude that a large investment is required in both law enforcement and rural development; that the success of the latter may be linked to investment in the former; and that without both of these, consumptive use of wildlife by local communities is likely to continue at a level that may prove in the long run to lead to the demise of much of Serengeti's wildlife.

ACKNOWLEDGEMENTS

We are grateful to the Tanzania Commission of Science and Technology for permission to conduct the study and the Director Generals of Tanzania National Parks, the Wildlife Division, Serengeti Wildlife Research Institute and the Conservator of Ngorongoro Conservation Area under whose auspices the study was carried out. We thank the Fritz-Thyssen-Stiftung and Max-PlanckGesellschaft (HH, MLE) and the Frankfurt Zoological Society (KLIC, SAH) for financial support. Special thanks are due to the SNP rangers who completed patrol record cards, interviewed hunters and filled in questionnaires. We also thank the management and wardens of the Serengeti National Park without whose dedication and hard work the Serengeti would not be the wildlife spectacle that it is today.

REFERENCES

Anderson, D. and Grove, R. (1987) *Conservation in Africa: people, policies and practice*, Cambridge University Press, Cambridge.

Arcese, P., Hando, J. and Campbell, K.L.I. (1995) Historical and present-day antipoaching in Serengeti, in *Serengeti II: Research, Management and Conservation of an Ecosystem*, (eds A.R.E. Sinclair and P. Arcese), University of Chicago Press, Chicago, pp. 506–533.

Barbier, E.B. and Markandya, A. (1993) Environmentally sustainable development: optimal economic conditions, in *Economics and Ecology*, (ed. E.B. Barbier), Chapman & Hall, London, pp. 11–28.

Batisse, M. (1986) Developing and focusing the Biosphere Reserve concept. *Nature and Research*, **22**(3), 1–10.

Blumenschine, R.J. and Caro, T.M. (1986) Unit flesh weights of some East African bovids. *African Journal of Ecology*, **24**, 273–286.

Campbell, K.L.I. (1989) *Programme Report, September 1989*, Serengeti Ecological Monitoring Programme, Arusha, Tanzania.

Campbell, K.L.I. and Borner, M. (1995) Population trends and distribution of Serengeti herbivores: implications for management, in *Serengeti II: Research, Management and Conservation of an Ecosystem*, (eds A.R.E. Sinclair and P. Arcese), University of Chicago Press, Chicago, pp. 117–145.

Campbell, K.L.I. and Hofer, H. (1995) People and wildlife: spatial dynamics and zones of interaction, in *Serengeti II: Research, Management and Conservation of an Ecosystem*, (eds A.R.E. Sinclair and P. Arcese), University of Chicago Press, Chicago, pp. 535–574.

Campbell, K.L.I., Huish, S.A. and Kajuni, A. (1991) *Serengeti National Park Management Plan 1991–1995*, Tanzania National Parks, Arusha, Tanzania.

Caughley, G. (1985) Harvesting of wildlife: past, present, and future, in *Game Harvest Management*, (eds S.L. Beasom and S.F. Roberson), Caesar Kleberg Wildlife Research Institute, Kingsville, Texas, pp. 3–14.

Clark, C.W. (1973) Profit maximization and the extinction of species. *Journal of Political Economics*, **81**, 950–961.

Clark, C.W. (1990) *Mathematical Bioeconomics: the Optimal Management of Renewable Resources*, John Wiley, New York.

Costanza, R. and Daly, H.E. (1992) Natural capital and sustainable development. *Conservation Biology*, **6**, 37–46.

Diamond, J. (1989) Overview of recent extinctions, in *Conservation for the Twenty-first Century*, (eds D. Western and M. Pearl), Oxford University Press, New York, pp. 37–41.

Dublin, H.T. and Douglas-Hamilton, I. (1987) Status and trends of elephants in the Serengeti-Mara ecosystem. *African Journal of Ecology*, **25**, 19–33.

Dublin, H.T., Sinclair, A.R.E., Boutin, S. *et al.* (1990) Does competition regulate ungulate populations? Further evidence from Serengeti, Tanzania. *Oecologia*, **82**, 283–288.

Edroma, E.L. and Kenyi, J.M. (1985) Drastic decline in bohor reedbuck (*Redunca redunca* Pallas 1777) in Queen Elizabeth National Park, Uganda. *African Journal of Ecology*, **23**, 53–55.

Ehrenfeld, D. (1988) Why put a value on biodiversity? in *Biodiversity*, (ed. E.O. Wilson), National Academy Press, Washington, pp. 212–216.

Geist, V. (1988) How markets in wildlife meat and parts, and the sale of hunting privileges, jeopardize wildlife conservation. *Conservation Biology*, **2**, 15–26.

Getz, W.M. and Haight, R.G. (1989) *Population Harvesting*, Princeton University Press, Princeton.

Hart, J.A. (1978) From subsistence to market: a case study of the Mbuti net-hunters. *Human Ecology*, **6**, 32–53.

Hofer, H. and East, M.L. (1993a) The commuting system of Serengeti spotted hyaenas: how a predator copes with migratory prey. I. Social organization. *Animal Behaviour*, **46**, 547–557.

Hofer, H. and East, M.L. (1993b) The commuting system of Serengeti spotted hyaenas: how a predator copes with migratory prey. II. Intrusion pressure and commuters' space use. *Animal Behaviour*, **46**, 559–574.

Hofer, H. and East, M.L. (1995) Population dynamics, population size, and the commuting system of Serengeti spotted hyaenas, in *Serengeti II: Research, Management and Conservation of an Ecosystem*, (eds A.R.E. Sinclair and P. Arcese), University of Chicago Press, Chicago, pp. 332–363.

Hofer, H., East, M.L. and Campbell, K.L.I. (1993) Snares, commuting hyaenas and migratory herbivores: humans as predators in the Serengeti. *Symposia of the Zoological Society London*, **65**, 347–366.

Hofer, H., Campbell, K.L.I., East, M.L. and Huish, S.A. (submitted) The cost-conscious poacher: models of the spatial distribution of illegal game meat hunting in the Serengeti.

Holling, C.S. (1993) Investing in research for sustainability. *Ecological Applications*, **3**, 552–555.

Howard, G.W., Jeffery, R.C.V. and Grimsdell, J.J.R. (1984) Census and population trends of black lechwe in Zambia. *African Journal of Ecology*, **22**, 175–179.

Ishwaran, N. (1992) Biodiversity, protected areas and sustainable development. *Nature and Research*, **28**(1), 18–25.

Kruuk, H. (1972) *The Spotted Hyena*, University of Chicago Press, Chicago.

Leader-Williams, N. and Albon, S.D. (1988) Allocation of resources for conservation. *Nature*, **336**, 533–535.

Lewis, D., Kaweche, G.B. and Mwenya, A. (1990) Wildlife conservation outside protected areas – lessons from an experiment in Zambia. *Conservation Biology*, **4**, 171–180.

Ludwig, D., Hilborn, R. and Walters, C.J. (1993) Uncertainty, resource exploitation, and conservation: lessons from history. *Science*, **260**, 17 and 36.

Maddock, L. (1979) The 'migration' and grazing succession, in *Serengeti, Dynamics of an Ecosystem*, (eds A.R.E. Sinclair and M. Norton-Griffiths), University of Chicago Press, Chicago, pp. 104–129.

Makacha, S., Msingwa, M.J. and Frame, G.W. (1982) Threats to the Serengeti herds. *Oryx*, **16**, 437–444.

Marks, S.A. (1973) Prey selection and annual harvest of game in a rural Zambian community. *East African Wildlife Journal*, **11**, 113–128.

Marks, S.A. (1977) Buffalo movements and accessibility to a community of hunters in Zambia. *East African Wildlife Journal*, **15**, 251–261.

Martin, R.B. (1994) Alternative approaches to sustainable use: what does and doesn't work. Paper presented at symposium on Conservation Through Sustainable Use of Wildlife, University of Queensland, Brisbane, 8–11 February 1994, Department of National Parks and Wildlife Management, Harare, Zimbabwe. Mimeo, 16 pp.

McDowell, L.R. (1985) *Nutrition of Grazing Ruminants in Warm Climates*, Academic Press, New York.

McNaughton, S.J. (1988) Mineral nutrition and spatial concentration of African ungulates. *Nature*, **334**, 343–345.

McNaughton, S.J. (1989a) Ecosystems and conservation in the twenty-first century, in *Conservation for the Twenty-first Century*, (eds D. Western and M. Pearl), Oxford University Press, New York, pp. 109–120.

McNaughton, S.J. (1989b) Interactions of plants of the field layer with large herbivores. *Symposia of the Zoological Society London*, **61**, 15–29.

McNaughton, S.J. (1990) Mineral nutrition and seasonal movements of African migratory ungulates. *Nature*, **345**, 613–615.

McNeely, J. (1989) Protected areas and human ecology: how National Parks can contribute to sustaining societies of the twenty-first century, in *Conservation for the Twenty-first Century*, (eds D. Western and M. Pearl), Oxford University Press, New York, pp. 150–157.

Michelmore, F., Beardsley, K., Barnes, R.F.W. and Douglas-Hamilton, I. (1994) A model illustrating the changes in forest elephant numbers caused by poaching. *African Journal of Ecology*, **32**, 89–99.

Milner-Gulland, E.J., Beddington, J.R. and Leader-Williams, N. (1992) Dehorning African rhinos: a model of optimal frequency and profitability. *Proceedings of the Royal Society London*, B **249**, 83–87.

Milner-Gulland, E.J. and Leader-Williams, N. (1992) A model of incentives for the illegal exploitation of black rhinos and elephants: poaching pays in Luangwa Valley, Zambia. *Journal of Applied Ecology* **29**, 388–401.

Murphree, M.W. (1993) *Communities as Resource Management Institutions*, IIED Gatekeeper Series No SA36, International Institute for Environment and Development, London.

Myers, N. (1972) National parks in savannah Africa. *Science*, **178**, 1255–1263.

Myers, N. (1979) *The Sinking Ark*, Pergamon Press, Oxford.

Newmark, W.D., Manyanza, D.N., Gamassa, D.G.M. and Sariko, H.I. (1994) The conflict between wildlife and local people adjacent to protected areas in Tanzania: human density as a predictor. *Conservation Biology*, **8**, 245–255.

Peterson, J.C.B. and Casebeer, R.K. (1972) *Distribution, population status and group composition of wildebeest*, Connochaetes taurinus, *and zebra*, Equus burchelli, *on the Athi-Kapiti plains, Kenya*, Project Working Document 2, Wildlife Management in Kenya, FAO, Nairobi.

Rasker, R., Martin, M.V. and Johnson, R.L. (1992) Economics: theory versus practice in wildlife management. *Conservation Biology*, **6**, 338–349.

Robinson, J.G. and Redford, K.H. (1991a) *Neotropical Wildlife Use and Conservation*, University of Chicago Press.

Robinson, J.G. and Redford, K.H. (1991b) Preface, in *Neotropical Wildlife Use and Conservation*, (eds J.G. Robinson and K.H. Redford), University of Chicago Press, pp. XV–XVII.

Schaller, G.B. (1972) *The Serengeti Lion: a study of predator prey relations*, University of Chicago Press.

Shaw, J.H. (1991). The outlook for sustainable harvests of wildlife in Latin America, in *Neotropical Wildlife Use and Conservation*, (eds J.G. Robinson and K.H. Redford), University of Chicago Press, pp. 24–34.

Sinclair, A.R.E. (1977) *The African Buffalo*, University of Chicago Press.

Sinclair, A.R.E. (1979) The eruption of the ruminants, in *Serengeti, Dynamics of an Ecosystem*, (eds A.R.E. Sinclair and M. Norton-Griffiths), University of Chicago Press, pp. 82–103.

Sinclair, A.R.E. (1985) Does interspecific competition or predation shape the African ungulate community? *Journal of Animal Ecology*, **54**, 899–918.

Sinclair, A.R.E. and Norton-Griffiths, M. (1979) *Serengeti, Dynamics of an Ecosystem*, University of Chicago Press.

Spinage, C.A. and Matlhare, J.M. (1992) Is the Kalahari cornucopia fact or fiction? A predictive model. *Journal of Applied Ecology*, **29**, 605–610.

SRCS (1991) *A Plan for Conservation and Development in the Serengeti Region, Phase II final report, Phase III action plan*, Unit. Rep. of Tanzania, Min. of Tourism, Natural Resources and Environment, Dar es Salaam.

Thirgood, S.J., Nefdt, R.J., Jeffery, R.C.V. and Kamweneshe, B. (1994) Population trends and current status of black lechwe (*Kobus*: Bovidae) in Zambia. *African Journal of Ecology*, **32**, 1–8.

Turner, M. (1987) *My Serengeti Years*, Elm Tree Books, London.

Wilkie, D.S., Sidle, J.G. and Boundzanga, G.C. (1992) Mechanized logging, market hunting, and a bank loan in Congo. *Conservation Biology*, **6**, 570–580.

10

Subsistence hunting and mammal conservation in a Kenyan coastal forest: resolving a conflict

Clare D. FitzGibbon, Hezron Mogaka and John H. Fanshawe

SYNOPSIS

Over 1000 local households hunt and trap a wide range of mammal species from Arabuko-Sokoke Forest, Kenya, including elephant-shrews, small rodents, bushpigs, duikers and two species of primate. Patterns of harvesting have changed over time: in particular the overall intensity has increased as a result of human population increases and a decline in wildlife habitat; different species are now targeted; and there has been an increase in trapping, particularly at the edge of the forest, and a decline in hunting. The available evidence suggests that primates and large ungulates are over-harvested, while current offtake rates of elephant-shrews, squirrels, bushpigs and duikers are probably sustainable. Species which are hunted, rather than trapped, are particularly vulnerable to overharvesting. While game meat harvesting is important for local people, both as a source of protein and to reduce crop raiding, over-harvesting will threaten mammal populations and reduce the biodiversity value of Arabuko-Sokoke. One possible solution may be to maintain the ban on hunting while allowing trapping on the forest periphery to continue, reducing the pressure on primates and large ungulates, and still allowing people to harvest the smaller species and control crop raiding.

10.1 INTRODUCTION

Subsistence hunting, particularly in national parks and protected areas, has been a source of contention for many years. The main issue is whether the hunting of wildlife can be compatible with conservation. Obviously people have hunted wildlife populations for hundreds of years, presumably sustainably, but as pressure on remaining wild areas increases, it is more and more likely that hunting will be having adverse effects. In general, the traditional view has been to ban subsistence hunting in national parks, in the belief that this is the best way to protect the remaining wildlife populations.

There are two problems with this approach. The first, a humanitarian one, is that a huge number of people in developing countries depend on game meat as their main source of protein, and in many areas the sale of meat and animal products is an important source of income (Asibey, 1974; Hart, 1978; Martin, 1983; Marks, 1989; Wilkie, 1989). The second, a more practical problem, is that the protection of threatened habitats depends on the continuing support of local people: unless these people benefit from the habitat and the wildlife it supports, there will be no long-term incentive to protect it (Bell, 1987; Ehrlich and Wilson, 1991; Bodmer *et al.*, 1994). The harvesting of wildlife products is one of the few benefits available to local communities, offsetting some of the direct and indirect costs of forest conservation (Balmford *et al.*, 1992). Thus, it is essential to determine whether controlled harvesting of wildlife populations can be compatible with conservation.

Currently the coastal forests of Eastern Africa are regarded as the most threatened forest type in Africa (Burgess *et al.*, in press) and Arabuko-Sokoke Forest, on the north Kenya coast, is the largest remaining fragment of such habitat. It is an important site for biodiversity conservation (Collar and Stuart, 1988; Burgess *et al.*, in press), providing habitat for a number of rare and threatened mammal species, including the Ader's duiker *Cephalophus adersi*, golden-rumped elephant-shrew *Rhynchocyon chrysopygus* (Nicoll and Rathbun, 1990; FitzGibbon, 1994), Sokoke bushy-tailed mongoose *Bdeogale crassicauda omnivora* and lesser pouched rat *Beamys hindei* (FitzGibbon *et al.*, 1995b), as well as a number of threatened bird species (Collar and Stuart, 1988). Although harvesting of wild animals is illegal in Kenyan forest reserves, local communities of Mijikenda and Sanya are known to harvest a wide range of mammal and bird species from the forest (Stiles, 1981; Mogaka, 1992). Harvesting of some species is probably sustainable, but the primates and large ungulates appear to be over-harvested (FitzGibbon *et al.*, 1995a). This chapter reviews the extent of current game meat harvesting from Arabuko-Sokoke, investigates how harvesting patterns have changed and the consequences for prey populations, and suggests ways in which harvesting may be regulated to minimize its impact on prey populations.

10.2   METHODS

The study was carried out in Arabuko-Sokoke Forest, a nature reserve covering 372 km^2 on the north Kenya coast (Figure 10.1), in 1991 and 1992. Data on the incidence of hunting were derived from a survey of 51 households living adjacent to the forest edge and 24 households living approximately 2 km from the forest boundary, between March and May

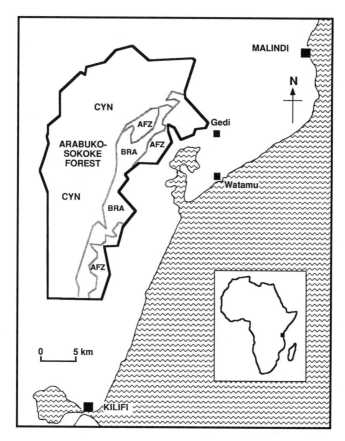

Figure 10.1 Map of Arabuko-Sokoke Forest and surrounding area, showing the three main vegetation zones (BRA, *Brachystegia* woodland; CYN, *Cynometra–Brachylaena* woodland; AFZ, *Afzelia* forest).

1991. Questions on hunting activity were included as part of a general socio-economic interview carried out by Hezron Mogaka. In addition, 16 regular hunters were asked to keep records of how often they went hunting and trapping, the number of traps set, trapping success rates and the prey species caught.

10.2.1 Trapping intensity and the abundance of mammalian prey

The forest consists of three main habitat types: (1) woodland dominated by *Brachystegia spiciformis*; (2) *Cynometra–Brachylaena* woodland; and (3) *Afzelia* forest (Kelsey and Langton, 1984; Figure 10.1). Distributed between these three main habitat types, 28 one kilometre transect lines were cut in the forest. In each habitat, half of the transects were on the forest edge where trapping and hunting intensity was predicted to be high, while the other half were positioned as far from the edge as possible. Immediately after cutting, the number of traps within 15 m of each transect line was recorded to provide an estimate of trapping intensity.

Line transects were used to determine the relative abundance of prey in the three main habitat types of the forest and to compare abundance between areas with high and low trapping intensity. The survey focused on a few key groups, namely primates (yellow baboon *Papio cynocephalus* and Syke's monkey *Cercopithecus mitis*), duikers (*Cephalophus* spp.), bush-pigs (*Potamochoerus porcus*), elephant-shrews (golden-rumped elephant-shrew *Rhynchocyon chrysopygus* and four-toed elephant-shrew *Petrodomus tetradactylus*) and squirrels (*Heliosciurius* and *Funisciurus* spp.). For some species (baboons, Syke's monkeys, golden-rumped elephant-shrews and four-toed elephant-shrews), it was possible to estimate population densities, while for others only relative abundance could be determined (for further details see FitzGibbon *et al.*, 1995a).

10.3   EXTENT OF HARVESTING IN ARABUKO-SOKOKE FOREST

Local communities harvest a wide range of mammal species from Arabuko-Sokoke Forest, ranging in size from small mice, rats and elephant-shrews to primates such as yellow baboons and large ungulates such as bushbuck, although these are now rare in the forest (Table 10.1). In the past, hunters used to take African elephant (*Loxodonta africana*) and black rhino (*Diceros bicornis*), but these animals are now rarely taken. Rhinos have disappeared from the forest, while the severe penalties imposed by the government make elephant hunting extremely risky. Both hunting and trapping take place in the forest. All species can be caught in traps but baboons, Syke's monkeys and ungulates, such as duikers and bushbuck, are also hunted.

Hunters use bows and poisoned arrows, spears, dogs and catapults. A wide range of traps are used, predominantly snares but also drop traps. While hunting is clearly species specific, traps are less so, but instead tend to catch animals of a particular size. However, trappers can target particular species by placing traps appropriately; for example, the small snares placed along four-toed elephant-shrew trails and the drop traps placed outside the burrows of giant pouched rats. However, since such a wide range of animal species is consumed, meat is rarely wasted. Traps have to be checked regularly (at least once per day) to ensure that carcasses are not stolen by other trappers or eaten by ants, baboons or other wildlife. This means that trapped animals are not left for exceedingly long periods, although the traps often kill animals outright.

A survey of households living close to the forest suggests that about 60 households hunt regularly in the forest, making an average of 2.4 hunt-

Table 10.1 Average number and biomass (kg) of mammals hunted (H) and trapped (T) by a regular hunter in one year (the occasional bushbuck and buffalo are taken, as well as several species of rodents, including the giant pouched rat *Cricetomys gambianus*)

	Number killed	*Biomass*	*Hunted/trapped*
Four-toed elephant-shrews	47	11.8	T
Golden-rumped elephant-shrews	26	14.3	T
Syke's monkeys	19	95.0	H & T
Yellow baboons	11	176.0	H & T
Bushpigs	6	480.0	T
Aardvarks[1]	5	275.0	T
Duikers	2	12.0	H & T
Squirrels	4	1.2	T
Mongooses[2]	5	7.5	T

[1] *Orycteropus afer.*
[2] *Herpestes, Ichneumia* and *Bdeogale* spp.

ing trips each week, while over 1000 people probably set traps in the forest at some time through the year. By determining the total number of people hunting and trapping in the forest, and the composition of their yearly catch, it is possible to estimate that over 5000 four-toed elephant-shrews, 3000 golden-rumped elephant-shrews, 1000 Syke's monkeys and 600 baboons are currently harvested each year (FitzGibbon *et al.*, 1995a). The total biomass of game meat removed by hunting/trapping each year is estimated to be 130 000 kg (350 kg/km^2).

Hunting takes place throughout the forest but trapping is concentrated around the periphery. Since traps have to be checked at least once a day, it is rarely worthwhile for trappers to travel far from their homes to set traps and traps are consequently rarely set more than 1–2 km from the forest edge. These time constraints are probably also the main reason that trapping intensity is not closely correlated with prey abundance and varies little across habitat, even though prey are far more abundant in some habitats than others (Table 10.2). Thus there were no significant correlations between the number of snares recorded along a transect and the relative abundance of bushpigs or duikers along that transect ($r =$ 0.168 and -0.207, $n = 14$, only using transects on forest edge), nor the number of four-toed elephant-shrew traps and the abundance of elephant-shrew trails ($r = 0.009$, $n = 14$). In addition, even though prey abundance is relatively low (Table 10.2) and trappers report low success rates in the *Brachystegia* habitat, the number of traps recorded along transects is not significantly lower in this habitat (Figure 10.2). Thus, it appears that people are prepared to put up with reduced success rates in

Table 10.2 Relative abundance of mammals in three habitat types (3 = highest density, 1 = lowest), as determined from transect surveys (see FitzGibbon *et al.*, 1995 (a) for methods)

Habitat type	Elephant-shrews		Duikers	Bushpigs	Syke's monkeys	Baboons	Squirrels	TOTAL
	Golden-rumped	Four-toed						
Brachystegia	1	1	1	1	1	3	1	9
Afzelia	3	2	2	3	3	3	2	18
Cynometra	2	3	3	2	2	2	3	17

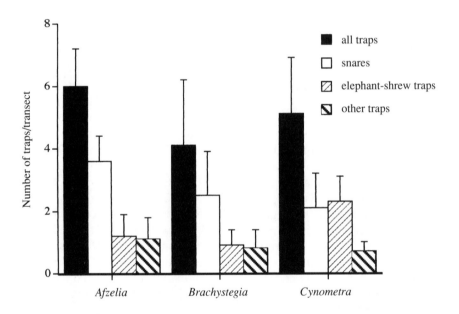

Figure 10.2 Relative abundance of traps (all traps combined; and snares, four-toed elephant-shrew traps and other trap types separately) in the three main habitat types.

areas with little prey, rather than not trap at all. Consequently, trapping is not density dependent and has the potential to reduce prey densities substantially around the edge of the forest.

10.4   CHANGING HARVESTING PATTERNS AND SUSTAINABILITY
       OF CURRENT OFFTAKE

Harvesting of wild animals from Sokoke has been taking place for many centuries, presumably sustainably. This section considers how harvesting patterns have changed and the consequences for prey populations.

10.4.1 Changes in the local community

The Sanya people are considered to be the indigenous inhabitants of the forest and were traditionally hunter-gatherers, harvesting fruits, medicinal plants and honey as well as wildlife. They lived in the forest until its gazettement in the 1950s, when they settled around the forest boundary and took up crop cultivation. They concentrated on the larger mammals, for example elephant and buffalo, and never hunted primates. The Mijikenda people arrived more recently, in the early 1900s, and were less discriminating, hunting a wide range of species, including both baboons and Syke's monkeys. While the Sanya people tended to travel around, moving on when game became scarce, the fact that the local people are now resident means that harvesting is more concentrated near areas of settlement. There is also some suggestion that the forest was divided into hunting territories, each group of households hunting a particular area. This system has broken down so that it is no longer worthwhile for hunters to protect wildlife stocks in their area.

10.4.2 Increase in human population density and loss of forest habitat

Over the last century, the whole coastal belt has experienced a rapid increase in population density, vastly increasing the demand for natural resources and for farm land. For example, in the coastal province of Kenya, population growth is estimated at 3.8% per year, partly due to migration into the area. The populations of Malindi and Mombasa have increased by up to 20% per year over the last decade, as people come in search of work in the tourist industry. Such population increases have put pressure on the supplies of natural resources, and the coastal forests have suffered heavily as they provide building materials, commercially viable timber, and a stock of unclaimed farmland. As a result, the area of forest has declined rapidly and harvesting intensity on the remaining wildlife populations has increased.

10.4.3 Changes in the use of harvested products

In the past, both hunting and trapping were carried out primarily to provide food for the family, although some game meat and other animal products such as elephant tusks and leopard skins were sold. Now, there is little commercial harvesting, and the majority of trapping is to provide food and to control crop pests. The forest is surrounded by agricultural land, households relying on their crops of maize, cassava, bananas and coconuts for most of their food and income. Mice, rats, Syke's monkeys, baboons, bushpigs and elephants cause extensive damage, many households losing between half and three-quarters of their crops. It is not surprising that trapping is now particularly focused on these species and it is the poorer people, who have few alternative sources of food, who do most of the harvesting (Figure 10.3). There is no hunting for sport, except by young boys who kill birds with catapults for fun, occasionally eating the bodies but generally feeding them to the dogs. Specific species are also killed as charms, for example the golden-rumped elephant-shrew.

10.4.4 Changes in traditional hunting practice

Hunting is becoming less common while trapping is on the increase, partly because hunting is more risky (both in terms of the risk of injury and the risk of being caught by forest guards) and also more time consuming. Although hunting is more successful overall, yielding an average of 5.5 kg meat per hour spent compared with 2.9 kg per hour from trapping, trapping is more reliable (34% of hunting trips are unsuccessful

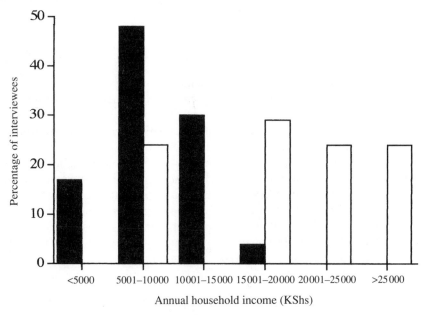

Figure 10.3 Annual cash incomes (KShs) for hunting (solid blocks) and non-hunting households (open blocks, n = 40 households).

while it is rare for all the traps set by a trapper to be unsuccessful – on average about a third of traps will catch per day). Hunting is also becoming less common because of a lack of men to participate. Traditionally only men hunted and set traps, and usually this is still the case, the father of the household doing most of the game harvesting, sometimes accompanied by his son(s). Because the men are now often employed elsewhere and the boys are at school, hunting parties now tend to be smaller and trapping, which can be done alone, is becoming the favoured form of harvesting. The change from hunting to trapping particularly affects the pattern of harvesting and the type of species that are harvested (see below).

10.4.5 Introduction of religion

The introduction of both Christianity and Islam into the region has not affected people's interest in game meat harvesting. It has, however, affected their belief in the power of the witchdoctors and as a result thieving of traps and their contents has become more common. In the past, for example, some people believed that anyone who stole a carcass would die from the spell cast on them by the trap owner. Now people have to set traps close to their homes so that they can check them regularly, increasing pressure on wildlife populations around the periphery of the forest.

(a) Consequences of changing harvesting patterns

So what are the consequences of the increased offtake levels and the changing harvesting patterns? A comparison of current offtake levels with very approximate estimates of maximum sustainable harvests suggests that the primates are over-harvested while current harvests of elephant-shrews, squirrels and duikers are probably sustainable (FitzGibbon *et al.*, 1995a) (Table 10.3). Bushpig harvests also appear to be sustainable. Over-harvesting has already reduced the densities of large ungulates, including bushbuck *Tragelaphus scriptus* and buffalo *Syncerus caffer*, to such low levels that hunting and trapping them is rarely worthwhile. The effect of harvesting clearly varies from one prey species to another.

What causes these differences in susceptibility to over-harvesting? One very important factor is the relative vulnerability of the species to trapping or hunting. Trapping has the ability to reduce population densities substantially – for example, reducing the abundance of four-toed elephant-shrews and squirrels by 41% and 66%, respectively (FitzGibbon *et al.*, 1995a) – but trapping is concentrated around the forest edge and consequently 60% of the forest is relatively unaffected. As a result, for

Table 10.3 Current population densities (1992), total annual harvest rates (from both hunting and trapping) and maximum sustainable offtake levels (all in no/km^2) for those mammal groups for which density estimates could be calculated. See FitzGibbon *et al.* (1995a) for calculation of maximum sustainable harvest rates.

	Current density	*Current harvest*	*Maximum sustainable harvest*
Four-toed elephant-shrews	391	15	274
Golden-rumped elephant-shrews	59	8	20
Syke's monkeys	58	3	1
Baboons	16	2	< 1
Duikers	63	1	3
Squirrels	11	1	6

those species which are only vulnerable to trapping, rather than hunting, the centre of the forest acts as a refuge. In contrast, hunting occurs throughout the forest and therefore species vulnerable to hunting, particularly the large ungulates and the primates, have no refuge. Thus the increase in trapping and the decline in hunting may actually be beneficial in terms of making harvesting more sustainable.

Several additional factors make the primates particularly susceptible: their low reproductive rates (Bodmer, 1994, in press); the fact that they live in groups and therefore more than one individual can be killed in one hunting foray; they are crop pests and are therefore targeted by farmers living adjacent to the forest; and they are easily attracted to bait and are therefore relatively easy to trap. The large ungulates were also targeted because their meat was so much in demand.

10.5   RESOLVING THE CONFLICT

Arabuko-Sokoke Forest is a very important site for mammal conservation. Current harvesting is apparently not affecting populations of the golden-rumped elephant-shrew. Its impact on the Ader's duiker and bushy-tailed mongoose populations is not known, but sightings of these species are now very rare. In addition, although none of the species negatively affected by harvesting is particularly endangered, loss of any one of them would reduce the conservation value of this forest and loss of the primates in particular would reduce the attraction of the forest for tourists. It is hoped that tourism will be a major source of income for forest management in the future and a nature trail and visitors centre have already been set up. It is clear that some sort of limitation of harvesting is required.

On the other side of the coin, Kilifi district (the administrative district encompassing the forest) is one of the poorest in Kenya. Despite a thriving tourist industry on the coast, average per capita income in 1992 was only 1500KShs (about £30), living conditions are poor and child mortality is high. Consequently, bushmeat is an essential source of protein for many families around the forest. The short-term benefits derived from the forest by people living locally are restricted, and are rarely sufficient to compensate them for the disadvantages, particularly the damage done to their crops by forest animals. Currently around 500 households directly benefit in cash from forest use or employment, while 700 households make significant savings through collecting bushmeat, firewood, building poles, medicinal plants, honey, fruits, mushrooms and water (Mogaka, 1991). It would be advantageous to allow people to benefit from game harvesting, but it is essential that harvesting should not be allowed to threaten the continued existence of mammal prey species.

The simplest solution would be to stop hunting, particularly of primates and large ungulates, while allowing trapping to continue around the forest edge. Since trapping affects only 40% of the forest, its impact is limited. This solution would reduce overall harvests by approximately 10% and still enable households living adjacent to the forest to control crop raiding and obtain supplies of game meat. A greater proportion of game meat would be derived from the smaller species, particularly elephant-shrews and rodents, which have higher reproductive rates and population densities and which can therefore withstand higher offtake rates than the larger species, such as primates.

Although allowing trapping to continue around the forest edge at current levels would probably be sustainable, there may be two problems with this approach. Firstly, as we have seen, the intensity of trapping is not dependent on prey abundance, and trapping can substantially reduce prey densities. Trapping around the edge of the forest may therefore drain the central portion of prey species, animals moving to the edge as densities there decrease. The second problem is that, if it became known that trapping in the forest was condoned, the intensity of trapping may increase, threatening prey populations. Furthermore, people may be tempted to set up temporary camps in the centre of the forest, where prey densities are higher, and start trapping there. However, more complex control systems – for example, licensing a restricted number of trappers and hunters or creating hunting zones – would probably be too difficult to implement and enforce, given the restricted resources currently available to forest managers. In the short term the loose arrangement suggested here is probably most likely to succeed, but an effective monitoring programme is clearly required in order to detect changes in harvesting levels and prey population densities.

ACKNOWLEDGEMENTS

We thank the Kenya Government for permission to undertake this study as part of the Kenya Indigenous Forest Conservation Project (KIFCON) and the joint National Museums of Kenya–International Council for Bird Preservation (ICBP) Arabuko-Sokoke Forest Project; Anthony Githitho, District Forest Officer, Kilifi, for allowing us to work in Arabuko-Sokoke Forest; and David Changawa, Francis Charo and David Ngala for extensive help in the field. The study was funded by the Overseas Development Administration, Frankfurt Zoological Society and BirdLife International (formerly ICBP).

REFERENCES

Asibey, E.O. (1974) Wildlife as a source of protein in Africa south of the Sahara. *Biol. Cons.*, **6**, 32–39.

Balmford, A., Leader-Williams, N. and Green, M.J.B. (1992) Protected areas of Afrotropical forest: history, status and prospects, in *Tropical Rain Forests – an Atlas for Conservation*, Vol. II Africa, (eds M. Collins, M. Sayer and J.A. Sayer), MacMillan, London, pp. 69–80.

Bell, R.H.V. (1987) Conservation with a human face: conflict and reconciliation in African land use planning, in *Conservation in Africa: people, policies and practice*, (eds D. Anderson and R. Grove), Cambridge University Press, Cambridge, pp. 79–101.

Bodmer, R.E. (1994) Managing Amazonian wildlife: biological correlates of game choice by detribalized hunters. *Ecol. Appl.*, **5**, 872–877.

Bodmer, R.E. (in press) Susceptibility of mammals to overhunting in Amazonia. *J. Wildl. Manage.*

Bodmer, R.E., Fang, T.G., Moya, L.I. and Gill, R. (1994) Managing wildlife to conserve Amazonian forests: population biology and economic considerations of game hunting. *Biol. Cons.*, **67**, 29–35.

Burgess, N., FitzGibbon, C.D. and Clarke, P. (in press) Coastal forests, in *Ecosystems and their Conservation in East Africa*, (ed. T. McClanahan), Longmans, Nairobi.

Collar, N.J. and Stuart, S.N. (1988) *Key Forests for Threatened Birds in the Afrotropical and Malagasy Realm*, Monograph 3, International Council for Bird Preservation, Cambridge, UK.

Ehrlich, P.R. and Wilson, E.O. (1991) Biodiversity studies: science and policy. *Science*, **253**, 758–762.

FitzGibbon, C.D. (1994) The distribution and abundance of the Golden-rumped Elephant-shrew *Rhynchocyon chrysopygus* in Kenyan coastal forests. *Biol. Cons.*, **67**, 153–160.

FitzGibbon, C.D., Mogaka, H. and Fanshawe, J.H. (1995a) Subsistence hunting in Arabuko-Sokoke Forest, Kenya and its effects on mammal populations. *Cons. Biol.*, **9**, 1116–1126.

FitzGibbon, C.D., Leirs, H. and Verheyen, W. (1995b) Population dynamics, distribution and habitat choice of the lesser pouched rat, *Beamys hindei*. *J. Zool.*, **236**, 499–512.

Hart, J.A. (1978) From subsistence to market: a case study of the Mbuti net hunters. *Hum. Ecol.*, **6**, 32–53.

Kelsey, M.G. and Langton, T.E.S. (1984) *The Conservation of the Arabuko-Sokoke Forest, Kenya*, International Council for Bird Preservation and University of East Anglia, International Council for Bird Preservation Study Report no.4.

Marks, S.A. (1989) Small-scale hunting economies in the tropics, in *Wildlife Production Systems*, (eds R.J. Hudson, K.R. Drew and L.M. Baskin), Cambridge University Press, Cambridge, pp. 75–95.

Martin, G.H.G. (1983) Bushmeat in Nigeria as a natural resource with environmental implications. *Environ. Cons.*, **10**, 125–134.

Mogaka, H.R. (1991) *Local utilization of Arabuko-Sokoke Forest Reserve.* Unpublished report to Kenya Indigenous Forest Conservation Project, Overseas Development Administration, Nairobi, Kenya.

Mogaka H.R. (1992) *A report on a study of hunting in Arabuko-Sokoke Forest Reserve.* Unpublished report to Kenya Indigenous Forest Conservation Project, Overseas Development Administration, Nairobi, Kenya.

Nicoll, M. and Rathbun, G. (1990) *African Insectivora and Elephant-shrews: an action plan for their conservation*, International Union for the Conservation of Nature, Gland.

Stiles, D. (1981) Hunters of the northern East African coast: origins and historical processes. *Africa*, **51**, 848–861.

Wilkie, D.S. (1989) Impact of roadside agriculture on subsistence hunting in the Ituri forest of northeastern Zaire. *Amer. J. Phys. Anth.*, **78**, 485–494.

11

The impact of sport hunting: a case study

David W. Macdonald and Paul J. Johnson

SYNOPSIS

The significance of sport hunting, as a source of rural revenue and infrastructure, a means of wildlife management and a catalyst for conservation, is often debated and less often quantified. A major difficulty in any such evaluation lies in comparing the different, often non-convertible currencies in which different factors may be measured. Such problems are at the root of many conservation dilemmas world-wide. As an example this chapter presents an analysis of some of the factors relevant to the functioning of hunting foxes with hounds in the UK. Although it has welfare implications, this example is not in itself a major conservation issue in terms of preservation of biodiversity, but it serves to illustrate in microcosm many recurrent issues in debates about wildlife. The results of a questionnaire survey of farmers (as users of, hosts to and participants in foxhunting) are presented along with data on the performance (e.g. annual tallies) of packs of hounds. These data are explored in terms of regional differences, and assessed in the context of conservation in the agro-ecosystem and in terms of fox population dynamics. The main aim is to disentangle and, where possible, to make a start at quantifying the threads of what has hitherto often been a muddled debate.

11.1 INTRODUCTION

Whether the goal be preservation of the imperilled, control of the pestilential, or any other of the diverse shades of relationship between people and the remainder of nature, there is one common characteristic of all issues in conservation. That characteristic is the entanglement of facts, figures and philosophies drawn from completely different ideological and scientific frameworks – we call this interdisciplinarity. Through the daunting algebra that is the conservationist's trade, the interdisciplinary equation must be solved, or at least exposed en route to an answer: the best practicable course of action. The consequences of hunting, whether for any or all of food, fur, control or sport, is just such an interdisciplinary issue.

Hunting, as one form of exploiting wildlife, may sometimes touch on major conservation issues. Decisions regarding the trophy hunting of big cats, ivory harvest of elephants or waterfowl crop of wetlands may affect global biodiversity, national economies and the recreational or employment profiles of large segments of societies. By comparison, debate about the consequences of some other hunting issues may be much more parochial, but nonetheless beset by the same daunting interdisciplinarity. In this chapter, as an exercise in disentanglement, we have chosen to explore some factors relevant to an evaluation of hunting foxes with hounds, not because it is a major conservation issue, but because it illustrates in microcosm the diverse facets of any such evaluation. Our approach is to identify, and calibrate, factors amenable to scientific scrutiny, and to separate those from aspects of the evaluation, such as ethical considerations, which may not be quantifiable. We seek to scrutinize scientific evidence which (although science may not be value-free) can inform but not necessarily dictate some ethical judgements. Our results are neither complete nor definitive, but nonetheless we hope they help to separate elements of a debate that, hitherto, has often appeared muddled.

Foxhunting on horseback with hounds has been popular since at least the fourth century BC. In the intervening years the role of the fox has varied (Longrigg, 1975; Macdonald, 1987; Cartmill, 1993). The Ancient Greeks hunted foxes and so, more enthusiastically, did the Romans. Aristotle interpreted hunting as a divine gift to mankind: if foxes were huntable, the gods had put them there to be hunted. In the Dark Ages in Europe, in a society reputedly besotted with hunting, foxes began as low status quarry: King Canute classified them as Beasts of Chase, a lesser category than Beasts of Venery. Nonetheless, foxes slowly climbed the ladder of merit as quarry and by the late thirteenth century King Edward I had a royal pack of foxhounds; by the Renaissance foxhunting was an indispensable part of a polite man's attainments, and by the eighteenth century the explosive popularity of foxhunting was affecting the British landscape dramatically. Between 1760 and 1797 more than 1500 Enclosure Acts took common land out of small holdings and converted it to parks and grazing land suitable for foxes and foxhunting. The proliferation of wild rabbits, originally imported to Britain by the Normans, also dates from around 1750 and doubtless increased the carrying capacity of the countryside for foxes. Between 1800 and 1850 it is said that the area of gorse in Leicestershire was doubled in an attempt to foster habitat suitable for foxes. The extent to which eighteenth century hunting shaped the countryside, or the new agricultural environment of the country suited hunting, may be debatable, but it remains that foxhunting has, historically, a major role in rural tradition. Of course, being a venerable tradition is evidence neither for nor against contemporary acceptability.

Clearly, during this golden era of English foxhunting, foxes were hunted unequivocally as quarry not vermin (indeed, hunts paid farmers to spare breeding earths on their land). Then, in the late eighteenth century, the advent of pheasant shooting brought radical change: the fox-hunter wanted many foxes and the gamekeeper wanted none. During this era, gamekeepers effectively eradicated such predators as polecats, martens and wildcats, and it is often said that foxes tended to survive this onslaught only where they were protected as a quarry for hunting. Simultaneously, agriculture began to intensify and in 1882 the arrival of barbed wire ended that golden age. Previously, advocates of foxhunting relied solely on unabashed singing of its praises as a sport, but by the twentieth century foxes shared the countryside with foxhunters, farmers, gamekeepers and an ever more politically enfranchised public, and the debate about foxhunting, always controversial (see More's *Utopia*), had become a minefield of quasi-scientific propaganda and dogma.

In an attempt to inform that debate we tackle five broad categories of question.

1. What is the organization, activity pattern and *modus operandi* of fox-hunting? This assessment of the scale of foxhunting gives some insight into its role in the rural socio-economic infrastructure (such as stimulating trade for farriers, saddlers, etc.).
2. What numbers of foxes are hunted and killed by fox hunts and, extrapolating from these tallies, what role may hunting play in fox population dynamics? This is essentially a consideration of the hunt as a predator, its foraging tactics and impact on its prey. However, aspects of this predatory behaviour, such as the length and duration of chases, and their varied outcomes, have some bearing on evaluating suffering endured by hunted foxes.
3. What impact does foxhunting have on the management of the countryside?
4. What are the perceptions of such interest groups as farmers, huntsmen and city-dwellers about foxes, their status as pests, and the role of foxhunting? Very importantly, we have tried to distinguish between what the facts are and what people think the facts are (and we highlight in the text danger points where we have no way of verifying information).
5. Aside from its possible roles in pest control and recreation, does foxhunting provide a useful measure for enumerating fox populations?

Our data come from a variety of diaries and questionnaires. Some data, for example on the opinions and behaviour of farmers, stem from the early 1980s. Some things may have changed in the meantime; in the case of the farmers' opinions they may have shifted with changes in the Common Agricultural Policy and behaviour altered in line with, for example, countryside stewardship schemes. Any such changes will be interesting, and amenable to quantification in subsequent studies, but do not affect our main purpose here, which is to expose the interwoven threads of biological, economic and sociological factors that characterize this, and all other, conservation issues. An important proviso throughout is that foxhunting differs greatly between the packs accompanied by mounted followers, often in the lowlands, and those followed on foot in upland, sheep-farming districts. Much of our account concerns the former.

11.2 METHODS

There are 206 registered hunts in Britain. Each one occupies a territory or 'country'. Between them, the hunt countries occupy a total area of approximately 145 000 km². In the south they are effectively contiguous. In upland sheep-farming areas the so-called 'foot packs' are followed on foot. A lowland hunt comprises mounted followers with a pack of hounds, and usually a number of followers in cars.

There are two parts to the foxhunting season: in September and October, 'cubhunting' takes place. This gives experience to young hounds and is said to promote the dispersal of foxes. It is also said by some hunters to be the time when greater emphasis is placed on killing, rather than chasing, foxes. Between November and late March, foxhunting proper occurs. On each hunting day, the hunt convenes at a particular district or 'meet'.

11.2.1 Questionnaire survey of hunt masters

We undertook a questionnaire survey of hunt masters in order to tackle such questions as:

1. What are the activity patterns of hunting packs of hounds?
2. What is the employment structure of hunts?
3. How many foxes do they move, hunt and kill?
4. By what means are these foxes killed?

In 1980, the masters of all 206 hunts in Great Britain were circulated with a questionnaire soliciting information on the size of their hunts in terms of the numbers and categories of followers, and whether the different trades of hunt servant were employed. They were also asked to supply, for as many years as possible during the previous decade, the number of foxes killed in each season's hunting, and the number of days on which the hunt was active in the season. The questionnaire was circulated with a letter from the Master of Foxhounds Association (MFHA) which urged masters to respond. A total of 80 masters provided responses to this initial questionnaire, and 60 yielded usable data on tallies.

In the following three years the masters were circulated with a daily diary sheet asking for details of each meet. They were asked to record, for each hunting day of the season, the numbers of foxes moved, hunted and killed (and the nature of the kill). Doubtless, the number of foxes reported as moved is more prone to error than the number reported killed, but the observations of professional hunt staff and scrutiny of large numbers of strategically placed followers certainly represents very intensive observer effort. Some masters volunteered to extract all or some of this information from their diaries for previous years. Obviously, a diminishing number of masters were able to provide comprehensive information for earlier years. Thus while 26 masters extracted data from their diaries for 1970/71, 53 completed diary sheets as the 1979/80 season progressed.

In 1994, we updated our data by circulating a further questionnaire, via the MFHA, to those hunt masters, or their successors, who had responded to the first questionnaire. Of 60 of these circulated, 39 were returned.

Data on the activity of upland foot packs in Wales was obtained from the Welsh Farmers Fox Control Association (WFFCA), which is outside the MFHA. With the cooperation of the WFFCA, a questionnaire was circulated in spring 1995 to 28 packs. A total of 18 were returned, giving data on the number of foxes killed in the 1993/94 season, and in the (incomplete) 1994/95 season. Also recorded were the number of lambs said to have been lost to foxes within the area they covered in 1993/94, and the number of farms affected.

We used a simple model to put the mortality levels estimated from these surveys into the context of what is known of the biology of fox populations. The VORTEX program (Lacy, 1993; Lacy *et al.*, 1994) was used for this exercise. VORTEX is a simple model and not suitable for exploring the intricacies of social organization in population processes. However, here we require an illustration of the likely impact of given mortality pressures on fox populations, and for this VORTEX is adequate. Thus, the intention was not to attempt a comprehensive analysis of all the factors likely to be influential, but to illustrate the likely impact of variation in total annual mortality in terms of population trends, given some realistic starting conditions. For this illustration, a population at a density and reproductive rate typical of the English Midlands was modelled.

In addition to the data derived from the questionnaire of masters, one pack in the English Midlands was followed for a total of nine days between mid September and early December 1979 to gather data on the duration of pursuits. During this period, 17 pursuits were observed.

11.2.2 Habitat structure and foxes moved

To explore the hunt's behaviour as a predator (albeit not one dependent for its survival on eating its prey), we investigated its success in finding foxes in different habitats. In doing so we had in mind that Macdonald *et al.* (1981) found an association between estimates of fox population density and habitat structure in contrasting regions of Britain. In tackling the question of whether, on a local scale, it was possible to relate numbers of foxes moved to these characteristics, we were able to investigate whether hunt records have potential as an index of fox numbers. The number of foxes moved by hounds was thought to be less affected by human intervention than was the number hunted or killed.

We concentrated upon two contiguous packs, the Bicester and Warden Hill, and the Heythrop, which hunted areas of *c.* 414 km² and 459 km², respectively (in Buckinghamshire, Oxfordshire and Northamptonshire), and had detailed records spanning 17 and 20 years. Mean numbers of foxes moved per day's hunting were calculated only for those meets at which there had been five or more hunts. The location of the meet is a variably good indicator of the area hunted.

Landscape characteristics were read from Ordnance Survey Maps (1:50 000) and, in the case of the Second Land Use Survey data (Coleman 1961, 1970; Coleman and Mags, 1965), from 6 inch:1 mile maps. Some of the topographical and climatological variables were selected so that we could use Bunce *et al.*'s (1981) system of land classification, whereby individual 1 km × 1 km squares were allocated to land classes. Other habitat features were selected because they seemed likely to be relevant on the basis of the natural history of the fox (Lloyd, 1980; Macdonald, 1987). Analyses explored the square kilometre around the meet and also the summed information from this and the eight surrounding squares. Consideration of these surrounding squares was relevant because the hunting pack often moves outside the square kilometre centred on a meet.

To compare the two hunt countries we used an ordination technique, DECORANA. To relate the fox abundance index to habitat characteristics we used Multiple Linear Regression analysis (MLR).

11.2.3 Questionnaire surveys of farmers and town dwellers

(a) Farmers' questionnaire

We used a questionnaire to tackle such questions as:

1. What are farmers' attitudes to, and levels of participation in, foxhunting?
2. How do these attitudes relate to their approach to conservation management of their farms?
3. How many foxes do they kill annually?

These questions were phrased within a questionnaire comprising a series of 130 questions soliciting information concerning all aspects of farming practices relevant to wildlife and was dispatched with a pre-paid reply envelope in 1980 to 2288 farmers in 10 regions of England (Macdonald, 1984; Macdonald and Johnson, submitted). We sought to represent each of the major agrarian regions of England. The regions were selected after examination of distribution maps in Coppock's (1976) *Agricultural Atlas of England and Wales*. Of the questionnaires, 867 (38%) were returned.

The questionnaire was designed after extensive consultation and pilot studies with farmers. It provided data for several projects, making it effectively impossible for farmers to anticipate how we would analyse their answers (only a small proportion of which concerned foxhunting). Questions of direct relevance to the current study included the farmer's estimate of how many foxes had been killed (by all methods) on their farm in the previous 12 months, the number of earths they knew to be present within the farm, and a series of questions about their attitudes to foxes (their perceived ranking as pests) and foxhunting, and their involvement with the local hunt. We have no means of verifying the various measures provided by respondents, and in interpreting the results we are mindful of the many potential biases associated with questionnaire surveys.

In 1992, a second questionnaire survey of farmers was carried out. This was devised primarily to assess farmers' perceptions of some common pests. With the aid of the National Farmers' Union (NFU), 460 questionnaires were distributed to a representative sample of farms in England, Scotland and Wales. Of these, 157 (34%) were returned. Responses concerning the amount of time and money lost because of foxes were used as an indication of the scale of economic damage attributable to them, as perceived by the farmers.

(b) Urban questionnaire

While mapping the distribution of urban foxes in Oxford in 1978 a questionnaire was delivered to each of 14 000 households and collected the next day. Most questions concerned sightings and experience of foxes (Macdonald and Newdick, 1980; Macdonald and Doncaster, 1985), but we also included some identical to those asked in the farmers' questionnaire. Our intention was to tackle the question: how does the background of the respondents influence their attitudes to foxes and foxhunting? Of the questionnaires delivered door-to-door, 3469 were returned.

11.3 RESULTS

11.3.1 The structure of a hunt

The following of hunts, in terms of participants, did not vary significantly in size with region, as reported by the masters who responded (Table 11.1). A typical day's hunting involved approximately 80 participants, including mounted followers and others. The area of land covered by hunt countries varied widely in all regions, though, on average, hunt countries in the North and in Wales and the West were smaller than hunts in the South and in the Midlands. The areas covered by the WFFCA foot packs were smaller than those of mounted hunts: the average size was 122.9 km², less than a quarter of the average size of mounted hunt countries in this region (Table 11.1).

Masters were asked about the numbers of hunt servants employed. In addition to a huntsman and one or more whippers-in, a hunt may employ terrier men, whose task is to dig out, with the aid of their dogs, a proportion of the foxes that have gone to ground. In traditional packs, 'earthstoppers' may also travel the vicinity of the day's hunt before dawn blocking foxholes to prevent the foxes escaping detection by hounds by spending the day underground, and to prevent hunted foxes seeking underground sanctuary. There was a tendency for hunts in the Midlands and Eastern region to have more earthstoppers than elsewhere (Table 11.2). Hunt countries in the North and in the Wales and West region cover smaller areas than in the South and East, and contained significantly fewer shooting estates and game keepers. An average of approximately one terrier man per hunt was employed (Table 11.2).

There was an average of 25 days cubhunting per season. Hunting proper occupied an average of a further 70 days per season. Hunts tend to convene two or three times in each week, and this did not vary significantly with region (Table 11.1).

Using 19 years of data from the Bicester and Warden Hill Hunt, there

Table 11.1 Hunt sizes and activity in 1981: there were no significant regional differences (MANOVA, Wilks lambda = 0.71, $F_{15,125.6}$ = 1.08, P = 0.38, all univariate ANOVAs ns)

Characteristic ($n = 80$)	Median	Mean (SE)
Mounted followers	50	52.2 (2.7)
Followers in cars	30	31.9 (2.2)
Subscribers in 1981	107.5	126.9 (11.6)
Subscribers in 1971	80	105.0 (12.1)
Subscribers in 1961	51	86.6 (13.1)
Days hunting per week	2	2.5 (0.08)

Table 11.2 Composition of hunts in 1981, and the pattern of regional variation (MANOVA, Wilks lambda = 0.28, $F_{33,100.9} = 1.64$, $P = 0.0325$)

Characteristic	Midlands and East	North	South	Wales and West	Univariate F	Overall mean (SE)
Hunt areas (km²)	860.7 (81.35)	589.2 (45.16)	839.4 (127.24)	634.4 (76.87)	$F_{3,197} = 2.53, P = 0.0532$	732.3 (43.1)
Percentage of hunt area owned by hunting participants	16.3 (3.43)	24.5 (4.99)	18.3 (2.92)	32.6 (9.37)	$F_{3,69} = 2.86, P = 0.0435$	21.5 (2.39)
Number of earthstoppers	12.9 (4.71)	5.4 (2.23)	6.2 (2.53)	0.8 (0.71)	$F_{3,74} = 2.64, P = 0.0559$	7.2 (1.79)
Number of terrier men	0.8 (0.14)	0.6 (0.20)	0.9 (0.17)	0.4 (0.17)	$F_{3,77} = 2.36, P = 0.0781$	0.7 (0.09)
Number of fence men	0.8 (0.37)	0.6 (0.21)	0.5 (0.12)	0.1 (0.14)	$F_{3,76} = 1.47, P = 0.2297$	0.6 (0.13)
Number of farmers in the hunt	642.9 (89.06)	321.5 (55.30)	442.1 (63.14)	488.5 (139.39)	$F_{3,69} = 2.86, P = 0.0435$	488.9 (43.32)
Number of hunting farmers	65.3 (28.48)	30.0 (4.82)	34.3 (7.74)	54.4 (16.36)	$F_{3,72} = 0.86, P = 0.4638$	46.28 (9.69)
Number of farmers banning the hunt from their land	6.7 (1.75)	2.6 (0.97)	5.1 (1.12)	4.7 (1.23)	$F_{3,75} = 2.23, P = 0.0972$	4.94 (0.71)
Number of farmers discouraging the hunt	1.68 (0.70)	1.7 (0.89)	2.5 (1.16)	0.4 (0.34)	$F_{3,71} = 0.61, P = 0.6085$	1.76 (0.47)
Number of shooting estates within the hunt	16.0 (2.67)	7.2 (0.92)	14.3 (2.93)	2.6 (0.74)	$F_{3,71} = 10.9, P = 0.0001$	11.2 (1.32)
Number of game keepers employed within the hunt	24.3 (4.49)	12.2 (3.02)	19.0 (2.50)	4.5 (1.09)	$F_{3,76} = 8.01, P = 0.0001$	16.9 (1.92)

was a significant relationship between the number of foxes killed per day at a meet and the frequency of visits there (excluding sites visited fewer than five times in 19 years: $n = 92$ meets, $r = 0.26$, $P = 0.0120$). The average proportion of hunting days that were at meets visited only once each season was 47.9% (32.3% twice, 13.1% three times, 8.1% four times). The pursuits observed ranged in duration between 4 and 64 minutes, with an average of 16.8 minutes ($n = 17$, SE = 3.88).

Scrutiny of the daily diaries from 36 hunts revealed that, of a total of 2062 kills for which sufficient detail was recorded, 1408 (67.5%) were effected by hounds catching a fox in a chase above ground. The rest went to ground, and were either dug out and shot, killed underground by terriers, or flushed out by the terriers and subsequently caught by the hounds. There were large between-hunt differences in the proportion of kills made above ground ($\chi^2_{[35]} = 385.2$, $P \ll 0.001$). Some masters recorded in the diaries that a particular fox was dug out and shot at the request of the farmer on whose land it was caught.

Some hunts made the majority of their kills after foxes had gone to ground and the maximum proportion of 'underground' kills was 74.8% (80 out of 107). There was evidence that hunts killing the highest proportion of foxes underground tended to be in the Wales and West region. The mean proportion in this region was 0.55 ($n = 10$, SE = 0.07), compared with between 0.21 and 0.26 elsewhere (ANOVA of the square root arcsine transformed proportions, weighted by the total foxes killed in each hunt, $F_{3,32} = 2.53$, $P = 0.075$). If Wales and the West is excluded, the proportion of foxes killed after going to ground is essentially constant across other regions at 0.25 ($n = 26$, SE = 0.06).

11.3.2 The numbers of foxes killed by hunts

Some hunts have long and fascinating records. Figure 11.1 shows the rate at which one typical Midlands hunt, encompassing almost entirely arable land (in Hertfordshire), killed foxes between 1838 and 1993. While in the majority of seasons the rate at which foxes were killed fell within the range 0.6–0.8 per day, some conspicuous peaks and troughs can be identified. Some of these are undoubtedly due to factors intrinsic to the hunt. For example, the number of hounds was reduced to a minimum by World War II, and efficiency was well down in the early 1940s. In 1970, the hunt amalgamated with a smaller neighbouring hunt, said to be less well 'foxed', and fewer foxes were caught in the years immediately following. Dry winters, such as those of the late 1980s, provide poor conditions for hounds to track the scent of the fox. Other fluctuations may reflect real changes in fox numbers. In the early years of the 1960s the use of poisoned pigeons was said by the master of the day to have reduced fox numbers (see also Lloyd, 1980).

Puckeridge Hunt

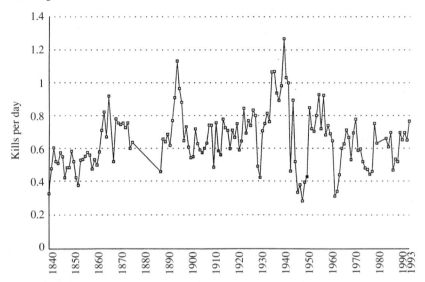

Figure 11.1 Mean number of foxes killed per day's hunting by the Puckeridge Hunt between 1838 and 1993.

Figure 11.2a shows the national average annual rate of fox kills by hunts for the last three decades, derived from the records of the hunt masters. There is an upward trend, particularly as far as the most recent data are concerned. Fluctuations due to harsh winters, such as those of 1962/63 and 1978/79, can be identified. (The latter season also coincided with a peak in the trade in fox skins, Macdonald and Carr, 1981). However, between 1960 and 1980, the rate did not deviate very far from *c.* 0.1 fox killed per km^2 per year. Extrapolated to the entire area covered by hunt countries, this suggests that approximately 14 500 foxes were killed annually by hunts during this period. In the latter years of the 1980s the rate was close to 0.15, corresponding to an annual total kill of 21 750. If a linear regression model is fitted to the tallies data for the 1980s (repeated measures approach, omitting the Wales and West data where the sample size was very small), there was a significant interaction between year and region ($F_{33,198} = 1.7$, $P = 0.015$), indicating that the regional trends were different. Inspection of these trends (Figure 11.3) suggests that this can be attributed to a more conspicuous upward trend in the South. The main effect of region was also significant in this model ($F_{3,18} = 3.3$, $P = 0.043$). Figure 11.2b shows the same tally data, weighted according to the size of the hunt country. These rates are lower indicating that more foxes tend to be killed per unit area in small hunts. On a local scale, tallies may be higher, as variable proportions of each hunt's

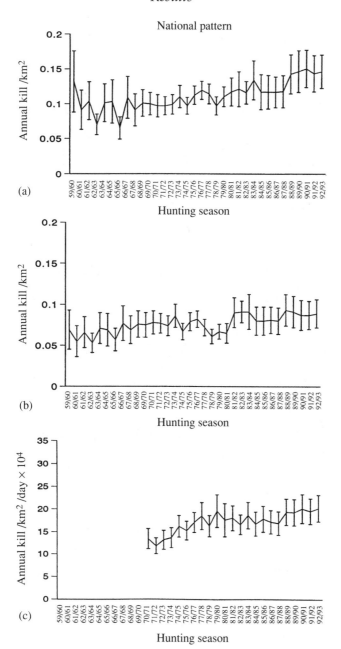

Figure 11.2 National pattern of fox kills by hunts between 1959 and 1993. (a) Simple mean density of kills over all the hunts responding; (b) mean density weighted by area of hunt countries; (c) foxes killed/unit area per day's hunting. (Bars are 1 × SE.)

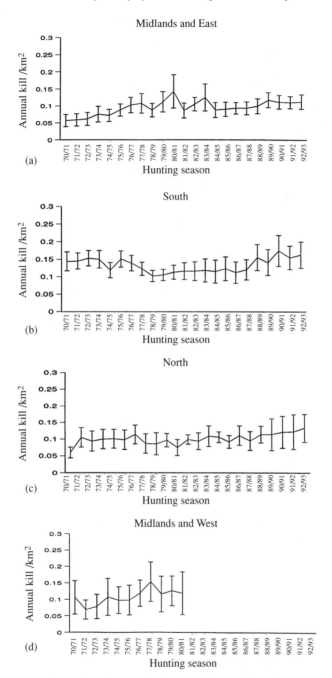

Figure 11.3 Regional breakdown of fox kills between 1959 and 1993. Simple mean densities for each season. (Bars are 1 × SE.) Means for Wales between 1980 and 1992 not plotted (*n* = 2).

country are inaccessible. The tallies for southern hunts tended to be higher; Macdonald and Newdick (1980) recorded a similar pattern with respect to urban foxes, which were found to be more prevalent in the South and Midlands than elsewhere.

The upward trend in numbers of foxes killed is not the result of a tendency for hunts to convene more often in recent years. Figure 11.2c shows the number of foxes killed per unit effort: the number of foxes killed per km^2 per day's hunting. The pattern for these data was similar to that for density of kills, and there was also a significant interaction between year and region ($F_{33,176} = 1.7, P = 0.016$).

The density of kills reported by the WFFCA packs were far in excess of those by mounted hunts anywhere. In the 1993/94 season, the average kill was 2.58/km^2 (SE = 0.80). The average weighted by the area covered by the pack was 1.16 (SE = 0.34). For the incomplete 1994/95 season, the figures were similar at 2.30 (SE = 0.76) and 1.00 (SE = 0.34), respectively. These densities are at least an order of magnitude higher than those reported by mounted hunts in the same region (Figure 11.3).

(a) Cubhunting

Over the entire period for which we have data on the extent of kills during the cubhunting phase of the season as well as overall kills (1971–1992), the national proportion of total seasonal kills which occurred in this phase did not fall below 0.43 (1988/89 season, SE = 0.047) or exceed 0.51 (1991/92 season, SE = 0.049). There was no evidence of any regional variation. Analyses of variance on the proportion killed during cubhunting (arcsine square root transformed) all yielded non-significant effects for region (all $P > 0.05$). However, our sample size for the Wales and West region is small ($n = 2$), and there is a widely held view that a higher proportion of foxes are killed during cubhunting in this region. The average number of days spent in cubhunting varied between 22.7 (SE = 1.29) in the 1979/80 season (an exceptionally harsh winter) and 28.3 (SE = 2.20) in the 1990/91 season. This compared with figures for the same seasons in hunting proper of 61.4 (SE = 2.68) and 74.6 (SE = 5.50).

In terms of the number of foxes caught per unit effort, the difference was not always statistically significant but foxes were consistently 'easier' to catch during cubhunting than during the hunting season proper for every season for which we have data. Figure 11.4 shows the pattern for the last decade.

We investigated whether there was any evidence that a hunt's efficiency, in terms of the ratios of moved foxes hunted or of hunted foxes killed, changed over the course of a season. For this, we examined the records of the Bicester and Warden Hill Hunt for which we have the

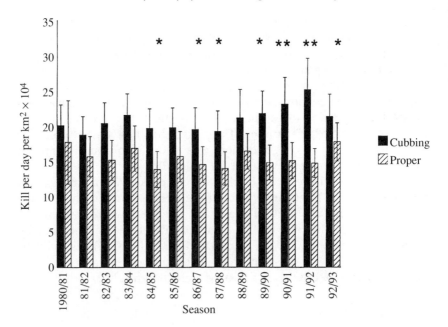

Figure 11.4 Comparison of rate of fox kills per unit effort during cubhunting and hunting proper during the 1980s. (Bars are $1 \times$ SE; $* P < 0.05$; $** P < 0.01$.)

longest sequence of hunt records detailing foxes moved, hunted and killed for each day's hunting. More foxes were moved at the beginning of the season ($F_{6,76} = 4.4$, $P = 0.0008$, Figure 11.5a), an observation which is consistent with the higher density of foxes at that time of year. There was, however, no difference between months in the daily mean number of foxes hunted ($F_{6,76} = 0.9$, $P = 0.5384$), suggesting that in September and October the hunt is in surplus with respect to its prey (this observation supports the use of numbers of foxes moved as a more reliable indicator of fox abundance than numbers killed). While the number hunted showed no variation across the season, the daily number killed did: more foxes were killed per day during cubhunting, as for the national pattern (above).

In terms of proportions, the number of hunted foxes relative to those moved was lower in the early stage of the season, while the proportion of hunted foxes that were killed was higher at this time (Figure 11.5b). The proportion of hunted foxes killed reached a low point at the end of the season. Non-exclusive and untested interpretations include those that later in the season less fit or less adept foxes have been weeded out, that survivors have learnt better escape tactics, or that the hounds hunt less effectively or persistently. The latter possibility is at odds with the belief that the hunting ability of a pack increases through the season.

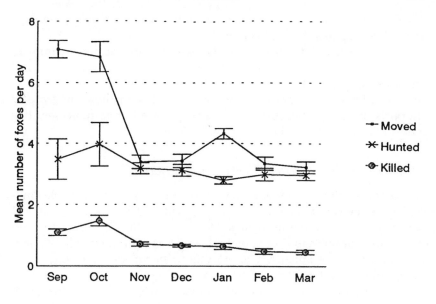

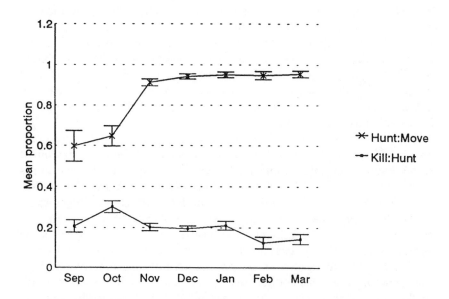

Figure 11.5 Mean seasonal activity pattern of the Bicester Hunt between seasons commencing 1964 and 1982. All days hunting within each month averaged, and the mean of these means presented; hence each point represents a mean for each month over all seasons represented. (Bars are 1 × SE.)

Another factor which may be relevant is that hunts are excluded from some shooting areas before February.

There were also differences between seasons in the mean daily number of foxes moved ($F_{17,76} = 6.5$, $P = 0.0001$), hunted ($F_{17,76} = 4.4$, $P = 0.0001$) and killed ($F_{17,78} = 2.0$, $P = 0.0195$). The relative smallness of the between-season difference in numbers killed compared with those moved also supports our suggestion that the numbers moved are a more sensitive index of fox abundance than are numbers killed.

(b) Number of foxes moved and habitat characteristics

We compared the Bicester and Heythrop countries in terms of the land-classes defined by Bunce *et al.* (1981), of which classes 1, 2, 4, 9 and 11 were represented in the study area. However, a DECORANA analysis of the habitat variables indicated that the landscapes of the two hunt countries differed substantially, with Heythrop being less hilly and more grassy than Bicester. These differences may explain why relationships between the mean number of foxes moved (MNM) and the five land-classes were inconsistent between the two hunt countries. The greatest MNM for Bicester (4.94 foxes moved/km^2 per hunting day) was recorded in Landclass 1, whereas the corresponding figure for Heythrop averaged 2.3 foxes. This landscape is characterized by gently rolling country, at medium to low altitude, comprising alluvial plain, low ridges or plateaux with little surface drainage. Hedges, mature trees and farm buildings are features of this varied lowland landscape, where the land use is chiefly cereals and good grasslands with limited native vegetation. The highest MNM for Heythrop was 4.3, in Landclass 4, which is characterized by virtually flat topography, almost entirely at low altitude, and comprises intensively farmed lowlands which are often built up.

Regression analysis of the whole Midlands data set (both hunts combined) revealed that half (50.8%) of the variance in MNM was explained by six habitat variables (Table 11.3). Separate analysis of the two hunts (each with about half the data sets and half the variance in MNM), showed that habitat variables were similarly effective predictors of MNM for both. However, the predictive habitat features differed between the two hunts. While for the Bicester Hunt sample, 50.7% of the variance in MNM was explained by six predictors and for Heythrop, 51.6% was explained, also by six predictors, only the area of parkland was a useful predictor in both cases (Table 11.3). This difference in useful predictors probably reflects the fundamental differences in the topography, ruggedness and land-use of the two areas, although both sets of predictors included attributes of hilliness. Indeed, the explanatory power of habitat variables dropped sharply on exclusion of those describing altitude and hilliness. Of the remainder, only 7.5% of the variance in MNM

Table 11.3 Stepwise linear regression analysis of mean number of foxes moved (MNM) against habitat variables

	Variable	*Students t*	*Variance explained*
Combined hunt areas	Height hill central square	−6.99***	50.8%
	Area park	4.80***	
	Area woodland	−3.19**	
	Water	−2.53*	
	Length 'B'-roads	−2.44*	
	Area fodder	2.43*	
Bicester	Area park	3.46**	50.7%
	Area woodland	−2.62*	
	Minimum altitude	4.67***	
	Altitude centre	−3.66***	
	Area agric. 1	2.50*	
	Slope	2.08*	
Heythrop	Area park	3.46**	51.6%
	Area open space	3.55**	
	Water	−2.74**	
	Area rootcrops	2.42*	
	No. contours	3.78***	
	Maximum altitude	−3.28**	

* $P < 0.05$.
** $P < 0.01$.
*** $P < 0.001$.

was explained for the Bicester Hunt (using area of grazed grass, area of fodder and agricultural landclasses 12 and 15). A single predictor (amount of urban area) explained 6.5% for Heythrop. The relationship between MNM and DECORANA scores confirmed the importance of altitudinal variables; MNMs were correlated principally with DECORANA axis 1, which was loaded heavily with these predictors. The analysis was repeated without altitudinal variables, to 'expose' other possible effects. When this was done, MNMs were predicted best by the 'new' DECORANA axis 4 – an axis characterized by wooded landscapes, with some built-up areas and minimal agricultural production.

In summary, considering the characteristics of the hunt country, there were consistent differences in the numbers of foxes found by hunts in different habitats. In particular, most foxes were found in relatively hilly landscapes and, perhaps because it encompassed more such landscapes, more foxes were found in the Bicester country than in the Heythrop country. Three categories of non-exclusive hypothesis might explain these results: foxes might be more numerous in these habitats, or easier to detect, or the two hunts may employ different tactics. Using numbers of foxes moved by the hounds, rather than numbers hunted or killed, may diminish any effect of the huntsman's tactics. Furthermore, the detectability of the foxes might be affected by both visibility (to on-lookers) and scent profile (to hounds); however, it was not obvious how either could explain the differences in MNM. Thus, although neither hypothesis can be rejected, the third possibility is arguably the most plausible, namely that MNM is a corollary of fox abundance which differs between local habitats. If so, then as suggested by Macdonald *et al.* (1981), there may be scope to use foxhounds in order to census fox numbers.

(c) Impact of foxhunting on the fox population

How do these numbers compare with other sources of fox mortality attributable to human activity? We can compare our data with the National Game Bag Census of the Game Conservancy (Tapper, 1992). These data include foxes killed by game keepers within shooting estates, derived from the game diaries. Nationally, approximately 2.0 foxes/km^2 per year were killed between 1980 and 1990. These data also show a clear trend upward since 1960, when the number killed was approximately 0.6/km^2. The data on hunting kills between 1978 and 1990 represent (nationally) between 6.0% and 7.5% of these kills (in 1977 the method of fox census was altered). It is possible that, before this, hunting kills were relatively higher. While the national rate of hunting kills did not increase very appreciably in the 1960s and 1970s, the game bag figures more than doubled. Regionally, the pattern is similar to that for hunting kills, with a much lower rate in the North. The game bag figures, however, are highest for the South West, while in our sample of hunts for this region foxes were killed at a similar rate to that in the North.

A completely independent estimate of foxes killed other than by hunt-ing is provided by our questionnaire survey of farmers for the year 1980. These figures (Table 11.4) are in close agreement with the Game Conservancy's game bag figures for that year. Nationally, the game bag average of 1.95 foxes/km^2 (SE = 0.33) does not differ statistically from the estimate weighted by farm size derived from the farmers (Table 11.4), despite the large sample size involved ($t = 1.30$, $P > 0.1$). The density of kills by farmers was highest in the Wales and the West region, and sub-stantially lower in the North. Comparing the hunt kills with the kills by farmers according to region (weighted data), hunting kills were relatively highest in the North (7.1%) and lowest in the South (3.2%). The equiva-lent figures for the Midlands and East and for Wales and the West were 7.8% and 4.5%, respectively. While we do not have data for the WFFCA packs for 1980, the 1993/94 density of kills would have represented 44.6%

Table 11.4 Mean number of foxes killed by farmers, number of fox litters produced annually on their farm and number of earths on their farms, estimated by the farmers (means weighted for area of farms (upper) and unweighted (lower); MANOVA on weighted data Wilks lambda = 0.88, $F_{9,1307}$ = 8.0, P = 0.0001)

	Midlands and East	North	South	Wales and West	Overall	Univariate F ratio
Mean annual kill/km²	1.8 (0.20)	0.9 (0.19)	2.3 (3.59)	2.6 (0.75)	2.0 (0.16)	$F_{3,648}$ = 5.6, P = 0.0009
	2.0 (0.24)	1.0 (0.24)	3.1 (0.45)	4.9 (1.67)	2.6 (0.28)	
Litter density/km²	0.9 (0.09)	0.6 (0.15)	1.4 (1.71)	1.6 (0.21)	1.1 (0.07)	$F_{3,607}$ = 13.2, P = 0.0001
	1.7 (0.32)	1.1 (0.53)	2.2 (0.37)	1.9 (0.25)	1.8 (0.20)	
Earth density/km²	1.0 (0.10)	0.8 (0.19)	2.1 (2.19)	3.3 (0.64)	1.4 (0.12)	$F_{3,658}$ = 41.0, P = 0.0001
	1.9 (0.31)	1.5 (0.60)	3.4 (0.41)	3.6 (0.61)	2.5 (0.22)	

of the density said to have been killed by farmers in the Wales and the West region in 1980.

Hunts usually move (i.e. put up) more foxes than are hunted or killed during the course of a day's hunting. The ratio of the number of foxes killed to the total number moved in a season can be used as a further index of the predatory success of hunts. However, because a proportion of meets are visited more than once, risking double counting of certain individual foxes, we calculate this index in two ways (Table 11.5). First, we assume that the foxes seen on repeat visits to a meet are different individuals. This gives a killed/moved index of 0.25 based on masters' estimates of foxes moved on an average day (1981 questionnaire) This is likely to be an under-estimate of the pack's predatory success. Second, we include in the moved tally only 65% of moved foxes (discounting, pro rata, 35% of hunting days, which is the proportion that were repeat visits to meets in our most complete data set); this gives a killed/moved index of 0.38, which may be an over-estimate of predatory success. Clearly, the extent to which the killed/moved index also measures the impact of the hunt on fox numbers depends on how precisely the numbers of foxes seen to be moved reflects the numbers present, which hinges on the effectiveness of the hounds at disturbing foxes, and this is likely to vary with, for example, scenting conditions.

In the three seasons between 1978 and 1981, more exact estimates of the numbers of foxes moved were taken from detailed examination of the hunting diaries. This enabled us to compare the figures on foxes moved arising from the diaries with those arising from the masters' estimates in the earlier questionnaire survey. Table 11.5 also presents the average number of foxes that were hunted but not necessarily killed in a day's hunting.

The estimates of the proportion of moved foxes killed were consistent; using the masters' estimates of the average number moved, most were between 0.20 and 0.25. Estimates using the diary data were lower; that for the 1979/80 season was the highest at approximately a fifth of all hunted foxes killed.

Does foxhunting have an impact on the numbers of foxes? It is possible tentatively to place these figures in the context of fox populations. Regional variation in the biology of the fox is an important consideration here. Radio-tracking studies have shown that there is enormous regional variation in the sizes of fox home ranges (e.g. Macdonald, 1981, 1987). For example, in the agricultural Midlands home ranges may be in the order of 250 ha. In the fells of Cumbria, ranges are closer to an average of 1000 ha. Much of this variation can be attributed to habitat differences, and there can be considerable local variation (see below).

To consider a typical Midlands example: the average home range size corresponds to a density of 0.4 territories/km^2 (if each territory is

Table 11.5 Hunting parameters extracted from daily diaries and masters' questionnaires, standard errors in brackets (figures for seasons 1990/91 and later are based on number of foxes that masters completing the 1994 survey said were moved in an 'averagely foxed' part of the country; adjusted proportion of killed/moved discounts foxes moved at repeated visits to a meet in a season)

Season	Diaries: mean moved/ day	Diaries: mean hunted/ day	Diaries: mean killed/ day	Diaries: p killed/ moved	All hunts: mean killed/ day	All hunts: total days hunting	All hunts: p killed/ moved adjusted	All hunts: p killed/ moved
1978/79	3.5 (0.25)	2.4 (0.17)	0.63 (0.070)	0.20 (0.029)	0.83 (0.055)	60.1 (2.68)	0.35 (0.046)	0.23 (0.029)
1979/80	3.4 (0.21)	2.5 (0.14)	0.71 (0.112)	0.22 (0.050)	1.00 (0.119)	61.4 (2.77)	0.50 (0.115)	0.33 (0.073)
1980/81	2.5 (0.32)	2.5 (0.19)	0.59 (0.129)	0.17 (0.018)	0.91 (0.080)	61.5 (2.89)	0.38 (0.048)	0.25 (0.031)
1990/91	–	–	–		1.16 (0.113)	74.2 (6.73)	0.34 (0.037)	0.23 (0.043)
1991/92	–	–	–		1.12 (0.126)	72.2 (5.67)	0.33 (0.031)	0.22 (0.039)
1992/93	–	–	–		1.22 (0.119)	73.8 (5.51)	0.36 (0.034)	0.24 (0.035)

250 ha). In such an area it is likely that groups of more than a pair will form, but if we assume conservatively that each spring a pair of foxes rears four cubs, then in the autumn the total density of foxes will be $2.4/km^2$. If by spring the density of foxes is again to be a stable $0.8/km^2$ (a conservative figure, particularly for farmland adjoining suburbia), it follows that the mortality due to all causes for all the foxes will, at equilibrium, be approximately 1.6 km^2, or two-thirds of the autumn density.

This conservative estimate of annual mortality is close both to the number of foxes that farmers in this region said were killed by them on their land (Table 11.4) and to the game bag estimate of foxes killed in this region (Tapper, 1992). This may imply that the greater part of fox mortality in the British countryside can be attributed to human activity (especially remembering the unquantified impact of road traffic accidents). It is important to remember that these estimates concern the censused farmland, and not the interstices of suburbia in which fox carrying capacity may be higher and mortality lower. The farmers also estimated that about one litter/km^2 was produced annually on their land, although we have no verification of this estimate. It is therefore likely that the number of foxes killed by hunts is a small fraction of total mortality, and probably less than 10% of man-made mortality. Furthermore, if the density of foxes is ultimately limited by food availability, then the loss may be compensated by increased survival of the remaining foxes, by the production of larger litters, or by previously non-breeding females coming into oestrus in the following season. The proportion of foxes removed by hunts each year would need to be in excess of this suppressed capacity to reproduce for there to be any impact on numbers. It is noteworthy, however, that the long term impact of any control method is not necessarily the principal concern of the wildlife manager, who may judge efficiency of control in terms of the timing of mortality (see discussion). Our main result here is that the magnitude of mortality due to hunting is relatively small.

To evaluate the likely total annual tally of foxes necessary to limit fox populations (in the long term) we used a simple model of a single hunt country. A notional area of 500 km^2 was used, around the size of a typical hunt country. Such an area would be expected to contain a minimum of approximately 400 adult foxes in the breeding season, and something like 1200 immediately after reproduction. We assume the carrying capacity (K) of this area is 600 foxes, and that mortality will increase to keep the density at or below this level, aside from some fluctuations due to random variation between seasons in both K and mortality. We explored two contrasting assumptions. In the first, 90% of vixens breed, regardless of population density (analogous to a situation where each territory is occupied by just one pair). In the second, the foxes start at the same density but only 40% breed when the population is at carrying capacity, rising to 90% when the population is low (analogous to a situation where territories at low population density are occupied by pairs, and accommodate additional but largely non-breeding vixens as the population increases). The proportion breeding in any season ($P_{(n)}$) was related to the number of foxes present in the same season according to a quadratic function (Lacy *et al.*, 1994):

$$P_{(n)} = P_{(0)} - \left[\left(P_{(0)} - P_{(K)}\right) * (n/K)^2\right]$$

where $P_{(0)}$ represents the proportion breeding at low density (0.9 in both models explored). In terms of probability, litter sizes were generated from a symmetrical distribution of mean 4.5 (Lloyd, 1980), and independent of population size.

Figure 11.6 shows how populations of foxes fared under these conditions. Each line represents the mean of 100 replicates, where each was simulated for 100 years. The lines show the effect of varying total annual mortality (applied evenly over age and sex classes). For the first model, the population is little affected by mortality as high as 65%. A critical value is reached between 65% and 70% when mortality exceeds the population's ability to reproduce, and each population is doomed to extinction. The same conclusion is reached by applying a purely deterministic model, and calculating the critical level of mortality as a function of mean litter size and fraction of breeding vixens. The critical level of mortality (d_c) is that level at which R_0 is forced below 1.0. If R_0 is defined as:

$$R_0 = \left[1 + (fL/2)\right]\left[1 - d\right]$$

where L is the mean litter size, f is the proportion of breeding vixens and d is mortality (proportion per year). Substituting $R_0 = 1.0$, and rearranging, we obtain the general result:

$$d_c = 1 - \left[1/\left(1 + fL/2\right)\right]$$

For the above example ($f = 0.9$, $L = 4.5$), this gives $d_c = 0.67$.

Fox populations under the second scenario were more vulnerable. But even at an annual mortality rate as high as 50%, the average population size was reduced by less than 20% relative to K. Even at 65% mortality, almost half of the simulated populations were extant after 100 years. Moreover, these estimates do not allow for any compensatory immigration. The foxes within our hypothetical hunt country were analogous to those of an island. In reality, unless all the bordering hunts adopted similar hunting pressures, the effect would be likely to be swamped by the arrival of dispersing foxes from outside, and possibly from the outskirts of human habitations.

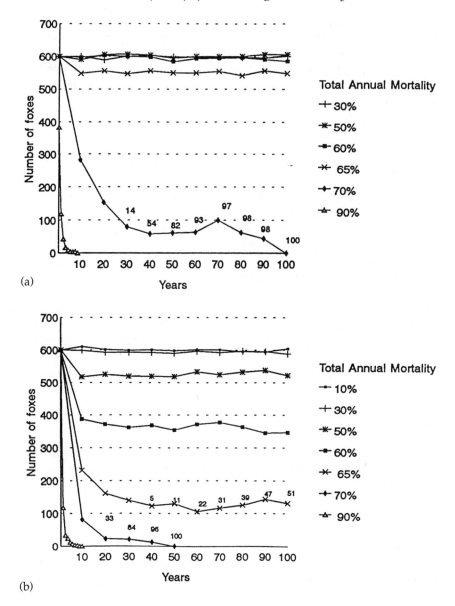

Figure 11.6 VORTEX simulations of effect of differing total annual mortality rates on a population of 600 foxes (at carrying capacity) in a hunt country of 500 km^2. (a) 90% of vixens breed every season, independent of density; (b) 40% of vixens breed when the population is at carrying capacity, rising to 90% at low density. Mean of 100 simulations. Numbers on lines represent number of 'extinct' simulations.

We investigated the extent to which the models were sensitive to variation in litter size. Litter sizes vary both regionally and between years (Lloyd, 1980). In the first model, a mortality level of 50% had no impact on the outcome (Figure 11.6a). If mortality was kept constant in simulations where litter size was varied, a critical value was reached at mean of approximately 2.5 cubs per litter, below which this mortality led to extinction. Similarly, a mortality level of 70% led to extinction where the mean litter size was 4.5. If the mean litter size exceeded approximately 5.5, this level of mortality had no impact on the simulated population.

The second model was sensitive to changes in the value of $P_{(0)}$. A reduction from 0.9 to 0.5 increases the impact of an annual mortality rate of 50% by reducing the population size from something fluctuating above 500 to one below 300.

11.3.3 Attitudes to fox control by hunting with hounds

(a) The rural view

Overall, 33.7% of farming respondents said they had at some time participated in hunting with hounds. This varied significantly between regions ($\chi^2_{[3]}$ = 12.2, P = 0.008), being highest in Wales and the West (46.7%) and lowest in the Midlands and East (29.4%). In the North and South samples the percentages were 38.2% and 34.1%, respectively.

A farmer's inclination to hunt did not stem from personally suffering damage by foxes. Thus there was no relation between whether the farmer said foxes had caused significant damage on their farm and their participation in hunting: 33.0% of farmers recording significant fox damage had been hunting while 36.4% where no fox damage was recorded had been hunting ($\chi^2_{[1]}$ = 0.89, P = 0.348). Also, there was no evidence that this pattern differed regionally (log-linear model on the three-way contingency table, SAS PROC CATMOD, $\chi^2_{[3]}$ = 2.05, P = 0.56). By contrast, there was a predictable relationship between fox damage and fox control in general (the answer to the question, 'Do you attempt to control foxes?'): 72.3% of farms where perceived damage was reported to occur made some effort to control foxes, while only 9.9% of the rest did so ($\chi^2_{[1]}$ = 315.1, $P \ll 0.0001$).

If they approved of the control of foxes, farmers were asked to state which of four reasons influenced this opinion (by placing a tick in a box or boxes corresponding to the relevant option). The four options, with the percentage of farmers citing the option, were: control of disease (47.2%); protection of domestic stock (69.8%); protection of game (46.1%); and simply because they were too numerous (66.1%). Although somewhat opaque, the latter category was included as an option because during pilot studies it was the reason most frequently voiced by farmers. It was not clear what disease they had in mind; at the time mange was infrequent and tapeworm *Echinococcus* little publicized, but rabies was much in the media. Only disease control was associated with participation in hunting; significantly fewer hunting farmers cited this reason than did non-hunting farmers (41.5% compared with 50.0%; $\chi^2_{[1]}$ = 4.5, P = 0.033). Regional differences (to some extent confounded with hunting participation) were more pronounced. Disease control was most frequently cited in the South (56.4%), and least frequently in Wales and the West (29.2%; $\chi^2_{[1]}$ = 21.2, P = 0.003). Protection of game was also cited relatively infrequently in Wales and West region (21.2% compared with > 45% in other regions; $\chi^2_{[1]}$ = 27.6, P = 0.0001). Neither protection of domestic stock nor the reason that foxes were too numerous differed regionally in their rates of response.

As well as being asked about motives for controlling foxes, the farmers were asked to give their attitude to different control methods. These are given in Table 11.6, broken down by region, by hunting participation and whether or not the farmers were directly affected by fox damage. (The relationship between these factors means that the results 'overlap' to some extent.)

Hunting farmers were less likely to approve of foxes being killed to improve pheasant shooting, or for their fur. They were also approximately twice as likely to approve of the active conservation of foxes. Damage attributable to foxes was associated with higher percentages of farmers who approved of foxes being killed to improve pheasant shooting and for their fur. Whether or not they believed that they had suffered damage by foxes had no influence on farmers' approval of hunting with hounds, or on the likelihood that they approved of the active conservation of foxes.

Hunting farmers were less likely than were non-hunting farmers to respond that other methods of fox control (shooting, snaring, poisoning and gassing – at the time poisoning was illegal but gassing was widespread although it has subsequently ceased) were humane. However, they were more likely to state that digging with terriers was humane. Snaring was the only method of control for which hunting participation made no difference to the percentage stating the method was humane (14.0%). Farmers who believed that they suffered damage due to foxes tended to consider all the common methods of control to be humane more frequently than did farmers who did not think they suffered damage from foxes. Hunting with hounds was the only method of control considered humane with similar frequency by farmers regardless of whether they sustained fox damage.

Table 11.6 Farmers' attitudes (%) to (a) motives for killing foxes and (b) different control methods, broken down according to region, whether the farm is subject to significant fox damage and participation in hunting with hounds (values of n in brackets)

	Region					Fox damage on the farm?			Farmers hunt?		
	Midlands and East	North	South	Wales and West	$\chi^2_{[6]}$	No	Yes	$\chi^2_{[2]}$	No	Yes	$\chi^2_{[2]}$
(a) Motive:											
Killed to improve pheasant shooting	46.3 (382)	39.0 (77)	42.6 (195)	28.6 (98)	12.9 $P = 0.043$	35.3 (496)	56.5 (214)	28.2 $P = 0.0001$	43.8 (488)	39.5 (253)	2.3 $P = 0.311$
Killed for fur	16.9 (378)	8.9 (79)	19.2 (193)	19.2 (99)	11.6 $P = 0.071$	12.8 (492)	25.1 (215)	26.6 $P = 0.0001$	18.9 (480)	12.2 (258)	9.0 $P = 0.011$
Hunted with hounds	66.7 (390)	68.6 (86)	65.8 (199)	61.2 (103)	2.2 $P = 0.902$	64.2 (514)	71.0 (221)	4.4 $P = 0.111$	51.0 (492)	93.4 (273)	169.4 $P = 0.0001$
Actively conserved	18.1 (376)	25.6 (78)	17.1 (193)	24.3 (103)	6.5 $P = 0.370$	20.7 (491)	17.6 (216)	1.3 $P = 0.523$	14.7 (483)	28.9 (253)	21.0 $P = 0.0001$
(b) Stating control methods are humane:											
Shooting	71.8 (411)	77.7 (85)	75.0 (212)	70.7 (99)	2.7 $P = 0.845$	70.6 (527)	77.3 (229)	28.2 $P = 0.001$	80.5 (524)	59.1 (269)	57.1 $P = 0.001$
Gassing	56.3 (407)	46.5 (86)	55.2 (210)	38.8 (98)	13.4 $P = 0.037$	48.8 (525)	62.0 (229)	3.9 $P = 0.144$	59.9 (519)	38.7 (269)	43.8 $P = 0.001$
Snaring	11.8 (400)	14.3 (84)	15.1 (206)	23.5 (98)	9.2 $P = 0.162$	9.1 (514)	24.6 (228)	11.9 $P = 0.003$	14.3 (509)	13.4 (267)	4.2 $P = 0.12$
Hunting (hounds)	58.8 (408)	64.7 (85)	57.1 (210)	57.1 (98)	4.8 $P = 0.570$	59.0 (525)	59.3 (226)	0.20 $P = 0.89$	41.4 (516)	91.2 (273)	210.9 $P = 0.001$
Poisoning	22.2 (401)	27.1 (85)	21.1 (204)	18.4 (98)	2.8 $P = 0.838$	17.4 (517)	32.6 (224)	24.4 $P = 0.001$	24.2 (512)	17.1 (264)	10.5 $P = 0.005$
Digging with terriers	23.9 (401)	22.6 (84)	21.9 (205)	39.2 (97)	14.9 $P = 0.021$	21.2 (515)	34.8 (224)	23.5 $P = 0.001$	19.1 (509)	36.5 (266)	28.1 $P = 0.001$

(b) The urban view

Table 11.7 presents a comparison of the urban and rural (specifically farming) responses. The majority of city dwellers thought that foxes needed to be controlled, but more thought they needed to be controlled in the city rather than in the countryside, the reverse of the farmers' view.

Farmers and city dwellers differed radically on what were thought to be acceptable motives for controlling foxes. Significantly fewer city dwellers approved of killing foxes to protect game, for their pelts or as sport with hounds, and significantly more approved of foxes being actively conserved (Table 11.7). The perceived reasons for fox control also differed, particularly with respect to fox numbers; a minority of city dwellers thought that foxes required controlling because they were too numerous.

There were also large differences between the urban and farmer samples in whether they thought hunting with hounds was effective and humane. Only 11.5% of the urban respondents thought that hunting with hounds was humane, compared with 58.4% of the farmers ($\chi^2_{[1]}$ = 686.2, $P \ll 0.001$); 11.9% of urban respondents thought it was effective, compared with 47.8% of the farmers ($\chi^2_{[1]}$ = 401.3, $P \ll 0.001$).

A separate question asked the respondents to state whether they were brought up in the city or in the countryside (Table 11.8). Country-bred respondents were significantly more likely to favour fox control in the countryside than were those brought up in the city. The proportion favouring control in towns could not be distinguished.

There were also differences between country-bred and city-bred city dwellers in terms of what they perceived to be the necessity for fox control. More city-dwelling respondents who had been brought up in the country thought that foxes required control to protect domestic stock and to protect game birds (Table 11.8). The proportions who thought that foxes were controlled because they were too numerous, or because they may spread disease, were similar.

11.3.4 Association of hunting with management of the environment

It is often alleged that one of the benefits to the rural environment which has to be considered in an appraisal of foxhunting is the management practices which accompany it. The argument is that hunting farmers are likely to place more value on such habitats as hedgerow and coverts, and be less prone to remove these for economic reasons. We examined whether there was any evidence for this link by comparing responses concerning past and present management strategies of these habitat features with the farmers' answers to the question concerning foxhunting participation.

Table 11.7 Comparison of attitudes of farmers with those of urban dwellers

Question	Urban dweller	Farmer	Statistics
Do foxes need to be controlled?	$n = 2100$	$n = 760$	
	% responding yes		
(a) in the country	47.7	73.9	$\chi^2_{[1]} = 457.1\ P \ll 0.001$
(b) in towns	61.9	70.7	$\chi^2_{[1]} = 26.3\ P \ll 0.001$
What are the reasons for fox control	$n = 1768$	$n = 751$	
To control disease	56.6	45.7	$\chi^2_{[1]} = 23.8\ P \ll 0.001$
To protect stock	48.7	67.6	$\chi^2_{[1]} = 76.7\ P \ll 0.001$
To protect game	14.4	44.5	$\chi^2_{[1]} = 269.0\ P \ll 0.001$
Foxes too numerous	21.1	65.1	$\chi^2_{[1]} = 454.2\ P \ll 0.001$
Do you approve of foxes being killed for these reasons?	$n = 2057$	$n = 760$	
To improve shooting	6.7	42.0	$\chi^2_{[1]} = 507.9\ P \ll 0.001$
For pelts	3.3	16.8	$\chi^2_{[1]} = 158.1\ P \ll 0.001$
For sport with hounds	11.8	68.4	$\chi^2_{[1]} = 340.2\ P \ll 0.001$
Do you approve of the active conservation of foxes?	46.0	19.3	$\chi^2_{[1]} = 165.6\ P \ll 0.001$

Farmers who said they had participated in hunting reported having removed less hedgerow than other farmers in the decade before the survey, both in absolute terms (metres per farm) and in terms of the density of hedgerow (m/ha) (Table 11.9). The different regional pattern of participation in hunting raised the possibility that these observed differences were attributable to land management practices varying with region, which were merely correlated with the pattern of hunting participation. However, the hunting parameter retained significant explanatory value when entered in a sequential model after region (Table 11.9). Residual variation in hedgerow loss adjusting for region was not much less than the 'raw' measures. Furthermore, the interaction terms were insignificant in these models, indicating that the association between hunting participation and hedgerow removal did not differ according to region. The association between hedgerow retention and hunting can be further strengthened by the farmers' responses to a question prompting them for their motives for hedgerow retention: farmers whose principal

Table 11.8 Comparison of attitudes of urban dwellers with different backgrounds

	Urban bred	*Country bred*	
Do foxes need to be controlled?	$n = 1260$	$n = 772$	
	% responding yes		
(a) in the country	45.6	53.1	$\chi^2_{[1]} = 10.9\ P \ll 0.001$
(b) in towns	62.3	64.0	$\chi^2_{[1]} = 0.9$ ns
What are the reasons for fox control	$n = 1077$	$n = 662$	
To control disease	55.2	58.1	$\chi^2_{[1]} = 1.3$ ns
To protect stock	45.6	53.6	$\chi^2_{[1]} = 10.6\ P < 0.01$
To protect game	12.3	17.8	$\chi^2_{[1]} = 10.3\ P < 0.01$
Foxes too numerous	19.5	22.6	$\chi^2_{[1]} = 2.9$ ns

Table 11.9 Farmers' participation in hunting and their management of hedgerows (estimated removal in the decade before the survey)

Hedgerow loss factor	*Participation in hunting with hounds*		*Hunting effect in two-way ANOVA region/hunting*
	Yes (n = 284)	No (n = 558)	
Density loss, m/ha, (SE)	2.84 (0.37)	5.15 (0.61)	$F_{1,758} = 9.90$, $P = 0.0032$
Mean absolute loss, m/farm, (SE)	413.0 (66.9)	519.7 (52.7)	$F_{1,785} = 6.81$, $P = 0.0092$

field sport interest was in hunting, or equally in hunting and shooting, were approximately five times more likely to tick 'to improve hunting' as a motive here compared with farmers either interested in shooting alone, or in neither hunting nor shooting (51% of hunting farmers and 51.5% of farmers equally interested in hunting and shooting, compared with 10.7% of farmers primarily interested in shooting and 7.5% of those citing an interest in neither; $\chi^2_{[3]} = 135.8$, $P = 0.0001$).

Hunting participation was not associated with the treatment of coverts and field corner spinneys. The proportions of farmers whose principal strategy had been encouragement, removal or leaving these alone did not differ between hunting and non-hunting farmers ($\chi^2_{[2]} = 2.60$, $P = 0.273$; and $\chi^2_{[2]} = 1.64$, $P = 0.440$, respectively).

11.3.5 Perceptions of agricultural damage by foxes

The major pestilential impact of fox predation is likely to be on game birds. There is evidence that fox predation limits wild partridge populations, and that fox control can reduce this effect (Potts, 1986); and mortality of pen-reared pheasants is a perennial cause of complaint. Regarding conventional agriculture, the loss of lambs to foxes is the most debated topic, but little quantified, where the impact of foxes on agriculture is concerned (e.g. BFSS Campaign for Hunting, 1994). Foxes are also increasingly cited as a significant pest of outdoor sow farming, but this has not yet been quantified.

We attempted to quantify the perceived damage to lambs from the responses of farmers. Of our sample, 649 kept sheep in a flock of any size. Flocks varied enormously in size, from only two sheep to several thousand (Table 11.10).

Of these farmers, 53.8% said that they had, at some time, lost lambs to foxes. The proportion was significantly higher in Wales and the West than elsewhere ($\chi^2_{[3]}$ = 14.4, P = 0.002). This sub-sample of farmers who reported that they had lost lambs were also asked what type of evidence had incriminated foxes. A high proportion offered the weak evidence that foxes had been seen in the area (82.7%), but many farmers also said that they had seen dead lambs at foxes' earths (46.1%). These observations do not distinguish between lambs taken before and after death, but 39.1% of farmers who had lost lambs also claimed to have seen, at some time in their lives, foxes attacking lambs. This varied significantly by region, with 53.2% of sheep farmers in the South claiming to have seen an attack, while only 17.7% of northern sheep farmers did ($\chi^2_{[3]}$ = 11.3, P = 0.010). It should be noted that these data may refer to observations accumulated over a number of years. Detailed accounts of foxes attacking healthy lambs are, unsurprisingly but frustratingly, rare in print.

The high percentages of farmers recording some loss does not necessarily imply the annual loss of large numbers of lambs. We also asked farmers exactly how many lambs they thought had been taken by foxes in the three years before the survey. The mean absolute number per annum was 1.74 (SE = 0.23), varying between 0.75 (SE = 0.17) in the Midlands and Eastern district to 4.45 (SE = 1.00) in Wales and the West. As a proportion of the flocks, perhaps a more reliable indicator of the economic importance of the perceived impact, the effect remained highest in the latter region (Table 11.10), and the regional difference was highly significant (MANOVA using the losses for the three years as a multiple response, Wilk's lambda = 0.93, $F_{9,1202}$ = 4.13, P < 0.001). As estimated by the farmers, losses of lambs which they thought were due to foxes did not exceed 1.0% (of their flock size) for any of the years for which we solicited data. The WFFCA data suggested a mean loss of 0.32

Table 11.10 Regional variation in perceived impact of foxes on sheep farming

Region	Mean flock size (SE)	Range	Median	Reported annual loss of lambs per sheep ×100		
				1976	1977	1978
Midlands and East (n = 341)	120.9 (12.5)	2–2000	200	0.62 (0.20)	0.34 (0.08)	0.39 (0.09)
North (n = 74)	177.6 (24.7)	3–1000	110	0.48 (0.13)	0.65 (0.18)	0.77 (0.19)
South (n = 127)	285.6 (53.6)	2–4000	400	0.36 (0.11)	0.30 (0.09)	0.45 (0.14)
Wales and West (n = 107)	523.5 (105.1)	2–7100	600	1.68 (0.84)	2.31 (1.37)	1.88 (1.07)

lambs/km^2 for the 1993/94 season over the areas covered by their foot packs.

Nationally, our sample of farmers questioned in 1991 about the total economic loss due to foxes in the previous year (both lost stock and the cost of protecting against foxes) reported an average cost of £58.90 (n = 151, SE = 14.60). This did not vary regionally ($F_{3,147}$ = 0.71, P = 0.548). The number of man hours said to have been expended annually in fox-related activity averaged 8.8 (n = 151, SE = 1.9). If those farmers who reported neither a financial loss nor any time spent controlling foxes are excluded (leaving farmers sustaining a cost of some kind, whether to stock or possibly to game) the equivalent figures were £211.80 (n = 42, SE = 45.0) and 20.8 hours (n = 42, SE = 4.5). Again, there was no detectable regional variation in this sample, though the figures were highest for Wales and the West (£381.90, SE = 183.50; 46.2 hours, SE = 15.9; n = 23).

11.3.6 Attitudes to foxhunting and conservation

As well as asking masters of hunts for quantitative details, we asked them a number of questions aimed at drawing out their opinion of the conservation role (if any) of their hunt. These prompted short essay- type responses, out of which broad categories of answer could be recognized.

We asked masters what factors they believed underlay differences they perceived within the hunt country between 'good', 'average' and 'poor' areas in terms of the numbers of foxes. The majority (more than 70%) said that predation by people was important, and over half specifically cited gamekeepers as an influence.

One question asked: 'What steps are taken by the hunt and by hunting people to produce wildlife (fox) habitat of which you have seen evidence in the last five years?' One intention behind this question was to estab-lish the master's perception of the primary function of the hunt: to get rid of foxes or to preserve them. Several types of activity were men-tioned: measures designed to favour or to hold foxes in an area, meas-ures designed to facilitate the access of the hunt to foxes in an area, and measures which either directly or indirectly benefit both foxes and other wildlife (notably game). Examples of this last category include the con-struction of artificial fox earths, and the creation of log piles and stick heaps, within which foxes may shelter. Some masters also reported that, in addition to the activity of hunt servants, hunting farmers built artifi-cial earths and log piles. Another conspicuous activity was the mainten-ance of 'covers', i.e. spinneys and thickets owned by the hunt. Just under half of respondents (45.5%) said their hunts maintained at least one cover, and a majority (52.6%) said they planned more. A majority of respondents also (58.7%) reported that they could identify differences in terms of land management between those farmers who participated in hunting (or shooting) and those who did not. Many masters said that steps taken to favour game also favoured foxes (preserving and creating cover), and 35.2% said they considered it to be part of a master's job to influence land management practices within the hunt country. This figure varied significantly across the regions ($\chi^2_{[3]}$ = 8.0, P = 0.046), with the proportion in Wales and the West (9.1%) being considerably lower than elsewhere (31.3–51.4%). This may reflect a difference in attitudes between lowlands, where we suspect that foxes are seen as a quarry, and upland sheep-farming areas where they are seen as pests.

Asked about the consequences for wildlife in general and for foxes in particular if there were no hunting in their country, masters were unanimous in stating that a reduction in fox numbers would result. Most argued that wildlife in general benefits from field sports. Another consequence frequently mentioned was a potential increase in alternative and less desirable methods of attempted control.

Finally, we asked masters about the role of the hunt in the rural community as they perceived it. Answers to this question were provided by 34 of the masters and showed marked consistency. They identified three roles: first, a recreational and social force embodying a traditional rural pastime; second, a humane method of controlling foxes keeping inhumane methods at bay; and third, a force for conservation of wildlife and preservation of the traditional landscape. The first of these was cited most frequently (82% of respondents). Only half mentioned fox control as part of the role of the hunt, and about a quarter cited hunting as a source of rural employment. A similar proportion used the word 'entertainment' in their replies. Almost a fifth cited conservation.

11.4 DISCUSSION

In the foregoing sections we have quantified some facts and perceptions, biological and socio-economic, of foxhunting and thereby reached some interim conclusions. Notable amongst these are:

- that foxhunting kills some 15 000 foxes annually and involves around 50 000 active participants;
- that in the early 1980s the reports of foxhunting farmers indicated that they were more inclined than others to retain hedgerows, and hunts stated that they maintained wildlife spinneys and coverts against the tide of agricultural intensification;
- that about a third of farmers in our sample hunted and about 65% of them approved of hunting (apparently as a sport more than a means of controlling foxes), whereas the great majority of city-dwellers in our sample, especially those with no rural roots, disapproved of hunting.

But how is the conservation equation, made up of these and other facts and figures, to be solved in pursuit of the best practicable way forward? We take a step towards this by summarizing evidence on three major issues which may weigh in the balance of acceptability:

- the need for fox control and the efficacy of foxhunting in satisfying that need;
- the wider contribution of foxhunting to the countryside;
- the evaluation of suffering caused by foxhunting.

One caution regarding generalizations, and stressed at the outset, is that packs supported by mounted followers differ greatly from those, generally characteristic of upland sheep farming country, where followers are on foot. Many of our findings are based largely on data stemming from the former (but see WFFCA, above).

11.4.1 Fox control

There are two questions at the heart of any discussion about fox control, and although both are amenable to scientific enquiry, they have been investigated with surprising infrequency. The questions are: do foxes need to be controlled, and does attempted control cost-effectively diminish the original problem. For example, in the control of rabies in European wildlife it is demonstrable that there is a strong case for fox control: reduced contact rate between infectious and susceptible foxes (and other species) limits the transmission of the disease. It is also demonstrable that gassing and shooting foxes generally failed to fulfil that aim, whereas oral vaccination generally succeeded in doing so.

In the context of rural Britain, the candidate requirement for fox control would be to reduce depredations on stock (e.g. lambs, piglets, poultry and game birds) or some rare ground-nesting birds. Existing evidence is that the general economic impact of fox predation on lambs and piglets is small, although particular cases can doubtless be severe. Fox predation on rare birds (e.g. capercaillie, hen harriers) is a highly localized problem, and the greatest hope of solving it may lie in non-lethal methods (Waugh *et al.*, 1993). There is minimal evidence relating the risk of locally severe depredation to particular strategies of prophylactic fox control; Hewson's (1990) comparison of lamb losses in areas with and without fox control illustrates a step in the direction required to provide such evidence. The practice of sheep farming varies greatly between regions, and so too may the efficacy of blanket prophylactic fox control, but in general there is minimal evidence that fox predation limits agricultural profit or that the effort put into prophylactic fox control is repaid by greater profits in the marketplace.

In contrast, there is mounting evidence that fox predation can limit both the shootable surplus and breeding stock of game birds (Lindström and Morner, 1985; Potts, 1986; Tapper *et al.*, 1991). There is thus the potential for fox control to reverse this impact, and evidence that there are circumstances under which it can do so (Tapper *et al.*, 1991). Obviously, whether promotion of game shooting is an adequate justification for the control of foxes (and other predators) is another intricate issue and an ethical question in its own right.

This question is made all the more interesting in that it conspicuously involves opposing effects on biodiversity and community ecology: habitat management associated with shooting may benefit species from butterflies to wood mice (Tew *et al.*, 1992) at the price of a spectrum of mammalian and avian predators (Potts, 1986). However, irrespective of any wider debate on game shooting, the relevant point here is that the harvest of game birds is a major perceived motive for fox control (and this perception gains at least some support from quantitative field studies).

We have shown that, over the entire area covered by the hunts, foxhunting kills about 0.10 foxes/km^2 per annum, whereas our simulations indicate that to limit fox populations in the English Midlands, a total annual mortality in excess of 0.60 foxes/km^2 would be required. While our data and those of the Game Conservancy's National Game Bag Census (Tapper, 1992) indicate that human predation may be the major mortality factor, perhaps even a limiting factor locally, on rural fox populations, it would appear that the contribution of foxhunting is small (see also Harris and White, 1994). Macdonald *et al.* (1981) estimated the British fox population at 250 000 from which an annual toll of *c.* 15 000 to hunting is, in itself, insignificant in terms of population dynamics. As a method of protecting game birds, arguably the main motive for fox control, foxhunting is not efficient. It seems similarly unlikely to offer general protection to lambs. However, the practice of 'on call' hunting, commonplace amongst the foot-packs of upland country and practised by some lowland hunts, may be more targeted at that minority of foxes which kill viable lambs (WFFCA reports 1.16 fox kills/km^2 in their *c.* 120 km^2 territories).

Furthermore, the timing of fox deaths is relevant to the efficacy of control, as well as the numbers of those deaths. While our VORTEX simulation shows that about 50% of the foxes alive in summer can die without diminishing the reproductive capacity of the population in the following spring, the absence of those foxes killed before pheasant poults are released in July can reduce the short-term risk of predation even if it does not limit fox populations. The extent to which that risk (itself largely unquantified) is reduced by the deaths of 0.10 foxes/km^2 through foxhunting during the preceding August to March seems likely to be small. In areas where fox density is low, the same absolute number of fox kills could be more influential.

While we found some evidence for regional differences in the number of foxes killed by hunts, it is not known how closely these are related to fox density on a smaller scale. The exact extent to which foxhunting (or any other method of control) diminishes fox damage can be measured unequivocally only by large-scale field experiments comparing statistically balanced treatment areas subjected to alternated control versus no control regimes. In the general absence of such data, the available circumstantial evidence suggests to us that it may be more realistic, at least outside the context of upland sheep-farming areas, to think of foxhunting more of a sport than as a method of fox control.

11.4.2 Hunting and the rural economy

Our data show that each hunt employs an average of one terrier man, and that some hunts employ several other hunt servants. It seems likely that these would be unemployed in the absence of foxhunting. Potentially, the impact is wider, as there may be many other employees who rely to some extent, and indirectly, on the existence of a hunt. Matson (1991, cited by Winter *et al.*, 1993) estimated that closing down a large pack of foxhounds could affect 57 jobs. However, the majority of these are jobs generated by horse husbandry (grooms, grain merchants, etc.), and the extent to which they would be affected depends on the extent to which these horses are kept solely for the purpose of foxhunting. Winter *et al.* (1993) carried out a detailed investigation of the economic impact of stag hunting with hounds in Devon and Somerset. They concluded that hunting played a relatively minor role in the economy, though on a local scale the impact could be important. Fewer than half the horses owned by hunt subscribers were said by their owners to be kept solely for hunting, and it cannot be concluded that these horses would not be kept for other purposes in the absence of hunting.

11.4.3 The wider countryside

The proposition that the activities of hunt personnel and, more especially, the supposed benign countryside management of hunting farmers is a particularly interesting one in terms of the debate about the desirable and undesirable features of foxhunting. Our results indicate that at least during the 1970s and early 1980s, hunting farmers had indeed been less inclined to remove hedgerows than the average farmer. Furthermore, many hunts owned or leased at least some spinneys which doubtless served as nature reserves at a time when the general agricultural trend was to destroy such habitats (Macdonald and Johnson, submitted). This result highlights a logical dilemma in evaluating hunting, namely that different factors are measured in different and generally non-convertible currencies. The relative weight given to the deaths and suffering of a number of foxes as opposed to the preservation of a length of hedgerow is, of course, an ethical matter on which judgements may vary. However, it is a decision that can be aided, if not fully determined, by knowledge of the sort we have sought to provide: the relevant numbers in the early 1980s were an average of 0.15 deaths/km^2 of all farmland (or about 10% of the summer fox population) versus about another 200 m of hedgerow/km^2 on the 33% of farms controlled by hunting farmers. Logically, the choice has become a false one in so far as hunting is nowadays far from being the only plausible motive for fostering hedgerows. This is an instance where the weight of factors in the balance has changed greatly for external reasons in the past decade.

Opinions and actions change with time, and our analysis from the early 1980s may no longer describe the current situation. In particular, our recollection of the case advanced by advocates of foxhunting in the 1970s was that an enthusiasm for hunting was the only realistic widespread motive for the then supposed (and here substantiated) tendency for some farmers to be benevolent custodians of the countryside. Clearly, it has always been logically possible that some other motive might cause farmers to nurture hedgerows, but in the 1970s the notion that society might pay them to do so probably did strike many as fanciful. Thus, the impact of hunting farmers, as quasi-conservationists, may genuinely have been greater then than now. A report by Cobham Resource Consultants (1992) indicates that providing coverts for foxes ranks low on the list of motives for retaining or planting small woods. Today, initiatives such as the Countryside Stewardship Scheme and the influence of Farming and Wildlife Advisory Groups typify sweeping cultural change which not only recognizes farmland as the arena for much conservation, but also acknowledges farmers as custodians of nature, and pays them for discharging that role. Of course, field sports may still influence farmers' perceptions and actions (indeed, some of the most influential conservation research is done by the Game Conservancy), but the promotion of foxhunting must surely have a lower rank today amongst factors promoting farmland conservation than it did formerly. Furthermore, habitat conservation may have climbed in the list of masters' priorities – 1994 saw the first inter-hunt conservation competition. It is also the case that many county wildlife trusts now ban hunts from their land. This is said to be at least partly a reaction to the illegal hard-stopping of badger setts of which hunts are suspected, an activity which has resulted in several recent prosecutions.

A different point to emerge from our surveys, and one with political implications, is the difference in opinion between the majority of the population, with no rural background, and those that have, or still do, live in the countryside. An estimated 20% of the population of the UK live in the countryside, while only 2% make a living from farming. Although we did not canvas opinions of rural-dwelling non-farmers, our results raise the strong possibility that opinion on foxhunting (and perhaps other rural issues) differed between town and country people. Only a small minority (11.8% in our sample) of city dwellers judged foxhunting to be acceptable, whereas a majority (68.4%) of farmers did so. If the great majority of the population nationally judges foxhunting morally unacceptable, one might doubt that a democracy will continue to tolerate rural defiance of this judgement, even if the majority of rural people took an opposite view. Farmers have day-to-day control of about 80% of Britain's landscape, but increasingly this role is seen, environmentally, as custodianship on behalf of the population as a whole.

11.4.4 Suffering and the philosophical context

Obviously, an overall evaluation of foxhunting would be incomplete without consideration of moral questions. We have no pretensions to the expertise of academic ethicists, but with the aim of presenting a rounded account we will mention ethical issues that we believe are relevant. In doing so, we are aware that there are those, both in the public and amongst professional philosophers, for whom a judgement on sport hunting in general, and foxhunting in particular, can be made on purely ethical grounds. For them, our attempt in this chapter to disentangle arguments and to quantify facts and perceptions may be irrelevant. Such a stance could derive from such philosophers as Jeremy Bentham and Henry Salt, who overturned biblical or Cartesian notions that nature existed only to serve mankind, and introduced concern for individual suffering. That school might judge, for example, that killing for sport is profoundly morally wrong, therefore questions concerning the numbers of fatalities and their impact on population processes become trivial. At another extreme, there are those for whom empathy with individual animals is irrelevant to discussions of wildlife management, but for whom respect for the integrity of ecosystems is paramount. Our stance lies between these pathocentric and ecocentric extremes. Indeed, the fact that debate about hunting so frequently turns on potentially quantifiable issues indicates that for many people the factors we have sought to clarify and quantify do have relevance as they seek to decide upon either or both of an ethical judgement or a pragmatic course of action. Naturally, those people will bear in mind that our study, as the first to attempt an interdisciplinary, non-partisan analysis, is declaredly not definitive but the exploration of an approach.

The crucial ethical question is whether foxhunting inflicts suffering and, if so, to what extent. Suffering is impossible to prove beyond one's own experience, but there are no scientific grounds for doubting that other mammals can suffer. Nonetheless, although the extent of suffering is difficult to measure, especially in a non-human species, the involvement of pain and fear are workable indicators of suffering and both are likely to be involved in the violent death of any mammal. In evaluating the extent of another creature's suffering, the existence of natural opiates which deaden pain complicates matters. However, such complications do not diminish the truth that the moral costs of an underestimate of suffering are usually deemed to be greater than those of an overestimate, and this truth is accepted even by those adopting a neo-behaviourist stance toward animal psychology (e.g. Kennedy, 1992).

Any discussion about suffering is complicated by humans' inconsistent moral attitudes to it, of which we illustrate just three. Clearly, human inconsistency towards suffering does not lessen its moral seriousness, and in illustrating inconsistencies we neither seek to justify them nor to argue that because something is widely tolerated it is morally correct. Nonetheless, we do think it helpful, when exploring an ethical judgement of foxhunting, to point out that related ethical inconsistencies are rampant.

- There are many cases where society tolerates the infliction of considerable suffering on mammals – for example, in the UK, the poisoning annually with anti-coagulants of millions of rats, or with strychnine of hundreds of thousands of moles, both leading to prolonged and apparently agonizing deaths (Atkinson *et al.*, 1994).
- Suffering is evaluated differently in different types of organisms (e.g. the death of billions of fish by suffocation attracts less public comment than those of the 15 000 or so foxes killed annually by foxhounds).
- Avoidable death by predation is not uniformly unacceptable. For example, millions of songbirds are killed annually solely because people choose to keep housecats. That this is an indirect effect of cat-keeping, whereas the deaths of foxes are a direct effect of foxhunting, has little bearing on the end result for those killed.

In trying to rank the deplorability of foxhunting amongst this morass of inconsistencies, it might help some people if we had a more familiar analogous model of the fox's capacity to suffer. Such a model is provided by the domestic dog: whereas most people are unfamiliar with foxes as individuals, most are familiar with individual dogs, which as members of the Canidae are close relatives of foxes. Our judgements about the depth and extent to which dogs can suffer are as unquantifiable as those about foxes, but they should at least be consistent.

A further complication is the often voiced prediction that if foxes were protected from hunting, the same number or even more would be killed by people using other methods. We do not know if this prediction (which is irrelevant in a judgement of the moral rectitude of killing foxes for sport, directly or indirectly) is correct, and it may not be very important to conservation in that fox populations do not seem to be threatened even by intensive control, but it is a reminder of the difficulty in assessing the net difference in suffering between each method of killing by people. From the human perspective, if not from the fox's, there is a distinction between man-made deaths and the 'natural' deaths which would otherwise achieve the mortality which stable fox populations suffer annually.

This is the ethical distinction between the actions of moral agents and non-moral agents. As ethicists classify non-human mammals as non-moral, the suffering they inflict cannot be evil in the way that suffering inflicted by people can be. However, from the point of view of the recipient, it is not obvious how suffering inflicted by hunting and other man-made deaths rank relative to that caused by starvation or disease (Scott Henderson, 1951). It is also the case that although the same number of foxes might die by different means, the individuals concerned could be different. On the topic of naturalness, it is often said that it is unnatural, and hence undesirable, for hounds to kill foxes, as both are predators. This is probably incorrect, as it is now widely apparent that inter-specific competition between similar predators is commonplace, and that competition often manifests itself in the larger predator killing the smaller. This phenomenon is especially marked amongst the Canidae, in which red foxes kill arctic foxes, coyotes kill kit foxes and red foxes, and wolves (effectively the same species as hounds), go out of their way to harass and kill red foxes (e.g. Hersteinsson and Macdonald, 1992; Macdonald, 1993). However, debate about the naturalness of a particular way of dying, or about the inevitability of dying one way or another, is potentially a distraction: the fact that we all die in the end does not diminish our applause for doctors who strive to postpone that inevitability.

How is the individual to arrive at an overall judgement on hunting foxes with hounds – or indeed, killing them with snares, bullets or poison? An ethicist might provide a logically robust answer solely in terms of the nature and meaning of wrongdoing by moral agents, arguing that it cannot be right to take sport in the avoidable infliction of suffering and death on sentient creatures, and that possible adverse and debasing effects on people of inflicting avoidable suffering are at least as relevant as is the suffering endured by the foxes. From this stance, to justify infliction of suffering and death for such trivial considerations as enjoyment or respect for tradition would bode poorly for progress in fostering animal welfare.

Others faced with the task of evaluating hunting might resort to a more utilitarian calculus which involves balancing negative factors (of which suffering is probably paramount) against positive ones. Such a calculus would not deny the responsibility humans may have to avoid inflicting suffering, but would seek to weigh that responsibility amongst as many other considerations as possible. Indeed, such a balance might diminish the relative importance attached to suffering but it would not affect the level of suffering itself.

Advocates of each of these two approaches could doubtless be unfairly parodied by the other: the pragmatist evaluator might seem ethically threadbare to the ethicist, who in turn might despair of unworldliness in ethical purism. However, in preparing this chapter we have found there was much to learn from advocates of both approaches, and much to gain from a blend of pathocentric and ecocentric philosophies.

It is easy to see, by comparison, society's current view of the balance of pros and cons in the parallel case of game shooting, which probably involves the deaths of at least five times as many foxes as does foxhunting. To a game manager, the prospects of doubling his partridge crop, and hence his profit, are amongst the factors which outweigh his remorse at the deaths, mostly by strangulation in snares, of foxes on his property. The game shooting industry is probably largely responsible for the deaths of in the order of 100 000 foxes annually in Britain (nowadays, most are lamped and shot with rifles, but a proportion, perhaps 25–30%, are snared and of these a proportion may strangle). Against these must be weighed the observations that this industry provides an incentive for habitat conservation on farmland, an estimated (in 1982) 12 000 jobs, more than £200 million worth of consumer expenditure and a game bag valued at £17 million. As two forms of farming, there are close parallels between the killing of foxes to protect reared phèasants and the killing of rats to protect stored grain.

In summary, the foxhunting debate defies unanimity of opinion since different people put different values on each of the relevant factors. There is no common currency with which to equate units of suffering versus units of employment versus units of cultural heritage and rural infra-structure and so forth. Scientists and economists can provide data to clarify these issues, but ultimately decisions will rest on values which are beyond the scope of science. Indeed, to the extent that fox control is deemed necessary, the data suggest that the lowland mounted packs make little contribution to it, and since farmland conservation is increasingly promoted by other sources, the debate increasingly focuses on the moral proprieties of taking recreation which inflicts suffering as opposed to the preservation of tradition and employment.

11.4.5 The future

Our intention has been to expose and, where we were able, to measure some factors that might inform a judgement on foxhunting. Lest there be any mistake, let us be explicit that we are advocates neither for nor against foxhunting and, indeed, as conservationists our main interest in this topic has been as a local illustration of the interdisciplinarity of larger conservation issues. Because the debate is only partly scientific, individuals armed with the same facts will make different judgements. Nonetheless, several salient findings of our enquiry suggest a probable outcome. In particular, in general (and possibly excluding upland foot-packs) the numbers of foxes killed by foxhunting have minimal impact on fox populations and therefore seem unlikely to contribute substantially to any perceived need for rural fox control. The most economically plausible case for fox control in much of Britain, and particularly throughout lowland England, is protection of reared gamebirds. Foxhunting is also seen as indispensable for commercial management of wild red grouse, and for the protection of rare species in some nature reserves. These activities are not invariably compatible with access by hunting hounds. Foxhunting is perhaps most straightforwardly seen, therefore, as a rural sport. As such, its practitioners have made a contribution to countryside conservation. This contribution could doubtless be increased, but seems destined to occupy a descending rank in the hierarchy of factors currently fostering conservation on farmland. In contrast, the role of foxhunting as a colourful pageant redolent with rural tradition, and fostering skill in horsemanship and dog-handling, is self-evident. Furthermore, it contributes directly and indirectly to employment, the rural economy and maintenance of rural crafts. Against these assets must be balanced the undeniable, if unquantifiable, cruelty inflicted upon foxes and the incompatibility of modern, highly technical farming practices with free access to the hunting field. Although the attitudes canvassed in our questionnaires might not be representative of the British electorate as a whole in the late twentieth century, they suggest to us that traditional foxhunting by mounted followers is unlikely to escape restricting legislation in the next few decades. Yet there is an increasing desire by urban-dwellers to participate in rural pastimes, a populous that is ever more knowledgeable about wildlife and eager to be out-of-doors, and a farming economy under mounting pressure to diversify. In such a climate there would seem to be a risk in banning foxhunting that the proverbial baby be thrown out with the bathwater.

Our impression is that those arguing for and against foxhunting have hitherto concentrated entirely on all-or-nothing outcomes, and these entrenched positions may have perpetuated the status quo. We raise the question of whether compromises exist which could minimize the cruelty involved in hunting.

First, it is arguable that a large proportion of the cruelty associated with foxhunting occurs when, having gone to ground, foxes are dug out using terriers. Banning the use of terriers would radically reduce, although certainly not eliminate, the cruelty in hunting. Such a ban would make particular sense if foxhunting, at least among the lowland mounted packs, abandoned the claim to fox control and focused instead on its role as a sport. It would also be relatively easy to police and could be readily compatible with legislation banning digging for badgers. In the same vein, foxhunters could be encouraged to seek ways to conduct their sport in which the aim was a skilful or challenging chase, rather than the death of the quarry. A potential difficulty with this option is that it might provide an incentive for more thorough earthstopping, possibly increasing the lengths of the chases that did occur. If earthstopping were also made illegal, foxhunting in Britain would be left in a similar position to that in the USA and Australia, where kills by the hounds are scarce and apparently there is little opposition from animal welfare societies. Second, another change which would seem likely to diminish the cruelty in foxhunting would be to curtail the season before vixens were heavily pregnant.

Clearly, neither of these proposals should be mistaken as making hunting 'cruelty-free'. Equally clearly, neither proposal would please the landowner who implacably wants foxes killed on his land, nor the person who implacably objects to them being killed. However, both proposals would seem highly likely to diminish the level of suffering that has prevailed for more than a century. The particular options we have raised may or may not be realistic, but our point is that if foxhunting is to continue, it is incumbent upon foxhunters to consider how suffering may be minimized.

Ultimately, to remove suffering from foxhunting it is necessary to remove the direct involvement of foxes, which leads us to the point that only the most superficial ingenuity appears thus far to have been devoted to making draghunting appealing. We gather that hitherto the excitement of draghunting has been perceived as a pale shadow of that provided by foxes. Indeed, one foxhunter explained to us that considering draghunting as a replacement for foxhunting was no more tenable than telling someone who loves Mozart that they should listen to the motorway because it is loud and rhythmic! Nonetheless, the opportunity to make draghunting more fulfilling is a challenge to be explored. Although we hear that drag hounds are hard to control, draghunting is less likely to lead to incidents of hunt trespass. With the opportunities offered by modern odour chemistry to synthesize scents of particular qualities, the opportunities for farmers to profit by diversifying the use of their land, and the great desire of ever more people to participate in benign country pursuits, there would seem to be very strong incentives, both cultural and economic, to explore with the greatest zeal and ingenuity ways of making draghunting attractive. This would seem the only course that is likely to preserve, and indeed potentially to enhance greatly, the traditions, skills, social infrastructure and employment associated with foxhunting.

ACKNOWLEDGEMENTS

We thank, particularly, Ian Lindsay for his participation in all parts of the project. We are grateful to Captain R.E. Wallace and A. Hart of the Masters of Fox Hounds Association, together with the many masters, farmers and others who participated. In a topic characterized, and perhaps caricatured, by acrimony between advocates of differing views, we would like to stress the open-minded generosity with which people of all shades of opinion volunteered their time to help us. Dr P.J. Bacon and F. Mitchelmore assisted with the habitat analysis and Dr C. Sillero assisted with VORTEX simulations (Professor R.M. May derived the deterministic equation for us). We thank Captain C.G.E. Barclay for permission to use the long sequence of data for the Puckeridge Hunt, and Mr Heddwyn Jones and Mrs Lynwen Evans of the WFFCA. We particularly thank the Reverend Professor Andrew Linzey whose critique of an earlier draft so deeply mistook our viewpoint that it prompted us to find, we hope, less ambiguous phraseology. Similarly, the paper benefited greatly from discussions with and comments by W. Andrewes, B. Appleby, C. Booty, J. Braeckman, J. Bryant, I. Coghill, Capt. I. Farquhar, Dr J. Ginsberg, Dr S. Holt, Dr G. Mace, Professor R.M May, Professor T. O'Riordan, and Dr S.Tapper.

REFERENCES

Atkinson, R.P.D, Macdonald, D.W. and Johnson, P.J. (1994) The status of the European Mole *Talpa europea* L. as an agricultural pest and its management. *Mammal Review*, **24**, 73–90.

BFSS [British Field Sports Society] Campaign for Hunting (1994) *Hunting: The Facts*, BFSS Publication, London.

Bunce, R.G.H., Barr, C.J. and Whittaker, H.A. (1981) *An Integrated System of Land Classification*, Annual Report for 1980, Institute of Terrestrial Ecology, Monks Wood, pp. 28–33.

Cartmill, M. (1993) *A View to Death in the Morning: Hunting and Nature Through History*, Harvard University Press.

Cobham Resource Consultants (1992) *Countryside sports: their economic and conservation significance*, Reading: The standing conference on countryside sports.

Coleman, A. (1961) The second land use survey: progress and prospect. *Geographical Journal*, **127**, 168–186.

Coleman, A. (1970) The conservation of wildscape: a quest for facts. *Geographical Journal*, **136**, 199–205.

Coleman, A. and Mags, K.R.A. (1965) *Land use survey handbook: an explanation of the second land use survey of Britain on the scale of 1:25,000*, 4th edn, Publications of the Second Land Use Survey.

Coppock, J.T. (1976) *An Agricultural Atlas of England and Wales*, Faber and Faber, London.

Harris, S. and White, P. (1994) *The Red Fox*, The Mammal Society, London.

Hersteinsson, P. and Macdonald, D.W. (1992) Interspecific competition and the geographical distribution of red and arctic foxes *Vulpes vulpes* and *Alopex lagopus*. *Oikos*, **64**, 505–515.

Hewson, R. (1990) *Predation upon lambs by foxes in the absence of control*. Report to the League Against Cruel Sports, London.

Kennedy, J.S. (1992) *The New Anthropomorphism*, Cambridge University Press, Cambridge, UK.

Longrigg, R. (1975) *The English Squire and His Sport*, Michael Joseph, London.

Lacy, R.C. (1993) VORTEX: A computer simulation model for population viability analysis. *Wildlife Research*, **20**, 45–65.

Lacy, R.C. Hughes, K.A. and Kreeger, T.J. (1994) *VORTEX Users Manual*, 2nd edn, Chicago Zoological Society.

Lindström, E. and Morner, T. (1985) The spread of sarcoptic mange among Swedish red foxes (*Vulpes vulpes* L.) in relation to fox population dynamics. *Revue d'Ecologie – la Terre et la Vie*, **40**, 211–216.

Lloyd, H.G. (1980) *The Red Fox*, Batsford, London.

Macdonald, D.W. (1981) A Report to the Sponsors of Our Investigation into Factors Affecting the Conservation of Wildlife on Farmland. (Unpublished report.)

Macdonald, D.W. (1984) A questionnaire survey of farmers' opinions and actions towards wildlife on farmlands, in *Agriculture and the Environment* (ed. D. Jenkins), ITE Publications, Cambridge.

Macdonald, D.W. (1987) *Running With the Fox*, Unwin Hyman, London.

Macdonald, D.W. (1993) Rabies and wildlife: a conservation problem. *Onderspoort Journal of Veterinary Research*, **60**, 351–355.

Macdonald, D.W. and Carr, G. (1981) Foxes beware: you are back in fashion. *New Scientist*, **89**, 9–11.

Macdonald, D.W. and Doncaster, C.P. (1985) *Foxes In Your Neighbourhood*, RSPCA pamphlet, Horsham.

Macdonald, D.W. and Johnson, P.J. (submitted) *Biological Conservation*.

Macdonald, D.W. and Newdick, M.T. (1980) The distribution and ecology of foxes (*Vulpes vulpes* L.) in urban areas, in *Urban Ecology* (eds R. Bornkamm, J.A. Lee and M.R.D. Seaward), Second European Ecological Symposium, Blackwells, Oxford.

Macdonald, D.W., Bunce, R. and Bacon, P.E. (1981) Fox populations, habitat characterisation and rabies control. *Journal of Biogeography*, **8**, 145–151.

Potts, G.R. (1986) *The Partridge. Pesticides, predation and conservation*, Collins, London.

Scott Henderson, J. (1951) *Report of the Committee on Cruelty to Wild Animals*, HMSO, London.

Tapper, S.C. (1992) *Game Heritage, an ecological review from shooting and gamekeeping records*, The Game Conservancy, Fordingbridge.

Tapper, S.C., Potts, G.R. and Brockless, M. (1991) The Salisbury Plain Experiment: the conclusion. *Game Conservancy Review of 1990*, **22**, 87–91.

Tew, T.E., Macdonald, D.W. and Rands, M.R.W. (1992) Herbicide application affects microhabitat use by arable field mice (*Apodemus sylvaticus*). *Journal of Applied Ecology*, **29**, 532–539.

Waugh, S.E., Macdonald, D.W. and Hirons, G.J.M. (1993) *A review of the effectiveness of anti-predator measures on nature reserves*. Report to the Royal Society for the Protection of Birds, Sandy, Beds.

Winter, M., Hellet, J., Nixon, J. *et al.* (1993) *Economic and Social Aspects of Deer Hunting on Exmoor and the Quantocks.* Centre for Rural Studies Occasional Paper No. 20, Cirencester.

12

Studies of English red deer populations subject to hunting-to-hounds

Jochen Langbein and Rory Putman

SYNOPSIS

Red deer (*Cervus elaphus*) around Exmoor and the Quantocks are the only red deer populations in the United Kingdom subject to traditional hunting-to-hounds. The history of deer hunting on Exmoor stretches back at least until the eleventh century and the establishment of the Royal 'Exmoor Forest'. The present style of hunting by riding to hounds is comparatively modern, revived on Exmoor around the middle of the eighteenth century; staghunting on the Quantocks dates back only to 1920, following reintroduction of red deer there specifically for this sport.

This chapter examines current numbers and distribution of red deer herds on Exmoor and the Quantocks and explores trends in population size over the last 20 years. Data are presented on condition (as body-weight) and fecundity. Based on these figures the authors attempt to extrapolate future changes in population size over the next 20 years given a range of hypothetical culling regimes. Against this context they then investigate estimates of the overall cull imposed on these deer populations in the recent past by differing methods, including both hunting and rifle shooting, to assess the impact of hunting-to-hounds on red deer numbers and its potential contribution in regulating population size. Finally, the effects of regular hunting on red deer distribution and social organization are addressed briefly.

The present studies suggest that these deer populations are highly productive. The annual cull taken by hunting, of around 2–4% of the autumn population, even when taken in combination with an estimated additional cull of 8–12% contributed by rifle shooting, has not prevented substantial population increases over recent decades.

12.1   INTRODUCTION

The history of deer hunting on Exmoor stretches back at least as far as the Norman Conquest and the establishment of the Royal 'Exmoor Forest' (based on the area still known today as the parish of Exmoor

Forest near the centre of the Exmoor National Park). Although deer hunting itself is so ancient, the present style of hunting by riding to hounds is comparatively modern (Burton, 1969). On Exmoor it was revived around the middle of the eighteenth century, where with some interruptions it has now had a 200-year history. Staghunting on the Quantocks dates back only to 1920, when red deer were reintroduced there specifically to revive this sport.

Three packs of staghounds operate in the West Country: the Devon and Somerset Staghounds (DSSH) based on the Exmoor National Park (NP) and environs; the Tiverton Staghounds (TSH) in the area immediately adjoining D&S country to the south and stretching to around Tiverton itself; and the Quantock Staghounds (QSH) hunting in and around the Quantock Hills. Each pack meets twice to three times a week during the hunting season, which extends from August to the end of April. Mature 'autumn' stags are predominantly hunted from the middle of August to the end of October. Thereafter hunting concentrates on hinds until the end of February, from which time younger stags (locally referred to as 'spring' stags) are hunted until the end of the season.

Throughout its long history, staghunting has primarily been practised for reasons of sport rather than as a means of control. While the need to ensure a supply of quarry assured the conservation of large areas of suitable habitat and tolerance of substantial populations of deer in such areas, hunting-to-hounds has never on its own been able to exercise a precise control over deer population numbers, in maintaining them within specified bounds. Thus, although the West Country is now the last stronghold of staghunting within Britain, hunting-to-hounds on Exmoor and the Quantocks accounts for only a proportion of the total annual cull imposed on the deer populations. As in other parts of the country, large numbers of red deer are also culled legally each year by marksmen with rifles or large-bore shotguns; an unknown but probably significant number of deer are in addition taken illegally by poachers, or die as a result of road traffic accidents.

This chapter examines the current status of red deer herds on Exmoor and the Quantocks and explores changes in numbers and distribution over the last 20 years. Data are presented on condition (as body-weight) and fecundity, and on the basis of these figures extrapolations of the likely future progression of population size are attempted. Predictions of population growth if existing culling levels are maintained are contrasted with growth predicted in the absence of any imposed cull; and estimate the level of culling required to maintain populations close to their current density with zero growth. Against this context we investigate culling levels actually imposed on these red deer herds in the recent past, as far as data available from both hunting and rifle-stalking allow, to consider the impact of hunting-to-hounds on red deer population numbers and its potential contribution in regulating population size. Finally, the effects of regular hunting on red deer distribution and social organization are addressed briefly.

Findings presented here arise foremost from an initial one-year study into the conservation and management of red deer on Exmoor and the Quantocks undertaken by the authors on behalf of the National Trust from October 1991 to September 1992.

12.2 METHODOLOGY

12.2.1 Study areas

The Exmoor National Park (692 km^2) extends over parts of West Somerset and North Devon, offering a diversity of upland, lowland and coastal habitats. Much of the area consists of improved hill pasture, but the Park (most of which remains in private ownership) also contains extensive tracts of heather and grass moors (34%) and woodland (10%); much of the latter is made up of tracts of ancient oak woodland growing along steeply sloping valley sides. The Quantocks Area of Outstanding Natural Beauty (AONB) (95 km^2) offers a similar mix of habitats located within five miles of the eastern boundary of the Exmoor NP.

12.2.2 Population numbers and structure

While population data were later extrapolated to offer estimates for the wider region, intensive assessments of deer numbers and age and sex breakdown were undertaken primarily within three specific subsections (35–55 km^2 each) within Exmoor: the National Trust's Holnicote Estate in the north-east; an area to the south of the National Park around Molland and Withypool; and an extensive moorland region within the former Royal Exmoor Forest. These sites were deliberately selected to offer the full range of different habitat types and conditions which may be experienced within the whole of the Moor, as well as giving representative geographical coverage. The Molland study area was in addition selected as one of the sites previously surveyed by Malcolm *et al.* (1984), so that direct comparison could be made between 1992 population levels and those recorded by Malcolm *et al.* in 1981/82.

Two independent but complementary methods were adopted at each site for determination of red deer numbers: direct census of animals seen during scan surveys from high vantage points overlooking a defined area of the site (vantage point counts) and assessment of population density from faecal accumulation in differing habitat types. For detailed methodology see Langbein and Putman (1992a), developed from methods established by Staines and Ratcliffe (1987) and Ratcliffe (1987) – vantage point counts; and Mitchell and McCowan (1983), Putman (1984, 1990) and Staines and Ratcliffe (1987) – faecal counts. Study sites were divided into several compartments of about 6 km^2 and vantage point counts were carried out on at least three or four different days from several points in each compartment between 10 January and 12 March 1992, giving data on minimum deer numbers and age/sex breakdown.

A second independent estimate of population density in each area was derived from dung counts carried out during March and early April. The time was chosen so that numbers of domestic livestock (especially sheep) on the moor would have been at their annual minimum for the three months preceding the assessment, minimizing risks of misclassifying sheep and deer dung. Dung decay rates and ground cover were also relatively low; hence the chances of finding dung were high. Similar data on dung accumulation for the Quantocks were made available for inclusion in the analyses by Winder and Chanin (personal communication; see also Winder and Chanin, 1992).

An additional overall estimate of minimum population size and distribution of red deer in both Exmoor and the Quantocks was by means of a large scale census over the entire area of each region. This was undertaken in March 1992 within the Exmoor area; a total of over 100 observers participated and were deployed simultaneously at different vantage points or survey routes throughout most of the Exmoor National Park area. A similar large-scale census was arranged within the Quantock Hills by the Quantock Deer Management and Conservation Group during May 1991 and February 1992.

12.2.3 Population change

While reference is made in the literature to various subjective and often widely varying estimates of past deer numbers in the West Country, no form of systematic annual census of the red deer populations over the whole of Exmoor or the Quantocks had taken place prior to 1991. There have been repeated small-scale censuses or surveys of the deer in certain subsections of the area over the last 20 years, which we have used to investigate patterns of change between years. Three main sources of objective past data were made available to us for such analysis:

- Records maintained by the Exmoor Natural History Society of numbers of deer sighted during replicate walks along a number of different nature trails in northern Exmoor, repeated several times per year since the mid 1970s.
- Data from a study of deer on Holnicote undertaken by extramural students of Exeter University during 1985 and 1986 under the

supervision of Dr P. Chanin, who recorded sightings from a number of regularly walked transect routes throughout the Holnicote area during 1985 and 1986. These records were compared against a repeat survey initiated by us in 1992, using identical routes and methods of survey, and carried out on our behalf by some of the same observers also involved during 1985/86.

• Estimates of Exmoor deer population numbers obtained by the Devon Wildlife Trust during 1981/82 (Malcolm *et al.*, 1984), based on dung counts and vantage point observations, including data on one of the main regions (Molland–Withypool) re-selected for intensive study by ourselves using comparable methods ten years later.

Each of these sources of information was analysed to determine changes in population size and structure between years, and to assess whether any consistent overall trends emerged for all regions, based either on comparisons of actual deer densities or else derived indices such as mean numbers seen on particular transect routes over the years. For each of these, changes in actual deer densities or trends suggested by derived indices were considered in terms of their implications for changes in relative population size, sex ratio, distribution and degree of aggregation between years. Trends detected in all three study sites may suggest more general trends for the region as a whole over this past 20-year period.

12.2.4 Population performance and body condition

Red deer carcass weights for the 1991/92 culling season, as well as some previous years, were provided by a number of stalkers culling in a range of sites within Exmoor, the Quantocks and also the Tiverton to Bampton area. During 1991/92 stalkers were also asked to extract the lower jaw from as many hinds culled as possible, to enable us to relate body-weights to known age, the latter being determined from tooth eruption patterns (after Lowe, 1967; Mitchell, 1967). As the time-scale of this initial study did not allow initiation of data collection to assess conception rates from cull material, reproductive performance was estimated mostly from observed hind:calf ratios.

12.2.5 Hunting records

Records of the number of deer accounted for by the three packs of stag-hounds in the recent past (up to 1991/92) were kindly made available by the Staghound Association. Details kept by each hunt, at least in recent years, usually include information on the number hunted of autumn stags, hinds and spring stags; a further category, referred to here as 'extras', gives a figure for the number of additional animals taken opportunistically through the season – usually injured deer accounted for either during hunting or as a result of a call to deal with a road casualty, fence injury or similar accident.

12.3 RESULTS

12.3.1 Population numbers

Visual census methods allowed us to establish minimum estimates of the end-of-winter population numbers for 1992 as 2100 red deer based on the Exmoor NP and a further 650 within the Quantocks AONB. All visual census methods, unless undertaken in entirely open country, are notorious for producing significant underestimates of the true population number. In view of the presence of extensive concealing habitats including woodland, large expanses of gorse and varied terrain, it was clear to us that many deer escaped detection during our vantage point counts.

Our second, indirect approach to estimation of deer density – from faecal accumulation – allowed more even sampling of all habitats potentially used by the deer. The results suggest that true winter population numbers in 1992 are likely to have been as high as 4750 red deer within the Exmoor National Park, with probably a further 1000 living within 10 km of its southern and eastern boundaries. Statistical confidence limits associated with our estimates of deer density in the various sample study areas were relatively wide, much of this variance arising from the high degree of aggregation of dung within the expanses of improved pastureland in the region (i.e. with heavy deer activity on pastures on some fields and estates but hardly any on others). If we make the assumption that these variance estimates may be extrapolated to the wider Exmoor region, this would suggest a 10% chance that true deer numbers lie below 3167, with conversely a similar 10% chance that numbers already exceed 6323.

Estimates derived from dung accumulation counts in the Quantocks AONB, based on the data collected by Exeter University (Winder and Chanin, 1992), suggest that red deer in that area numbered at least 800–900.

This estimate for the Quantocks AONB (95 km^2) indicates an average density of 8–9 red deer/100 ha. Deer density estimates for Exmoor based on dung counts for Holnicote in the north-east (14/km^2) and Molland in the south (12/km^2) are equally high, although rather lower densities were recorded in much of the more exposed, less wooded western half of the Park (c. 3/km^2). The relatively high density estimates resulting from our dung counts methods have recently been further substantiated by replicate visual surveys using spotlight counts at night in the Exe valley near Dulverton, where up to 620 deer were recorded within just 2650 ha (21/km^2) (Langbein, 1993). For comparison, density estimates recorded in forestry as well as open hill land in Scotland range from 0.5 to 31.5/km^2 (Stewart, 1985; Staines and Ratcliffe, 1987) with up to 40/km^2 noted in particularly favourable areas (Ratcliffe, 1984). Thus, although our estimates of numbers are considerably above those previously recognized for the Exmoor and Quantock region, they still suggest overall deer populations of only moderate density, reaching the higher levels above 15/km^2 in only a small proportion of their range.

Density comparisons of red deer populations in specific areas within the Exmoor NP (e.g. Holnicote, Molland–Withypool) during 1992, with earlier estimates in the same areas, are very consistent across all our study sites and suggests that populations within the Park have increased significantly over the last 10–15 years. The same seems true in the case of the Quantocks. General consideration of increasing deer populations within Exmoor and the Quantocks must however be viewed against a background of probable under-estimation of numbers in the past. The most widely accepted figures of around 500–800 deer within the Exmoor NP during the 1970s (Lloyd, 1970, 1975) and 1500 during the 1980s (Allen, 1990) were very probably well below the true numbers in the Park at that time. The analyses presented here, based on data collected in recent years in the Holnicote and Molland–Withypool study areas, suggest average net rates of increase (as the balance of births over natural mortality and cull deaths) of between 5% and 10% per year since the 1970s.

12.3.2 Body-weights and size

While red deer in the West Country are generally thought to be substantially larger than their counterparts in Scotland, few published data are available on their body size. Carcass data collected during the present study (Table 12.1) provided average dressed carcass weights for adult hinds (> 1 yr) culled during the 1991/92 season of 48.9 kg on the Quantocks (n = 15), 49.9 kg on Exmoor (n = 41) and 49.5 kg for animals from the Tiverton/Bampton area (n = 20). Comparable data on weights of carcasses at the same stage of dressing-out are available from Ratcliffe's study in Scottish forests (Ratcliffe, 1987). This shows weights of adult hinds to average 43.8 kg taken across 12 different woodland regions throughout Scotland, and 52.4 kg at Grizedale in north-west England. Weights in the West Country are thus seen to be well above (*c.* +12%) averages attained across commercial forests in Scotland, and compare well with those in the very high performance population at Grizedale.

Table 12.1 Average body-weights (dressed carcass weights after removal of head, viscera and lower limbs) of red hinds culled in the West Country (data pooled from a number of stalkers in each region)

	Quantocks AONB	*Exmoor NP*	*Tiverton/Bampton*
Hinds culled 1991/92	48.9 kg	49.9 kg	49.5 kg
(nos. weighed)	(15)	(41)	(20)
Pooled data 1987/92	48.9 kg	48.1 kg	52.8 kg
(nos. weighed)	(15)	(70)	(31)

12.3.3 Pregnancy and calving rates

Reproductive success in red deer, and particularly the proportion of yearling hinds which conceive in any population, has been shown to be closely related to body-weight and early growth rates (Hamilton and Blaxter, 1980; Clutton-Brock *et al.*, 1982; Ratcliffe, 1987). Our data suggest that performance of red deer in the West Country would be much higher than those of deer on the Scottish moorlands, and should lie near the high end of the range found among Scottish forest deer and populations elsewhere.

The time-scale of our original study did not enable us to initiate collection of data on conception rates from cull material, although such research is now under way (Langbein, in progress). For assessment of reproductive performance we must at this stage rely mostly on calf:hind ratios observed in the field (expressed as the number of calves recorded per 100 hinds). Observed ratios of calves to hinds in the field have been used widely to indicate approximate reproductive rates but must be interpreted with great caution to make appropriate allowance for the proportion of yearling hinds present in autumn (which could not have calves at foot except in the rare instance of conceiving as calves themselves) and, if ratios are assessed after the beginning of the cull in autumn, to allow for any differential culling of hinds and calves.

Calf:hind ratios were assessed during all vantage point counts, whenever sufficiently clear views permitted calves and hinds to be clearly distinguished. A total of 532 sightings of red deer on Holnicote between January and March 1992 could be confidently differentiated into hinds and calves, resulting in a calf:hind ratio of 52 calves per 100 hinds; figures for Molland–Withypool based on 322 sightings gave a ratio of 59 calves per 100 hinds for that area. Calving rates projected on these figures, after allowing for yearlings in the hind groups, approach 69% and 79% for the two sites, respectively. These estimates, based on direct observations, are not dissimilar to earlier findings by Malcolm *et al.* (1984) and Winder (unpublished) who independently both suggested pregnancy rates of 73% derived for 11 hinds culled at Arlington (western

Exmoor fringe) and 15 culled on the Quantocks, respectively. Conception rates in excess of 70% are also supported by a more extensive study of carcass data (Langbein, in progress), with provisional results indicating 88% for adult ($n = 68$) and 79% for yearling hinds ($n = 27$).

12.3.4 Population growth rates

Specific comparisons of current red deer numbers within particular sub-sections of Exmoor, with estimates available for those same areas 15–20 years before, give net rates of increase (as the balance of reproductive recruitment over natural mortality and cull deaths) of between 5% and 10% per year since the 1970s. An independent estimate of potential growth rate may be provided by analysis of data presented here for population condition and fecundity. Deer populations in the West Country continue to be subject to culling by various means; thus we do not present our projections of population trend as an estimate of the maximum natural potential for increase given only natural schedules of fecundity and mortality. Instead, projections are made for population growth under a number of different culling regimes (varying in overall proportion of population culled and sex/age structure of that cull). Simulations, of which some examples are shown in Figure 12.1, were developed from an autumn (post-breeding) starting population of 5630; based on our winter (pre-breeding) estimates of red deer numbers and population structure within the Exmoor NP. Basic parameters used were derived as follows.

- **Calving rate**: 75% of the adult hind population at the end of spring (based on conservative estimates of conception rates as 60% among yearling hinds, 80% among older hinds).
- **Natural mortality**: calves 10%; hinds 2%; stags 5% (as percentage of summer population).
- **Imposed cull**: simulations show model culling at 18%, 20% or 22% (as percentage of summer population and taken off in addition to natural mortality). Culling is assumed to be unselective, taking the same percentage level for juveniles and adults, and carried out in direct relation to the prevailing sex ratio; that is, if there are two females to every male in the population, the total cull should reflect that ratio.

Assumptions of calving rates were based on observed calf:hind ratios and pregnancy rates determined during this study; natural mortality rates are equivalent to those employed by Ratcliffe (1987) for studies of high performance populations in Scotland.

The simulations indicate that, given our assumptions of calving rates and natural mortality, a cull of below 18% would allow populations to continue to increase rapidly, with a likely doubling of total numbers

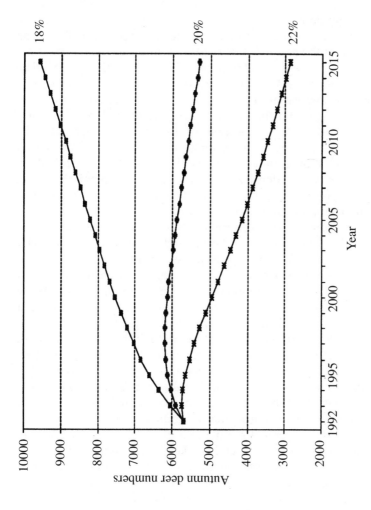

Figure 12.1 Projected autumn deer numbers within the Exmoor National Park given annual cull levels of 18%, 20% or 22% of the autumn population. (For other parameters used in the simulations, see text.)

within 30 years. In contrast, a cull at 22% might be expected to cause a reduction in population numbers at about the same rate. Population projections based upon an intermediate cull figure in the region of 20% suggest that such a culling level would be required to maintain the total population numbers close to those currently present. It emphasizes also the narrow margins between a consistent cull which may effectively control numbers and one which allows populations to rise or fall sharply.

The above simulations are presented to provide a relatively simple illustration of some of the effects of alternative approaches to culling. While we feel that the assumptions of calving rate and natural mortality are appropriate with regard to the populations in question, such values would need to be assessed in rather greater detail and with due consideration to age-related variation of mortality, to allow detailed modelling of population changes. Furthermore, we have not included any density-dependent factors in our projection of future population size (i.e. changes in the rate of recruitment or mortality resulting from changes in deer density itself). Such density-dependent factors are not thought to start to have an effect at a density of less than half the ecological carrying capacity of the population range – a level that is likely to be relatively high on Exmoor, and perhaps higher still on the Quantocks, due to ready access to improved pastures and agricultural crops for the deer.

Data available to date on the condition and fecundity of red deer in the West Country suggest that these populations are highly productive and likely to have an annual recruitment rate (reproduction minus natural mortality) in excess of 20%. Observed average rates of increase below such levels over the past 20 years or so thus reflect the impact of legal culling (both hunting and rifle-shooting), poaching and road casualties. However, the overriding trend of a continuing increase in population size of between 5% and 10% per year (section 12.3.1) suggests that the total 'imposed' mortality from all these sources is insufficient to maintain stable population levels – and has on average only accounted for about half to two-thirds of the annual recruitment. Based on the simple simulations presented here, we suggest that an annual cull of just over 20% would be required to maintain populations close to the current population numbers, assuming that this cull is taken in direct proportion to the sex ratio among the autumn population. Given our 1992 estimate of population numbers, this translates into a total cull requirement of 1139 animals within the Exmoor NP, or 1259 animals if including the entire DSSH country, and a further 192 on the Quantocks.

Analyses carried out on the hunt statistics provided to us are presented in Table 12.2. The overall number accounted for by each of the packs of staghounds since 1960 is shown in Figure 12.2. Numbers killed are seen to have increased slightly between 1960 and the early 1970s, since which time total numbers taken by all three packs combined have remained close to but rarely exceeded 200 per season. Highest numbers are taken by the DSSH, averaging about 95 since 1980, with averages of 53 and 38 by the QSH and TSH, respectively, over the same period. Whether or not our estimates of current population numbers are very accurate, it is widely acknowledged that red deer numbers have risen significantly within both the Exmoor and Quantock regions since the 1970s. Despite that rise, the numbers killed by the hunts have shown little change. The percentage of the annual increment taken by the hunt has thus gradually fallen in real terms; in the case of the Exmoor population, this would suggest that hunting (including dealing with injured deer) accounted for about 6–8% of the population within D&S country during the mid 1970s, but for only *c.* 2.5% by 1991/92 (that is, falling from a contribution of around two-fifths to one-eighth of the total cull requirement if the aim is to maintain numbers close to existing levels). If figures for a winter deer population on the Quantocks of the order of 850 deer are accepted as accurate, this implies that the average hunt kill per season may still account for some 5% of the autumn population in this area (or a quarter of the cull required for zero growth).

To summarize, hunting would seem to account currently for 2.5% of the red deer population of Exmoor, and nearly 5% in the Quantocks. Such a cull level is insufficient on its own to impose any form of regulation on numbers in the area. Even in the highly unlikely event that our population estimates are twice as high as actual numbers present in the area, animals accounted for by the D&S would still amount to only 4% out of an estimated total cull requirement of 20%. In neither area is hunting-to-hounds able to impose a high enough cull to exercise control over population numbers, without substantial additional culling undertaken by other means.

12.3.5 Effects of hunting on deer behaviour

A number of contrasting claims are commonly made regarding the direct effects of the hunt on deer behaviour. Some opponents of hunting claim that it 'terrifies' the deer to such an extent that a number of animals may die later as a result of trauma-induced myopathy even if they manage to escape from the hounds (Henshaw and Allen, 1989). Staghunters maintain that, unless hounds are actually hunting their 'line', 'other deer in the area pay no more heed to the hunt than a grazing wildebeest does to a pride of lions lunching off a mate nearby' (Lloyd, 1990). At the same time, staghunters aver that 'the presence of hounds in an area disperses the herds of deer and encourages them to move to other areas'. Clearly these conflicting claims cannot all be true, and a brief résumé of our impressions in this regard may be useful here.

While the behaviour of the deer was not directly investigated during this study, it has already become clear from previous sections that deer

Table 12.2 Numbers of red deer accounted for by the three packs of West Country staghounds over the last 10 seasons

Year	Autumn stags	Spring stags	Total stags	Hinds	Extras	Total
Devon and Somerset Staghounds						
1982/83	24	14	38	23	23	84
1983/84	21	15	36	65	29	130
1984/85	20	12	32	20	10	62
1985/86	13	15	28	34	22	84
1986/87	21	12	33	40	21	94
1987/88	18	18	36	41	26	103
1988/89	13	13	26	36	28	90
1989/90	12	16	28	37	29	94
1990/91	21	17	38	20	33	91
1991/92	16	22	38	39	37	114
Total 1982–1992	179	154	333	355	258	946
Tiverton Staghounds						
1981/82	10	9	19	14	4	37
1982/83	7	11	18	14	4	36
1983/84	9	14	23	18	5	46
1984/85	10	10	20	17	4	41
1985/86	7	12	19	12	3	34
1986/87	6	8	14	15	3	32
1987/88	4	10	14	15	4	33
1988/89	5	6	11	13	2	26
1989/90	7	11	18	9	5	32
1990/91	7	8	15	9	5	29
1991/92	5	9	14	22	2	38
Total 1982–1992	77	108	185	158	41	384
Quantock Staghounds						
1981/82	9	14	23	6	12	41
1982/83	9	13	22	5	20	47
1983/84	9	7	16	12	14	42
1984/85	6	6	12	7	20	39
1985/86	?	?	?	?	25	25
1986/87	6	11	17	10	20	47
1987/88	10	10	20	10	17	47
1988/89	9	10	19	26	14	59
1989/90	9	9	18	9	22	49
1990/91	9	6	15	9	14	38
1991/92	9	13	22	20	56	98
Total 1982–1992	85	99	184	114	234	532

In addition to deer killed during the autumn-stag, hind and spring-stag seasons, those shown as extras are animals of either sex dealt with by the hunts, such as fence or road casualties.

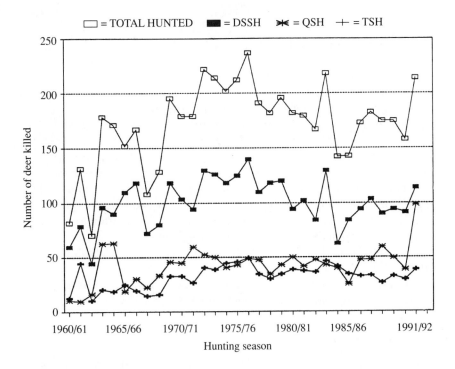

Figure 12.2 Total number of deer accounted for by each of the three packs of West Country Staghounds since 1960 (DSSH, Devon and Somerset Staghounds; QSH, Quantock Staghounds; TSH, Tiverton Staghounds).

on Exmoor and the Quantocks are relatively easy to view in comparison with many other red deer herds, in as far as many regularly join large groups in very predictable feeding or resting areas, with groups of in excess of 40 hinds or 20 stags commonplace. The formation of large groups in open areas may in part be explained by an intrinsic property of social organization in ungulates in general: that on moving from closed habitats to more open ones, smaller social units tend to coalesce into larger aggregations (e.g. Putman, 1988); this allows them to benefit from 'safety in numbers' and reduce vigilance rates.

Large areas of ground are covered during the course of any one hunt, leaving few predictable refuges. Such disturbance is likely to increase the tendency to aggregate still further, possibly resulting in formation of larger aggregations rather than in fragmentation or dispersal. A similar increase in group size among red deer in response to disturbance has also been recorded in deer parks at times of heightened levels of public disturbance (Langbein and Putman, 1992b), as small groups disturbed in cover join up into larger ones in the open.

Thus, despite over 20 hunt meets on the estate during that time, several large herds of deer (30+) were observed on almost every visit to Holnicote between October 1991 and May 1992; similarly, a group of 12–14 stags was observed regularly on a specific set of fields in the south of Exmoor on several days directly following as well as directly preceding a hunt during March, which had drawn its quarry from that particular group of stags. In general, it would appear that, rather than causing more even dispersion of deer, hunting may result in somewhat tighter aggregation. Even groups actually targeted or displaced by the hunt seem to return to their regular home ranges within a very short period. On the other hand, such observation would also suggest that hunting does not result in any major disruption of the immediate behaviour pattern of the local deer herds. Indeed, it might be argued that the pattern of behaviour observed may itself have developed in adaptation to the regularity of hunting activity; hunting itself may contribute to a more aggregated, visible population than might otherwise be found. Stalking with rifles, in the absence of hunting, by contrast is often observed to lead the deer to become rather more secretive, as they increasingly seek the relative safety of woodland or other concealing cover (e.g. Challies, 1985).

12.4 DISCUSSION

Red deer populations subject to hunting-to-hounds in south-west England, as well as many other deer populations throughout England that are not subject to such hunting, have increased significantly over the last 20 years. The total number of deer accounted for by the three remaining packs of staghounds in the West Country has changed little over the same period, totalling between 175 and 225 deer in most years. Hunting has rarely been able to control deer numbers on its own; additional culling by other means has been practised in the area for many years, originally using shotguns, though increasingly replaced by stalking with high velocity rifles since restrictions regarding permissible weapons against deer were imposed by the Deer Act 1963. On the basis of our current estimates of population size and annual recruitment rates (births minus natural mortality) a cull of close to 20% of the autumn population numbers is likely to be required to prevent further increases. While during the mid 1970s hunting may have contributed up to a third of such cull requirements, its net contribution today is unlikely to exceed one-eighth or a quarter of the cull needed on Exmoor and the Quantocks, respectively, in order to prevent further significant population growth. Indirect assessment of the cull taken by other control methods, based on estimated population growth over recent years, suggests that legal shooting and poaching accounts annually for 8–13% of the population (two-fifths to three-fifths of the probable cull requirement to achieve zero growth).

The unpredictable though frequent disturbance of deer by hounds and riders in hunted areas, is likely to contribute to an observed tendency of deer in the region to aggregate in large groups even during the day; such groups are commonly formed in open moorland and farmland areas providing good visibility rather than maximum concealing cover. Hunting may also contribute to a somewhat patchier deer distribution than might otherwise occur, with artificially high concentrations of deer in 'safe' areas provided partly by estates supportive of hunting, to ensure a suitable supply of quarry – but ironically also by landowners directly opposed to hunting, in order to provide sanctuaries or refuges where deer may not be hunted.

Our studies of the deer around Exmoor and the Quantocks were initiated specifically to assess the current status, conservation and management of these populations, rather than addressing the welfare of hunted deer or the ethics of differing means of culling. Red deer have undoubtedly thrived over recent years, with numbers probably higher now than at any other time during this century, and body condition and fecundity indicative of a highly productive population. The current, relatively small contribution of hunting-to-hounds to the total annual cull does not suggest that hunting threatens the continued conservation of this species in the region. However, neither is the red deer distribution in the West Country limited by any means to those areas where hunting of deer continues to be practised: significant and increasing herds are present also throughout most other areas of west Somerset, north Devon and south into Cornwall. As such, according to Prescott-Allen and Prescott-Allen's scheme of evaluation (Chapter 4), the present impact of hunting on red deer should probably be viewed as being neutral with regard to conservation of the species in this region. The continuing controversy in this country surrounding hunting-to-hounds as a legal means of culling deer will thus need to be decided largely on the grounds of ethics, animal welfare and social considerations, rather than on the basis of the overall effect of hunting on the deer populations concerned.

ACKNOWLEDGEMENTS

This investigation was undertaken on behalf of and funded by the National Trust. In addition we should like to express our thanks to the numerous other conservation and field sports organizations, landowners, deer managers and naturalists who have assisted us during this study through provision of information as well as samples from deer culled in the West Country.

REFERENCES

Allen, N. (1990) *Exmoor's Wild Red Deer*, The Exmoor Press, Dulverton.

Burton, S.H. (1969) *Exmoor*, 3rd edn, Westaway, London.

Challies, C.N. (1985) Establishment, control and commercial exploitation of wild deer in New Zealand, in *Biology of Deer Production, Bulletin 22*, (eds P.F. Fennesy and K.R. Drew), Royal Society of New Zealand, pp. 23–26.

Clutton-Brock, T.H., Guinness, F.E. and Albon, S.D. (1982) *Red Deer: Behaviour and Ecology of Two Sexes*, University of Chicago Press, Chicago.

Hamilton, W.J. and Blaxter, K.L. (1980) Reproduction in farmed deer. I. Hind and stag fertility. *J. Agric. Sci. Camb.*, **95**, 261–273.

Henshaw, J. and Allen, R. (1989) A case of suspected capture myopathy in a West Country red deer. *Deer*, **7**(9), 466–467.

Langbein, J. and Putman, R.J. (1992a) *Conservation and Management of Deer on Exmoor and the Quantocks*, National Trust, London.

Langbein, J. and Putman, R.J. (1992b) Behavioural responses of park red and fallow deer to public disturbance and effects on population performance. *Animal Welfare*, **1**, 19–32.

Langbein, J. (1993) Studies of red deer on Exmoor: red deer activity in the Bridgetown to Dulverton region of the Exmoor National Park, and feasibility assessment of coloured feed markers in investigations of deer movement patterns. Ministry of Agriculture, Fisheries and Food, Project CSA 2279. Unpublished report.

Lloyd, E.R. (1970) *The Wild Red Deer of Exmoor*, Exmoor Press, Dulverton.

Lloyd, E.R. (1975) *The Wild Red Deer of Exmoor*, 2nd edn, Exmoor Press, Dulverton.

Lloyd, E.R. (1990) *The Staghunting Controversy: some provocative questions with straight answers*, Masters of Deer Hounds Association, South Molton, Devon.

Lowe, V.P. (1967) Teeth as indicators of age with special reference to red deer of known age from Rhum. *J. Zool. (London)*, **152**, 137–153.

Malcolm, S., Piatowski, A., Morgan, D. *et al.* (1984) *Exmoor Red Deer Survey, 1981–82: A comparison of lowland and upland habitats*, Devon Trust For Nature Conservation, Exeter, Devon.

Mitchell, B. (1967) Growth layers in dental cement for determining age in red deer. *J. Zool. (London)*, **36**, 279–293.

Mitchell, B. and McCowan, D. (1983) The faecal accumulation method for estimating and comparing population densities and site occupation by red deer. NCC NERC contract HF3 03 269 ITE Project 528.

Putman, R.J. (1984) Facts from faeces. *Mammal Review*, **14**, 79–97.

Putman, R.J. (1988) *The Natural History of Deer*, Christopher Helm.

Putman R.J. (1990) Patterns of habitat use; an examination of the available methods, in *Methods for Study of Large Mammals in Forest Ecosystems*, (eds G.W.T Groot Bruinderink and S.E. van Wieren), Riksinstituut voor Natuurbeheer, The Netherlands, pp. 22–31.

Putman R.J. and Langbein J. (1990) Factors affecting performance of deer in parks. Report PECD 7/2/65 to Department of Environment.

Ratcliffe, P.R. (1984) Population dynamics of red deer in Scottish commercial forests. *Proc. Roy. Soc. Edinburgh*, **82B**, 291–302.

Ratcliffe, P.R. (1987) Red deer population changes and the independent assessment of population size. *Symp. Zoo. Soc. London*, **58**, 152–166.

Staines, B.W. and Ratcliffe, P.R. (1987) Estimating the abundance of red deer (*Cervus elaphus* L.) and roe deer (*Capreolus capreolus* L.) and their current status in Great Britain. *Symp. Zoo. Soc. London*, **58**, 131–152.

Stewart, L.K. (1985) Red deer, in *Vegetation Management in Northern Britain*, Monograph No. 30, (ed. R.B. Murray), British Crop Protection Council, Croydon, pp. 45–92.

Winder, F. and Chanin, P. (1992) Red deer on the Quantocks: summary of preliminary results. Report to STNC, July 1992. Unpublished report.

Part Four

Wildlife Trade and Conservation

13

Zimbabwe: a model for the sustainable use of wildlife and the development of innovative wildlife management practices

Michael D. Kock

SYNOPSIS

In the African context, Zimbabwe has been one of the most aggressive promoters of the sustainable use philosophy. In contrast to the fact that the world's biodiversity is shrinking daily, Zimbabwe's wildlife management practices, both in the Parks and Wildlife Estate (PWLE) and the communal/private land sector, are expanding. More than 30% of the country's land mass is now under some form of wildlife use. Few countries in the world can match this trend. There are greater numbers of several wildlife species, including the elephant, than at any time in the country's history, despite human population growth and land tenure problems. A marked exception to this is the black rhinoceros. It is possible that if innovative wildlife management practices were introduced into the conservation of this species (including consumptive use) several years ago, the black rhinoceros's status would have been very different today.

Why are land areas for wildlife increasing? Zimbabwe boasts several progressive and innovative conservation initiatives, including the Communal Area Management Programme for Indigenous Resources (CAMPFIRE), the Conservancy concept, Intensive Protection Zones (IPZs) for the rhinoceros and detailed evaluation of multi-species (cattle and/or wildlife) production systems. Within these initiatives, sport hunting and live sales of wild animals are generating considerable income to both the private sector and communal wildlife programmes. Other innovative research programmes include use of electric fencing in control of problem animals. This chapter presents information on these initiatives and challenges the view, from an African perspective, that exploitation (sustainable use) of wildlife is negative and may only achieve short-term economic objectives.

13.1   INTRODUCTION: ZIMBABWE – AT THE LEADING EDGE OF CONSERVATION

As we approach a new century, the problems facing humanity are enormous. This is reflected by unchecked population growth, continued ethnic strife, political upheavals and civil wars. There is a distinct lack of vision by many political leaders and politicians in terms of an understanding of environmental problems and the value of renewable natural resources in contributing towards sustainable development. Humanity must adopt life styles and development paths that respect and work within nature's limits (IUCN/UNEP/WWF, 1991): to do otherwise would be foolish and short sighted and will court disaster in time.

Africa is a continent of great biological diversity, but with the considerable problems that have been outlined above it faces significant constraints in supporting and maintaining this diversity. If Africa's problems with the environment and management of natural resources are to be addressed, solutions uniquely African need to be found. It is within this context that Zimbabwe is a model for developing, initiating and evaluating resource management programmes that are innovative and address key issues. These initiatives have been developed by individuals, government and university departments, and conservation organizations with a unique understanding of Africa. This understanding is the key to successfully unlocking the door to a better life for Africa's rural poor and moving towards integration of conservation with development needs.

Through many of these initiatives, Zimbabwe has come into conflict with conservation and animal welfare organizations, many of whom believe that the solution to conserving natural resources lies in eliminating or banning any form of consumptive use. It is apparent that many of these organizations who purport to be the saviours of wildlife fail to appreciate the link between economic and social factors concerning resource management and the conservation of biodiversity in Africa. The trend in Zimbabwe is for accelerated integration of a rapidly expanding wildlife industry into the mainstream economy of the country. In contrast to the fact that the world's biodiversity is shrinking daily, Zimbabwe's wildlife management practices, both in the Parks and Wildlife Estate (PWLE) and the communal/private land sector, are expanding. More than 30% of the country's land mass is now under some form of wildlife use (Table 13.1) (Zimbabwe Trust, 1992; Martin 1993; Prescott-Allen and Prescott-Allen, 1996).

Following the decimation of wildlife by both disease (rinderpest pandemic) and uncontrolled hunting in the nineteenth and early twentieth centuries, policies have evolved in Zimbabwe that have resulted in a steady expansion of wildlife of many different species. There were more

Table 13.1 Wildlife as a land use in Zimbabwe, 1994 (source: Department of National Parks and Wild Life Management)

Land category	Total area (km²)	Area under wildlife (km²)	Percentage (%)
State land	57 010	49 000	86
Communal land	162 790	30 000	18
Commercial farms	170 961	37 000	22
Totals	390 761	116 000	30

elephant, crocodile, ostrich, buffalo and other species in Zimbabwe in 1994 than at any time in the country's history, and these populations continue to increase through innovative wildlife management practices (Wildlife Producers' Association, personal communication). A marked exception is the black rhinoceros. Individuals who have been operating at the sharp end of rhino conservation over the last decade believe that if a broader based approach to conserving these animals had been adopted several years ago, including consumptive use such as sport hunting and legal horn trade, the status of this charismatic megaherbivore would have been very different today (R.B. Martin, personal communication).

13.2 INTERACTION BETWEEN HUMANS AND WILDLIFE

Habitat loss, legal (uncontrolled) and illegal hunting represent the greatest threat to Africa's biodiversity. Habitat loss is a direct result of population pressures, civil wars and land tenure problems. During and after the scramble for Africa, colonial governments arbitrarily defined national borders (Pakenham, 1991) and established game reserves within strict boundaries. These boundaries created a 'hard edge' between those indigenous people who may have lived within these reserves historically, but now live on the border; that is, in communal lands (Figures 13.1 and 13.2). There were no benefits to indigenous people from these game reserves, and any exploitation of wildlife within (other than by the colonial powers) was by illegal hunting (snaring and shooting), and illegal harvesting of wood and grazing of livestock within park boundaries. This 'hard edge' is indefensible in the light of Africa's population growth and hunger for land; any conservation initiatives that fail to recognize this are doomed. The conflict generated by the 'hard edge' can vary, but invariably results from resentment, including a history of racial conflict, crop damage and potential for loss of life and property.

Within Zimbabwe, significant numbers of wildlife are found in communal land areas (Table 13.1) and in many instances these animals will move between communal lands, safari areas and national parks. These areas are usually in Natural regions IV and V which receive minimal

PEOPLE AND WILDLIFE

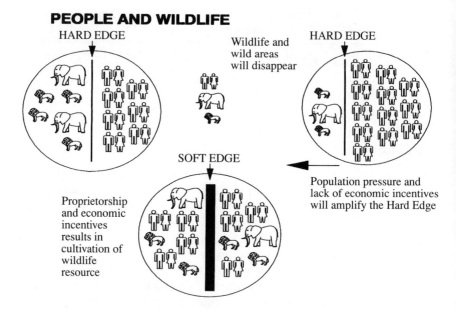

Figure 13.1 The 'hard edge': a division between wildlife and people, an historical fact and further constrained by population pressures. Solution: softening of the 'hard edge' through programmes such as CAMPFIRE.

rainfall and are unsuitable for most agricultural pursuits. Communities are often at barely subsistence levels and survival (both historically and in some instances currently) does not take in the necessities of preserving wildlife and considering animal welfare (Hutton, 1994).

13.3   TO EXPLOIT OR USE SUSTAINABLY: A POSITIVE CONCEPT OR A QUICK FIX?

It is becoming increasingly apparent that wildlife conservation in Africa cannot be viewed in isolation from other larger economic and social factors (Makombe, 1994). The dilemma is to balance conservation with development. Exploitation, by definition implies that something is 'turned to advantage', that one makes use of for one's own ends, and this carries the connotation that exploitation is negative. Many conservation initiatives in southern Africa 'exploit' wild animals, but certainly not negatively. We must be careful not to allow this term to be used when describing conservation programmes such as CAMPFIRE or sport hunting.

An alternative term, as adopted by IUCN/UNEP/WWF (1991) in the *World Conservation Strategy: Caring for the Earth,* is sustainable use. This applies to natural or renewable resources, and implies using these

Figure 13.2 Graphic representation of the 'hard edge': fence line with Matobo National Park on the right and communal land on the left. Some softening of 'hard edge' with controlled seasonal grass cutting for thatching and sale by communal land people. Note thatching grass pile in background in National Park.

resources at rates within their capacity for renewal. In the context of Zimbabwe's conservation initiatives, this term will replace 'exploitation'.

In understanding wildlife conservation in Africa it is essential to comprehend the basics of conservation, preservation and sustainable use. These are key concepts that in the author's opinion have been manipulated and misunderstood in the developed world. Conservation implies that a resource should be used (this can be for either non-consumptive or consumptive use) (Passmore, 1974) and inherently recognizes that the resource has a value, both aesthetic and economic. Preservation implies that the resource should not be used, but preserved for future posterity, and is valued solely for its aesthetic appeal (Passmore, 1974). Within the African context, and particularly the southern African context, preservation has no major role to play in the future of wildlife and wild areas, although it may be important in other geographical areas and with particular species (Robinson, 1993). This is because of preservation's protectionist message, which is inappropriate in view of Africa's social and economic problems. Conservation's major driving force under African conditions is utilization, which is synonymous with the maintenance of biological diversity, and the key to the success of this philosophy is sustainability (Zimbabwe Trust, 1992; Makombe, 1994). This emphasizes the need for people to manage biological diversity as an essential foundation for the future, to maintain wildlife populations for their benefit and to use species sustainably to enhance quality of life (Makombe, 1994).

13.4 SPORT HUNTING

Exploitation of wild animals through sport hunting has existed for centuries throughout Africa. In the nineteenth and early twentieth centuries hunting pressures were considerable, with little understanding of sustainable offtake. In southern Africa, particularly South Africa, many species of plains game were brought close to extinction by the end of the nineteenth century. Over the last several decades there has been an increase in understanding by the hunting fraternity of the need for monitoring offtake and trophy size, and for setting hunting quotas. Hunting has become more closely linked with conservation and to the future of wild animals and rural communities.

Sport hunting in Zimbabwe is a growth industry. The key to this was the Parks and Wildlife Act 1975, which promoted the sustainable utilization of wildlife by conferring proprietorship of wildlife resources on the 'owners or occupiers of alienated land' (Murphree, 1991). Indeed, the growth of the wildlife industry in Zimbabwe had its impetus in this act. Between 1985 and 1990 there was a 42% increase in the number of days of sport hunting; this represented an increase in revenue of 117% (Table 13.2) (Bond, 1994). Within the CAMPFIRE programme incomes from hunting quotas almost tripled between 1990 and 1993 (Prescott-Allen and Prescott-Allen, 1996).

In the developed world there are individuals and organizations who see no link between sport hunting and conservation. Hunting is often perceived as purely killing animals for sport, thus creating an emotive smokescreen that has clouded the perceptions of many individuals. That sport hunting and conservation are inexorably linked can be amply demonstrated by the following examples.

Table 13.2 Growth in sport hunting in Zimbabwe, 1985 to 1990 inclusive (source: Price Waterhouse and Environmental Resources Limited, 1992)

Year	Hunting days	Income (US$)	Income (Z$)
1985	7 966	4 313 343	7 079 177
1990	11 338	9 368 171	24 718 129
% change	42%	117%	249%

13.4.1 Sport hunting of white rhinoceros in South Africa

Since 1968, sport hunters have shot 820 adult white rhino bulls. During the same period, the population has increased from 1800 to 6370. Revenue earned from these hunting activities has exceeded US$22.3 million (Adcock and Emslie, 1994). This represents a clear indication of hunting's conservation component, with many ranching and game park enterprises moving towards profitability, and has promoted the continued existence of their rhino populations. Trophy hunting has been highly sustainable and has generated significant revenue.

13.4.2 Sport hunting and CAMPFIRE

Sport hunting constitutes the main source of income to most districts in Zimbabwe implementing CAMPFIRE. Safari operators lease hunting concessions from district councils by paying dues, in the form of either trophy fees, a percentage of gross revenue, or a lump sum payment. Of the wildlife-based revenue earned through CAMPFIRE, 90% was derived from sport hunting between 1989 and 1992, representing over US$1.8 million (Z$10 million) (Bond, 1994; Child and Bond, 1994). Zimbabwe earns approximately Z$1 billion from tourism and the sport hunting contribution is 40%. This is a clear demonstration of the conservation and sport hunting link in addressing ecological and rural development needs in southern Africa (see section 13.8 for further discussion).

Sport hunting in Zimbabwe is a valuable industry but it is also acknowledged as a 'high risk' one and necessarily carries higher profits to cushion against unforeseen problems, such as civil unrest. The safari hunting market is one where demand exceeds supply, but there is limited scope to increase it in Zimbabwe since in most areas the offtake is close to the limit of sustainability (Martin, 1993, 1994a,b). It is likely that a multi-faceted approach will be adopted with eco/adventure tourism, the increased profitability of which will replace sport hunting in many areas.

13.5   CAPTURE AND LIVE SALES

With the trend in land use indicated in Table 13.1, trade in live animals amongst farmers within southern Africa has escalated. This has been further enhanced by the ability to move various species across borders and by a more rational approach adopted by veterinary authorities (C. Foggin and E. Anderson, personal communication) towards restrictions on animal movement – buffalo in particular. Many key species, including elephant, rhino (black and white), buffalo, sable, roan, nyala and Lichtenstein's hartebeest, are being traded profitably. Wildlife auctions have been held regularly over the last few years (Table 13.3). For example, a recent auction of 120 sable antelope generated US$318 000 (Z$2 640 000), and the first Natal Parks Board auction of five black rhino in 1992 netted 2.2 million rand (approximately US$600 000) (Wildlife Producers' Association, personal communication). Success in this aspect of the wildlife industry requires the maintenance of professional standards in game capture and minimum standards with holding facilities, whilst closely monitoring animal welfare.

13.6   CONSERVATION AND COMMERCIAL CONSUMPTIVE USE OF WILDLIFE

In Zimbabwe, a successful conservation initiative (albeit non-mammalian) has been achieved by placing a high economic value on the Nile crocodile.

Table 13.3 Some recent prices from wildlife auctions in Zimbabwe, 1993 (source: Wildlife Producers' Association, Harare, 1993; exchange rate (approx.) US$1 : Z$6.5)

Species	Unit price (US$)	Unit price (Z$)
Sable (female)	3077	20 000
Elephant	2667	17 000
Sable (male)	2308	15 000
Giraffe	1102	7000
Eland	1038	6700
Zebra	615	4000
Bushbuck	500	3250
Ostrich	423	2800
Blesbok	415	2700
Tsessebe	385	2500
Reedbuck	249	1600
Oribi	234	1500
Wildebeest	231	1500
Impala (female)	81	525
Impala (male)	62	400

The population of Nile crocodile is noticeably healthy if one spends any time in the wild areas of Zimbabwe, and even in some of the not so wild areas. Some say it is too healthy, and this makes for interesting anecdotal accounts of encounters with crocodiles, although some unfortunately have ended in tragedy. In the 1960s Nile crocodile populations were severely depleted throughout most of the range of the species (Pinchin, 1994). A major crocodile conservation programme was initiated in Zimbabwe, and by 1983 was showing promise. Through CITES, Zimbabwe's population was downlisted to Appendix II, allowing the legal export of crocodile products and expansion of the industry. The numbers of crocodile in Zimbabwe are estimated to be > 50 000 in the wild, with 150 000 in captivity. The success of this conservation programme has filtered through to CAMPFIRE communities with crocodile ranchers paying local communities for egg-collecting rights and also providing employment. This is another example of 'exploitation' providing long-term benefits to a species and to rural development.

13.7 MULTI-SPECIES PRODUCTION SYSTEM PROJECT

In evaluating the success or otherwise of various conservation initiatives within the private sector in Zimbabwe, several conservation organizations are involved in implementing research programmes. One such organization is the World Wide Fund for Nature (WWF). The WWF Multi-species Production System Project was implemented to evaluate critically various farming enterprises involving cattle, or cattle with wildlife, or wildlife alone (D. Cumming, personal communication). The survey focused on the question of which of these three alternative land-uses can best exploit semi-arid rangelands in a sustainable manner (Jansen *et al.*, 1992; Price Waterhouse, 1994). With serious questions being asked concerning the viability of cattle farming in certain areas of Zimbabwe (Natural regions IV and V) and the impact on the environment by cattle in these marginal areas, the need for a detailed economic and environmental evaluation of these various enterprises was long overdue. Politically, replacing cattle (a traditional animal in African society) with wildlife is a sensitive issue, and wildlife utilization has been seen as a white-dominated industry. It has become critical, therefore, that facts and figures are made available demonstrating the economic performance of wildlife versus cattle and the value of wildlife to the economy of Zimbabwe.

The results of the survey of 89 ranches with either cattle or wildlife reinforced the notion that well-managed wildlife enterprises are potentially more financially viable than cattle enterprises, specifically in Natural regions IV and V. On average, wildlife had a return on investment of 8.6% compared with cattle at 2.5%, with respective net revenues per hectare of US$1.11 compared with US$0.60 (Bond, 1994). It was made clear that, in order for a wildlife ranching enterprise to be successful, experience in marketing was essential. Caution was expressed in advocating wildlife as the panacea for economic and environmental problems and a key constraint to both is the negative effects of a number of government policies.

There are examples of this type of study carried out elsewhere in southern Africa (Price Waterhouse, 1994) but there is little doubt that the historical prejudices surrounding wildlife in relation to traditional agricultural enterprises (fostered heavily by colonial administrations) are being eroded significantly, allowing the sustainable use philosophy to flourish and the environment to recover.

13.8　CAMPFIRE: COMMUNAL AREA MANAGEMENT PROGRAMME FOR INDIGENOUS RESOURCES

Many modern African governments are almost as distant from rural land use practices as former colonial regimes (Metcalfe, 1993). A critical issue here is policy, and as Murphree (1991) states: 'Unless policy on tenure and natural resource management seriously considers the third option of communally-based resource management regimes for much of our land (Zimbabwe) there is little reason, either from the historical record or from analysis of the factors and dynamics involved, to be optimistic about our environmental future.' There is little doubt that pressure on land will continue to grow, and any posturing from the West concerning the need to maintain biodiversity purely for aesthetic reasons will be engulfed by short-term political and subsistence/survival factors. With a 'hard edge' philosophy to managing our wildlife resources came resentment from many local communities who believed that government, through the national parks, cared more for wildlife than for people.

In Zimbabwe a key principle in resource management that attempts to answer the 'hard edge' problem is that 'people seek to manage the environment when the benefits of management are perceived to exceed its costs' (Murphree, 1991; Martin, 1993). CAMPFIRE addresses this and vests proprietorship of wildlife resources in the local people: it advocates that rights of access to wildlife must be based on distinctive community and resource boundaries (Metcalfe, 1993; Prescott-Allen and Prescott-Allen, 1996). This programme operates on the belief that wildlife utilization as a form of land use will only be endorsed by rural communities when individual members receive direct benefit (Murphee, 1991; *CAMPFIRE Newsletter*, 1994). CAMPFIRE also operates on the principle that 'a communal resource management regime is enhanced if it is small enough (in membership size) for all members to be in occasional face-to-face contact, enforce conformity to rules through peer pressure and has a long-standing collective identity' (Murphree, 1991). The principles of CAMPFIRE ensure that more than 50% of any revenues earned from wildlife resources should accrue to local communities and to the lowest level. In Zimbabwe, 25 rural districts have the 'appropriate authority' to manage their wildlife resources. Within these, 12 districts with 70 wards (82 000 households) and 600 000 people have active CAMPFIRE programmes. This is carried out on approximately 1 million hectares and has generated Z$10 million in revenue (*CAMPFIRE Newsletter*, 1994; B. Child, 1995, personal communication).

There are several key issues within CAMPFIRE that are linked to sustainable use and a few examples are presented here.

13.8.1 Economics and accountability

In accepting CAMPFIRE, a community has decided that wildlife will be a major form of economic activity in the community and will be important for the livelihood of the people. Dividends earned through wildlife activities are disbursed to households within the community, but there are also collective community concerns. These are addressed through peer and community pressure. A percentage of each household's dividend is allocated, for instance, to building a new schoolroom or purchasing a new grinding mill. There is accountability and transparency in this process as dividends are handed out with the community present.

Comments that have been recorded demonstrate the value of grassroots conservation with an economic base, and the following is a quote from Murphree (1991) concerning a statement by one of the councillors at a CAMPFIRE meeting in which revenues were shared: 'This money comes to you from your wildlife. It is your money. The decision is yours. You cannot wait for government. You can develop your community according to how you decide.' The rural African is exploiting a resource sustainably and recognizes wildlife as being crucial to development and providing a better living for their family.

13.8.2 Elephants and sport hunting

In the continuing debate over the 1989 CITES ivory trade ban, Zimbabwe has always argued strongly that the ban would adversely affect the success of community-based resource programmes. There is concern that any pressures to stop elephant sport hunting (trophies are still allowed to be exported back to countries such as the USA) would further jeopardize CAMPFIRE programmes, resulting in a reduction in economic incentives for sustainable wildlife utilization in crucial communal land areas in southern Africa. The elephant plays a vital role in revenue generation in these programmes (Taylor, 1993). The value of elephants to the rural poor does not come just from sport hunting: culling provides much needed protein and hides can be processed and sold.

Pressures to stop sport hunting of elephants are real, an example being a moratorium proposed in 1992 by the US Fish and Wildlife Service (USFWS, undated) to ban or restrict elephant sport hunting. Bond (1994) states that the total value of the 1992 communal lands sport-hunted quota of 89 elephants and other animals is over Z$3 million (US$352 000) (Figure 13.3). It is estimated that the 1989 CITES ban has cost rural communities more than Z$4 million in lost revenues over four years (B. Child, personal communication). The ban has resulted in the inability of rural communities to sell, legally, ivory that has been collected from elephants shot during problem animal control (PAC). Approximately 15 tons (or half of Zimbabwe's ivory stockpile) represents PAC ivory.

13.8.3 Fencing

In several of the CAMPFIRE areas, problems related to negative interactions between wildlife and community members, particularly associated with crop damage, have been solved by appropriately placed electric fencing (Murphree, 1991; Hoare and Mackie, 1993). For example, in the planning of resource management by one CAMPFIRE community a wildlife committee was formed and they soon moved into land-use planning. They set aside 20 km² for fields and settlement, to be surrounded by an electric fence. The rest of the ward was set aside for wildlife with the development of sport hunting and eco/adventure-based tourism.

13.9 CONSERVANCIES: CONSERVATION AND THE PRIVATE SECTOR

Conservancies offer one of the most exciting wildlife developments in southern Africa with the potential of contributing significantly to the maintenance of biodiversity.

There is no formal definition of a conservancy. The term can be applied to any number of privately owned properties which are amalgamated into a single complex in order to enable more efficient management, utilization and protection of some or all of the natural resources in the area (du Toit, 1992; Price Waterhouse, 1994). Zimbabwe is not unique in adopting the conservancy concept – South Africa has established a number of large conservancies in Natal and bordering the Kruger National Park (Penzhorn, 1994). The uniqueness of Zimbabwe's large conservancies stems from the fact that they were developed to provide a safe haven for the black rhinoceros, outside of the PWLE. The three major conservancies were formed in 1991 in the lowveld (other smaller conservancies exist elsewhere in Zimbabwe). The main focus of conser-

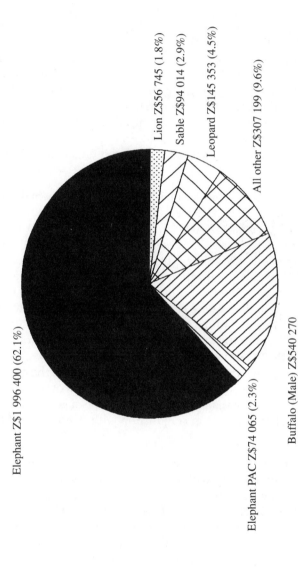

Elephant Z$1 996 400 (62.1%)

Lion Z$56 745 (1.8%)

Sable Z$94 014 (2.9%)

Leopard Z$145 353 (4.5%)

All other Z$307 199 (9.6%)

Buffalo (Male) Z$540 270

Elephant PAC Z$74 065 (2.3%)

**Total income
Z$3 214 046**

Figure 13.3 The value (Z$) of elephants and other animals to CAMPFIRE in eight districts in Zimbabwe, 1992 (PAC, problem animal control). Source: Bond (1994).

vancy agreements is cooperative management of the wildlife resource (Price Waterhouse, 1994).

The key to the success of a conservancy lies in its constitution, and this is built around four main principles:

1. That internal game fencing is limited, in order not to divide a conservancy into compartments and thereby interfere with the natural movement and breeding of animals.
2. In the event that a conservancy property passes into the hands of an agency whose land use and wildlife management practices are not consistent with the conservancies, that property will be excised from the conservancy, and conservancy assets retrieved.
3. The members are jointly responsible for meeting recurrent management costs.
4. Management of the conservancy wildlife is based on sound scientific principles.

Within Zimbabwe, three large conservancies have been established. Save Valley (321 000 ha) and Chiredzi River (89 000 ha) are located in the south-east lowveldt; Bubiana (128 000 ha) is located in the south-west lowveldt. Among or within them, these conservancies hold a large number of Zimbabwe's black rhino. The most exciting aspect of the conservancies has been their realization of the enormous potential for conservation programmes other than rhinos, including high quality tourism and limited sport hunting. These conservancies have much potential (in the light of government failure to finance adequately conservation efforts in the PWLE through the Department of National Parks) for competing with some of the major national parks in Zimbabwe.

In the context of mammal exploitation and sustainable use, the conservancy concept amply demonstrates the long-term potential of utilization. Three examples will be discussed briefly. These have been generated in a report by Price Waterhouse (1994) in conjunction with the owners of the three conservancies.

1. In a short-term ranch study, an evaluation was performed comparing cattle and wildlife on a ranch within one of the conservancies. Based on an annual stocking rate of 1 livestock unit (LSU)/20 ha for cattle, gross revenues were likely to be Z$429 000. At 1LSU/10 ha, revenues increase to Z$858 000, or Z$18–36/ha, but environmental degradation would be significant. In contrast, a wildlife production model would generate (in the short term) revenues of Z$1 478 000, or Z$61.31/ha. With the latter, more people would be employed and more revenue would be spent in the local economy.
2. A medium-term study involved the running of a 40-bed safari lodge and an eight-bed exclusive camp within a conservancy. The safari

lodge generates annual gross revenues of Z$9 million and the small camp Z$1 million. Profits from these operations are significant, as are the number of local people employed from surrounding communal lands.

3. The third example is based on the Londolozi operation in South Africa, which is part of the Sabie Sand Wildtuin near Kruger National Park and demonstrates the long-term potential of a former cattle ranch converted to wildlife. In 1992, Londolozi generated over Z$18.6 million, with net profits of Z$4.64 million. In addition this operation has a very strong socio-economic component with support for surrounding local communities.

It is likely that the major conservancies in Zimbabwe will have a strong component of eco/adventure tourism in their operations, but will be diverse in supporting limited sport hunting, live sales of wildlife and controlled culling with meat for the local communities. The establishment of these conservancies has added significantly to land area under wildlife use in Zimbabwe. Politically, they are acceptable because they are located in Natural regions IV and V where communal land degradation is significant.

The conservancies, therefore, offer some hope of assisting these communities to climb out of the spiral of subsistence living and poverty; they have the potential to create increased environmental awareness and to impact on individual perceptions of conservation/sustainable use as much as CAMPFIRE. It is hoped that they will operate hand-in-hand.

13.10 NATIONAL PARKS CONSERVATION PROGRAMME: INTENSIVE PROTECTION ZONES FOR BLACK AND WHITE RHINOS

The decline of the black rhinoceros in Zimbabwe has been dramatic with the first indications of a poaching onslaught in 1986. Over the last eight years, the population has declined by more than 90%, with significant loss of human life due to a very aggressive anti-poaching stance adopted by the Zimbabwean Government (ZBRCS, 1992). Despite this effort, the population now rests at approximately 260 animals.

The black rhino is an example of a mammal that has not paid its way in conservation terms (Martin, 1993) and there are individuals who question why it should. Stark facts from the last decade on the decline of the black rhino provide some of the answers. Many conservationists in Zimbabwe believe that this decline has been accelerated by a strict protectionist policy. There is no doubt that the rhinos that lived in communal lands in the 1980s were significant in number. Many of these communal lands are located on the northern border and adjacent to many national parks. The loss of these rhino can be attributed not just to illegal hunting, but also to a failure to place a value on these animals by the local communities.

The belief is that if a small percentage of older adult males were selected and offered on quota to be sport hunted within a CAMPFIRE programme, the enormous earnings (estimated at US$100 000–200 000, or Z$800 000–1 600 000) would have stemmed the decline of this species in Zimbabwe. It has been amply demonstrated in CAMPFIRE that poaching virtually stops when wildlife assumes a role in the local economy (B. Child, personal communication; see also Chapter 4). In the case of the rhino, there would have been less collaboration and more reporting of the presence of illegal hunters.

At present, Zimbabwe's black and white rhinos are protected in Intensive Protection Zones (IPZs) in the PWLE and private land conservancies. The IPZs represent areas within national parks or safari areas that have viable populations of rhinos, with increased scout density, better training, more equipment and vehicles. In addition, the majority of rhinos have radio-collars which serve to provide information for law enforcement and behavioural data. All these populations have been dehorned (Plate 1) and in many instances dehorned again. The collection of data on a recent operation indicated a calf survival rate (calves less than 6 months old when mothers were dehorned) of more than 71% and probably close to 100%, and 71% for calves older than 6 months. This is in sharp contrast to a report by Berger and Cunningham (1994) who indicated that survival rates may be close to 0%. No black rhinos have been known to have been poached in the last 26 months in Zimbabwe and this has been attributed to dehorning and improved law enforcement.

13.11 CONCLUSIONS

The question of whether sustainable use of natural resources will succeed will depend on many factors and the perfect mix and match may be hard to attain. Despite the positive results seen in CAMPFIRE in Zimbabwe, there have been some early failures and problems still exist (Chapter 4; B. Child, personal communication). There is no doubt that a balance must be attained between non-use, consumptive use and non-consumptive use and that dialogue must be established between proponents of each of these. In this chapter, arguing for a philosophy of use with natural resources, several examples have been given that support the concept of sustainable use as a long-term strategy in preserving biological diversity in Africa. In considering sustainable use there are several important factors that need to be taken into account and summarized:

- political, social and economic considerations;
- philosophy of use;
- cash and sustainability;
- a shifting conservation axis;
- maintenance of biodiversity into the next millennium.

13.11.1 Political, social and economic considerations

Whether wildlife should pay its way depends on many factors but the most important are: where the wildlife is in the world; political and historical considerations; land tenure issues; and the status of species (viable, threatened or endangered). Conservation cannot operate in a vacuum and too many issues are driven by environmental considerations without taking into account both economic factors and socio-cultural backgrounds (Martin, 1994a,b). The answer must lie at the regional and country level, or even at the village level, in a given country. It is vital to resurrect the community as the unit of social and economic analysis with respect to natural resources (Bromley, 1993).

Historically, governments have paid lip service to conservation, with inadequate amounts of their budgets going towards protected areas. In 1994, Zimbabwe's budget allocated Z\$1.67 billion to defence and only Z\$44 million to the custodians of wildlife on state land: the Department of National Parks and Wild Life Management. Therefore, without political support nurtured by conservation education, any conservation initiative will be doomed in the long term.

13.11.2 Philosophy of use

In the debate about preservation versus sustainable use, it is recognized by both sides that preserving habitats is an important component of any conservation philosophy. But positive feedback results from the promotion of a philosophy of use. The Parks and Wildlife Act 1975 in Zimbabwe was predicted to result in the decline of wildlife throughout the country, leading to local extinctions (Martin, 1994b). In fact the exact opposite became true. Not only have wildlife numbers increased, but also the land available to wildlife in the private and communal land sector now approaches 18% of Zimbabwe's total land area. As well as resulting in the preservation of habitats, this has promoted a far-reaching conservation philosophy in which all indications point towards long-term sustainability.

13.11.3 Cash and sustainability

In southern Africa an economic/ecological partnership exists with interest (extra wildlife) accumulating in the bank despite some rigorous initial spending (exploitation). The capital (habitat/ecosystem and wildlife) has been protected by healthy returns (earnings and interest accumulated) due to economic incentives to protect. Cash is one of Africa's most effective development extension agents (Murphree, 1991) and as Child states (Child and Peterson, 1991): 'Real and immediate benefits, graphically illustrated by cash, cement the relationship between wildlife and economic development. These incentives are crucial to encourage communities to cultivate their wildlife resources.'

13.11.4 A shifting conservation axis

The issue of sustainable use has polarized the conservation world. Those groups that preach preservation and animal welfare organizations that vehemently oppose any use, especially hunting, have gained the high ground in recent years, adversely influencing organizations such as CITES. It has been a clear indication of 'convenience'conservation operating by remote control. This has recently changed: the axis has shifted and rightly so. During the recent 1994 CITES meeting, sustainable use, instead of being perceived as extreme, has moved on to centre stage with more overlap with the preservationist groups (R. Martin, personal communication; Martin, 1993). The more extreme animal welfare groups have become marginalized, but it is imperative that this should not detract from the need to consider animal welfare within wildlife management practices. Indeed, it is vital to support those animal welfare groups who occupy the middle ground.

All too often recipes supplied by the West are based on a belief in the infallibility of science for matters which are primarily socio-economic (Martin, 1994a). To quote Clarke (1992): 'Modern western science, moreover, is being offered to (imposed upon?) cultures globally, as a problem-solving device, dragging along with it its particular set of assumptions and its selected, and biased vision.' It is ironic that Zimbabwe is regarded as a pariah in the eyes of many conservationists and organizations in the West, when more land is being put under wildlife management. Many Zimbabwean conservationists believe that proprietorship and economic benefits are the key initiators of sustainability (Martin, 1994b) and that ecological criteria have been too weighty.

13.11.5 Maintenance of biodiversity into the next millennium

Under the current world 'climate' and political leadership, a decline in biodiversity world-wide will continue despite all efforts by governments, international organizations such as CITES and others. This is an inescapable fact and is made more stark by events in Africa, a continent with such enormous potential but so many depressing failures, politically, socially and environmentally. Is there any cause for optimism? Only if, as the World Conservation Strategy 1991 states, we 'adopt life styles and development paths that respect and work within nature's limits'. The onus and focus of effort to achieve this differs between the developed and developing world but the ultimate goal is the same: a biologically diverse and ecologically healthy planet. To achieve this many individuals, organizations and policy makers must remove their 'blinkers', show more tolerance towards alternative viewpoints (especially those from the developing world) and be driven by flexibility, creativity and innovation in addressing environmental and conservation issues.

Every creative act in science, art or religion involves a new innocence of perception liberated from the cataract of accepted beliefs.

Arthur Koestler

REFERENCES

Adcock, K. and Emslie, R.H. (1994) The role of trophy hunting in white rhino conservation, with special reference to Bop parks, in *Proceedings of a Symposium on Rhinos as Game Ranch Animals*, (eds B.L. Penzhorn and N.P.J. Kriek), South African Veterinary Association, Onderstepoort, South Africa, pp. 35–41.

Berger, J. and Cunningham, C. (1994) Phenotypic alterations, evolutionary significant structures, and rhino conservation. *Conservation Biology*, 8(3), 833–840.

Bond, I. (1994) The importance of sport-hunted African elephants to CAMPFIRE in Zimbabwe. *TRAFFIC Bulletin*, 14(3), 117–119.

Bromley, D.W. (1993) Common property as metaphor: systems of knowledge, resources and the decline of individualism. Presidential address to the Fourth International Conference of the International Association for the Study of Common Property, Manila, June 1993.

CAMPFIRE Newsletter (1994) November, CAMPFIRE Association, PO Box 661, Harare, Zimbabwe.

Clarke, M.E. (1992) Worldviews, science, and the politics of social change. Paper presented at the Third Annual Conference of the International Association for the Study of Common Property, Washington DC, September 1992.

Child, B. and Bond, I. (1994) Marketing hunting and photographic concessions in communal areas, in *Safari Operations in Communal Areas in Matabeleland, Proceedings of the Natural Resources Management Project Seminar and Workshop*, (ed. M.A. Jones), Department of National Parks and Wild Life Management, Harare, Zimbabwe, pp. 37–55.

Child, B. and Peterson, J.H. (1991) *CAMPFIRE in Rural Development: The Beitbridge Experience*, Harare, DNPWLM/CASS Working Paper 1/91.

du Toit, R.F. (1992) Large-scale wildlife conservancies in Zimbabwe: opportunities for commercial conservation of endangered species. Paper presented at 3rd International Wildlife Ranching Symposium, Pretoria, October 1992.

Hoare, R. and Mackie, C.S. (1993) *Problem animal assessment and the use of wildlife management fencing in communal lands of Zimbabwe*, WWF (MAPS) Project Paper No. 39, Harare, Zimbabwe.

Hutton, J. (1994) A 'war' between humans, wildlife. *CAMPFIRE Newsletter*, 8. CAMPFIRE Association, PO Box 661, Harare, Zimbabwe.

IUCN/UNEP/WWF (1991) *Caring for the Earth: a Strategy for Sustainable Living*, Gland, Switzerland.

Jansen, D., Bond, I. and Child, B. (1992) *Cattle, Wildlife, Both or Neither: Summary of Survey Results for Commercial Ranches in Zimbabwe*, WWF Multispecies Animal Production Systems Project, Paper No. 30, WWF Harare, Zimbabwe.

Makombe, K. (ed.) (1994) *Sharing the Land: Wildlife, People, and Development in Africa*, IUCN/ROSA Environmental Issues Series No. 1, IUCN/ROSA, Harare, Zimbabwe and IUCN/SUWP, Washington, USA.

Martin, R.B. (1993) 'Should wildlife pay its way'? Paper presented in Perth, Australia. Zimbabwe Government Publication, Department of National Parks and Wildlife Management, Harare, Zimbabwe.

Martin, R.B. (1994a) The Influence of Governance on Conservation and Wildlife Utilisation. Paper presented at a conference on Conservation Through Sustainable Use of Wildlife, University of Queensland, Brisbane, Australia. Zimbabwe Government Publication, Department of National Parks and Wildlife Management, Harare, Zimbabwe.

Martin, R.B. (1994b) Alternative Approaches to Sustainable Use: What does and doesn't work. Paper presented at a conference on Conservation Through Sustainable Use of Wildlife, University of Queensland, Brisbane, Australia. Zimbabwe Government Publication, Department of National Parks and Wildlife Management, Harare, Zimbabwe.

Metcalfe, S. (1993) CAMPFIRE: Conservation can succeed. *Wildlife Watch*, **1**(3), 28–29.

Murphree, M.W. (1991) Communities as Institutions for Resource Management. CASS Occasional Paper series, National Conference on Environment and Development, Maputo, Mozambique. CASS, University of Zimbabwe, Harare, Zimbabwe.

Pakenham, T. (1991) *The Scramble for Africa*, Abacus Books, London.

Passmore, J. (1974) *Man's Responsibility for Nature*, Duckworth, London and Scribner, New York.

Penzhorn, B.L. (ed.) (1994) *The Future Role of Conservancies in Africa?* Wildlife Monograph No. 1, Ondersterpoort 1994, Du Toit Game Services, Sunnyside, Pretoria, South Africa.

Pinchin, A. (1994) Conserving the Nile crocodile in Zimbabwe – the value of sustainable yield utilisation. *International Zoo News*, **251** (41/2), 19–24.

Prescott-Allen, R. and Prescott-Allen, C. (eds) (1996) *Assessing the Sustainability of Uses of Wild Species. Case studies and initial assessment procedure*, IUCN, Gland, Switzerland.

Price Waterhouse (1994) The Conservancies: New Opportunities for Productive and Sustainable Land-Use. Save Valley, Bubiana and Chiredzi River Conservancies with Price Waterhouse Wildlife, Tourism and Environmental Consulting, Harare, Zimbabwe.

Robinson, G.R. (1993) The limits to caring: sustainable living and the loss of bio-diversity. *Conservation Biology*, **7**(1), 20–28.

Taylor, R.D. (1993) Elephant management in Nyaminyami District, Zimbabwe: turning a liability into an asset. *Pachyderm*, **17**, 19–29.

USFWS (undated) *US Endangered Species Act. Proposed Guidelines on African Elephant Sport Hunted Trophy Permits*, US Fish and Wildlife Service.

Zimbabwe: At the Leading Edge of Conservation. 1993. A report in response to the Humane Society of the United States. Department of National Parks and Wild Life Management of Zimbabwe.

Zimbabwe Trust (1992) *Wildlife: relic of the past, or resource of the future*, Environmental Consultants (PVT) Ltd, Harare, Zimbabwe.

ZBRCS (1992) *Zimbabwe Black Rhino Conservation Strategy*, Zimbabwe Government Publication, Department of National Parks and Wild Life Management, Harare, Zimbabwe.

14

Sustainable utilization: the lessons of history

David M. Lavigne, Carolyn J. Callaghan and Richard J. Smith

SYNOPSIS

The concept of sustainable utilization of wild life[1] has taken on new connotations in recent years, placing increased emphasis on trade in dead wild life in the marketplace. The authors question whether this change is progressive and in the best interests of maintaining wild populations and biodiversity, or whether it is regressive and designed to achieve objectives that are quite different from those implied by the concept.

During the nineteenth century, the exploitation of animals, including mammals, was largely unregulated and markets in dead wild life were rampant. Eventually, numerous populations were depleted, local populations were extirpated and some species were rendered extinct. This trend was reversed in some parts of the world during the twentieth century as safeguards were successively added in an attempt to reduce human impacts on wild life populations.

The decline in biodiversity, nonetheless, continues. The number of endangered and threatened mammals is increasing, many because of direct exploitation. Any solution to this crisis requires the promotion only of those wild life uses that have proven to be truly sustainable. Promoting the general view that wild life must have value in the marketplace in order to be conserved and relaxing safeguards, which were put in place to enhance the continued existence of exploited populations, may well achieve some short-term economic objectives. But it will do so, history suggests, only at the further expense of the populations we are trying to conserve.[1]

---

[1] Although this volume is concerned with the exploitation of mammals, the fundamental principles involved in sustainable utilization apply far more broadly. Here, we use the original term **wild life** (two words, e.g. Hewitt, 1921), following Anon (1990), to mean 'all wild organisms and their habitats – including wild plants, invertebrates, and microorganisms, as well as fishes, amphibians, reptiles, and the birds and mammals traditionally (i.e. in recent decades) regarded as wildlife' (one word).

14.1 INTRODUCTION

There was a time when the meaning of the term 'sustainable utilization of wild life' was clear and unambiguous. Simply put, 'a species of animal must not be destroyed at a greater rate than it can increase' (Hewitt, 1921). Consistently, the World Conservation Strategy stated that when we use renewable natural resources, including wild life, they 'should not be so heavily exploited that they decline to levels or conditions from which they cannot easily recover' (IUCN/UNEP/WWF, 1980).

Today, however, sustainable utilization has come to mean something quite different. Now often subsumed under the mantra of sustainable development (IUCN/UNEP/WWF, 1980, 1991; WCED, 1987), it embodies the view that if wild life is to be conserved – especially in the developing world – it 'must pay its own way' (Child and Child, 1990; Eltringham, 1994). Central to this 'use it or lose it' philosophy (Baskin, 1994) is the proposition that markets for dead wild life, including parts and derivatives, must be promoted on a global scale (UNCED, 1992; Rasker *et al.*, 1992; Robinson, 1993). In addition, it is often suggested that wild life management should be de-centralized, giving control to local communities or to private individuals, e.g. landowners and game ranchers (Mossman, 1975; Child and Child, 1990; Teer, 1993). Only when these communities or individuals begin to reap economic benefits from 'their' wild life – the argument goes – will they have reason and motivation to ensure its continued survival.

Such views have been widely expressed, for example, by the International Union for the Conservation of Nature and Natural Resources (IUCN) and the World Wide Fund for Nature (WWF) (IUCN, 1992; Thomsen, 1992) and were, to a large extent, reflected in the discussions at the Earth Summit in Rio (Johnson, 1993). They continue to be enthusiastically embraced by many biologists, wild life managers and journalists around the world (e.g. Asibey and Child, 1990; Block, 1990; Bonner, 1993). Those who disagree with such views are often branded as emotional, liberal protectionists who have not only lost touch with the natural world but who also do not understand the realities of living in the developing world, especially in Africa (see for example Adams and McShane, 1992; Bonner, 1993).

One does not have to be an emotional protectionist, however, to recognize that sustainable utilization, in its current formulation, is a seriously flawed argument. One only needs to know something about human nature and the history of wild life management.

The present chapter examines the central arguments put forward in favour of sustainable utilization. It concludes that any short-term gains that might arise will go largely to the North and to a small, elite minority in the South; they will not generally accrue to the masses who are supposed to be the main beneficiaries. In the longer term, sustainable utilization, as it is now being promoted, is simply a recipe for sustaining over-exploitation and reducing options for future generations.

14.2   FIRST, SOME CLARIFICATION

In any discussion of sustainable utilization, it is first essential to distinguish between biologically sustainable activities and economically sustainable activities. Biologically sustainable activities ensure that wild life populations are maintained at sufficiently high levels that their future is not placed in jeopardy. Economically sustainable can mean something quite different. It is not beyond the realm of possibility, for example, that an extinct population could provide sustainable economic benefits in perpetuity, assuming that the capital raised by killing the animals in the first place was invested wisely. Indeed, for those natural resources where the growth rate is less than the prevailing interest rate, e.g. large whales, elephants and old growth forests, it will be more profitable for the exploiter to over-harvest the resource as quickly as possible – even to biological extinction – and to invest the proceeds, than to harvest the resource in a biologically sustainable manner (Clark, 1973a,b, 1989; Caughley, 1993).

For the purposes of the present chapter, the discussion is limited primarily to the question of whether placing a value on dead wild life in the marketplace is a biologically sustainable activity.

In the modern literature on sustainable utilization, there is a tendency to equate sustainable use with non-consumptive use (e.g. Hoyt, 1994), the implication being that any consumptive use is not sustainable. This is incorrect. Hunting of numerous North American species, for example, has proved to be both biologically and economically sustainable throughout most of this century (e.g. Geist, 1988, 1993, 1994). Conversely, there are other situations where 'non-consumptive' use, e.g. some cases of ecotourism, may be neither biologically (Chapter 4) nor economically sustainable (Place, 1995).

The key distinction, which must be made when discussing sustainable utilization of mammals, is between the use of living animals – including hunting for personal consumption – and the sale of dead animals, their parts and derivatives in the marketplace (Geist, 1988). With adequate regulation, the former can be a sustainable activity; the latter almost always leads to unsustainable exploitation (e.g. Geist, 1988, 1994; Norse, 1993).

14.3   THE LESSONS OF HISTORY

Given the widespread and growing acceptance that promoting the marketing of dead wild life – thereby giving it value – is in the best interest of conservation, one might presume that it was based on past examples where such commercialization had made a positive contribution to the conservation of wild life populations. Yet, as recently as 1970, Ehrenfeld (p. 129), in describing the 'hypothetically most endangered animal', noted that among the mammals that are particularly susceptible to over-exploitation, are those with 'valuable fur, hide, oil, etc. ... [which are] hunted for the market or hunted for sport' in the absence of 'effective game management.'

If Ehrenfeld were correct, it would suggest that proponents of sustainable utilization today are promoting something that only 25 years ago was considered to increase the likelihood of extinction. This raises at least three questions. Was Ehrenfeld wrong in his assessment or, alternatively, are the proponents of sustainable utilization ignoring the lessons of history? Or are they are simply pursuing the 'new world deception' that is associated more generally with the concept of sustainable development (Willers, 1994, p. 1146)? Some insight into these questions can be gained from a brief review of the history of wild life management in North America.

14.3.1 The North American example

Over the past 70 years, North Americans have not only used their wild life in a variety of ways but they have done so in a manner that, argu-ably, has been both biologically and economically sustainable (Geist, 1988, 1993, 1994). This was not always the case, however, so the methods by which sustainable utilization was achieved should provide an instruc-tive model as we strive to achieve this objective on a global scale.

Prior to the arrival of Europeans, North American aboriginals used mammals – both marine and terrestrial – as sources of food, clothing and shelter for more than 1100 years (e.g. Wright, 1987). Although trade net-works were in existence well before the first contact with Europeans, commercial activities among aboriginals are believed to have been ancil-lary to subsistence (Wright, 1987).

When Europeans arrived in the New World in the sixteenth century, they found an abundance of wild life, both in the sea and on the land (Mowat, 1984). These resources formed the basis of North America's first two industries: fisheries, which included marine mammals, and the fur trade (Bliss, 1987). Both played key roles in the settlement of the conti-nent and in its economic development and, even to this day, the fishery and the fur trade in Canada continue to play a significant role in the national psyche.

The commercial fur trade, for example, became well established dur-ing the seventeenth and eighteenth centuries, providing exports to Europe, where fur-bearing mammals had been over-exploited and in decline since the 1400s. This time was characterized by a fight for monopolistic control of the industry, primarily between the French and the English (Innes, 1956). By then, the motivation behind wild life exploitation in North America had changed. Wild life was viewed largely in terms of its economic value as a commodity and the objective was to put as many wild life products as possible into the international marketplace, to satisfy a seemingly insatiable demand. In the process, the cultural traditions of native people were changed for all time, from a subsistence-based economy to one dependent on global mercantilism (McGee, 1987).

Because of the high market demand and unmitigated competition between the two largest trading companies, the Hudson's Bay Company and the North West Company (Giroux, 1987), many furbearer populations became over-exploited by the early 1800s. With continuing European settlement, market hunting and hunting for 'sport' intensified through the late 1800s, a time that is now remembered as the 'era of over-exploitation' (Shaw, 1985). This era was characterized by widespread, unregulated exploitation of wild life for the marketplace. The continuing high demand for wild life products overseas, the unregulated nature of the harvest and intense competition between traders, especially for fur-bearers, led to further over-exploitation and depletion of many populations (Chapter 3).

By the end of the nineteenth century, the list of fur-bearing mammals that had undergone marked population declines or local extirpation due to over-exploitation was long and growing (Geist, 1989). It included: beaver (*Castor canadensis*) (Obbard *et al.*, 1987), American bison (*Bison bison*) (Rorabacher, 1970), sea otter (*Enhydra lutris*) (Riedman and Estes, 1988), wolverine (*Gulo gulo*) (Hash, 1987), river otter (*Lutra canadensis*) (Melquist and Dronkert, 1987), American marten (*Martes americana*) (Strickland and Douglas, 1987), fisher (*Martes pennanti*) (Douglas and Strickland, 1987) and lynx (*Lynx canadensis*) (Monk, 1980).

Sea mammals fared no better (Mowat, 1984; Busch, 1985). The walrus (*Odobenus rosmarus*), for example, was extirpated throughout the southern extent of its range from Cape Cod to the northern Gulf of St Lawrence by 1800 (Lavigne and Kovacs, 1988); polar bears (*Ursus maritimus*) disappeared from the north shore of the St Lawrence (Mowat, 1984); and grey seals (*Halichoerus grypus*) were reduced to such low levels that they were only re-discovered in the late 1940s (Mowat, 1984). Elsewhere in the North-West Atlantic, humpback whales (*Megaptera novaeangliae*) were heavily exploited and depleted, particularly between 1866 and 1912 (Mitchell and Reeves, 1983). In the western Arctic, commercial whaling for bowhead whales (*Balaena mysticetus*), which began in 1848, rapidly depleted the population and the industry collapsed by 1914 (Bockstoce and Botkin, 1983).

Other taxa were similarly affected (reviewed in Mowat, 1984). Most notable, were the extinctions of the great auk (*Pinguinus impennis*), the Eskimo curlew (*Numenius borealis*) and the Labrador duck (*Camptorhynchus labradorius*).

There is no doubt that the precipitous decline in North American wild life in the late nineteenth century was caused by over-exploitation (Hewitt, 1921; Mowat, 1984), and the reasons for it were apparently understood. Twenty years later, C. Gordon Hewitt, the Dominion Entomologist and Consulting Zoologist, would write, 'It is universally recognized now by sportsmen and conservationists that the free marketing of wild game is one of the greatest factors tending rapidly to exterminate our native game resources' (Hewitt, 1921, p. 331).

Hewitt's use of the word 'universally' now seems ironic or in error, given that some conservationists still promote the 'free marketing of wild game' as a means of stopping the depletion of game resources. None the less, it is noteworthy to reflect on where the profits gleaned from the exploitation of North American wild life, largely for foreign markets, went. The lion's share, of course, went overseas with the pelts, blubber, feathers and other products derived from dead wild life; they did not go in large part to the local people, in particular to the aboriginal peoples or even to the European trappers, the fishers, sealers and whalers, who did the killing. Fortunes were made by merchants, such as those on Water Street in St John's, Newfoundland (Lavigne and Kovacs, 1988), and by the owners of large companies, such as the Hudson's Bay Company in England (Bliss, 1987). The legacy of that reality can still be seen today, for example, in the isolated outport communities of Newfoundland and in impoverished native communities across Canada.

At the turn of the century, it was evident that something had to be done to reverse the precipitous declines in North American wild life. Thus dawned the 'era of protection' (Shaw, 1985) and with it the birth of wild life conservation and management in North America, which over the past 70 years has created, arguably, the most successful and economically productive system of sustainable wild life use anywhere (Geist, 1988, 1993, 1994).

Without going into great detail, suffice to say that North America's success in wild life conservation and management has been based on four policies (summarized in Geist, 1988, 1989):

1. the public ownership of wild life;
2. the elimination of markets in the meat, parts and products of vulnerable wild life: large mammals, waterfowl, pigeons, shore-birds and song birds (but not of less vulnerable furbearers);
3. the allocation of the material benefits of wild life by law (i.e. centralized wild life management) – and not by the marketplace, birthright,

land ownership or social position – which effectively makes every citizen a 'shareholder', generating a sense of proprietorship in those who choose to use 'their' wild life;

4. a prohibition on the frivolous killing of wild life.

While there are some local and regional exceptions to these policies, they generally ensure that the killing of wild life is an economic liability (Geist, 1989). It costs money and time to kill animals and in the process this generates a demand for goods and services and makes a major contribution – some US$70 billion annually – to the North American economy. For perspective, this figure is several times the value of the global trade in wild life products, both legal and illegal, estimates of which range from US$8 billion (Geist, 1994) to about US$20 billion (Prescott-Allen and Prescott-Allen, 1989, cited by Swanson, 1992; also see Chapter 15).

14.3.2 Is the twentieth century North American fur trade an anomaly?

One notable exception to the ban on market hunting of North American mammals is the fur trade. Yet many North American furbearer populations have also recovered in numbers and have reoccupied areas from which they were previously extirpated, even in the face of continued commercial exploitation. Large numbers of pelts are still put into trade annually and no species has been driven to extinction. By most criteria, the twentieth century fur trade must be judged to be a biologically sustainable activity. An obvious question, therefore, is whether the sustainable marketing of North American fur products nullifies the generalization that placing value on wild life products increases the likelihood that species will be exploited in an unsustainable manner. The available evidence would suggest that it does not.

As was the case for other North American wild life, a number of conservation measures were introduced for furbearers early in the twentieth century, in response to the over-exploitation of the previous century. For example, in most regions of Canada, registered trap-lines were assigned exclusively to individual trappers (Anderson, 1987), a form of semi-privatization that helped to curtail competitive trapping, allowed trappers to participate in the management of the resource and reduced the potential for poaching (DeAlmeida and Cook, 1987). Other conservation measures initiated to prevent over-harvesting included the introduction of quotas, seasons, pelt sealing (a means of tracing pelts from trapper to finished product), restocking depleted populations, establishing game preserves (refugia) and, as with other wild life, the creation and enforcement of conservation laws.

Nonetheless, the nature of the economic factors controlling the fur trade continues to harbour inherent risks of over-exploitation. Most

North American furbearers are, however, buffered from over-exploitation by their life history traits and by market conditions that differ from those of many other wild life species currently in trade. The majority of North American furbearers currently traded, for example, possess characteristics typically associated with so-called r-selected species (Pianka, 1970). They mature quickly and produce a large number of offspring over a relatively short life span. Such populations or species, therefore, have a high intrinsic growth rate and are able to sustain intense trapping pressure (Geist, 1989).

In contrast, furbearing species that exist naturally at low densities and have low reproductive rates (K-selected species; Pianka, 1970) remain more sensitive to price-induced over-exploitation in the marketplace (Todd, 1981; Bailey *et al.*, 1986; Douglas and Strickland, 1987). Examples of increased susceptibility to over-exploitation in relation to reproductive strategies is a general phenomenon (Ehrenfeld, 1970), as exemplified by numerous species – including some furbearers – currently in the marketplace: North American lynx, fisher and sea otter, the large cats (e.g. Bengal tiger, *Panthera tigris tigris*), African elephants (*Loxodonta africana*), black rhinos (*Diceros bicornis*), and many species of parrots (Hoyt, 1994).

The fur industry is also similar to other trade in wild life in that market forces control hunting effort. Fur harvesting pressure is primarily determined by economic returns per unit effort (Usher, 1971; Todd and Boggess, 1987; Siemer *et al.*, 1994). But the North American fur industry differs from many other wild life industries in that economic returns have generally decreased in value relative to societal earnings since the turn of the century. In the early 1900s, for example, a prime silver fox (*Vulpes vulpes*) pelt was worth the equivalent value of a labourer's yearly income and, as recently as the 1930s, a mink (*Mustela vison*) pelt was valued at more than an average week's salary. Currently, however, a fox or mink pelt is worth the equivalent of a few hours of a labourer's wage (Novak, 1987). It is questionable whether compliance with conservation regulations and, therefore, sustainable use of these resources would have resulted if economic returns from the fur trade had remained high relative to other societal earnings.

There is one additional reason for the apparent sustainability of the North American fur trade this century. The trapping of furbearers requires initially an investment in technology (e.g. traps, snares) that is not readily available to the general public. Trapping also requires specialized skills and is labour intensive, often under inhospitable conditions. Together with the relatively low economic return, such constraints effectively reduce the incentive to poachers, which is not the case when more valuable wild life can be taken with readily available technology, e.g. guns.

So, the North American fur trade does not nullify the generalization that placing a market value on wild life promotes over-exploitation. It does, however, begin to illustrate the biological and economic conditions, and legal requirements, under which market hunting might actually prove to be sustainable.

14.4 DISCUSSION

The legacy of wild life in the marketplace is well documented and remains at the root of many conservation issues today. Generally, putting a price on dead wild life almost invariably leads to over-exploitation and increases the 'extinction potential' of target species (Ehrenfeld, 1970). Consider, for example, the history of commercial whaling (Gaskin, 1982) and commercial sealing (e.g. Busch, 1985), the effects on elephants of the trade in ivory (Pagel and Mace, 1991), the decline of rhinos exploited for their horns (Cumming *et al.*, 1990), and the precarious status of tigers, killed for their skins, bones and penises (Chapter 16).

The conclusion is not restricted to trade in mammals. Current issues of concern include the rapidly developing commercial fishery for sea cucumbers in the Galapagos Islands, the effects of the exotic bird trade on wild populations (Hoyt, 1994), the depletion of temperate old growth and tropical rainforests (Johnson, 1993) and, perhaps most significant, the present state of the world's commercial fisheries (e.g. Ludwig *et al.*, 1993; Anon, 1994). These examples simply reaffirm the conclusion of several recent commentators that exploitation, generally, and market hunting in particular, can have serious consequences for wild populations, including extinction (Table 14.1) and the attendant loss of biodiversity (Wilson, 1992). Norse (1993, p. 81), for example – editing a volume, ironically co-sponsored by a number of organizations including IUCN and WWF, which elsewhere support the marketing of wild life – stated: 'Species that people use as commodities are inherently at risk of population reduction or elimination.' Consistently, Talbot (1995, p. 7) noted: 'Virtually all species and stocks of wild living resources ... which are being harvested commercially have been or are being depleted.'

From the foregoing, it seems evident that those who today advocate generally the marketing of dead wild life are actually promoting a return to nineteenth century attitudes and practices, which usually lead to unsustainable exploitation. In this regard, the re-defining of sustainable utilization to include marketing of dead wild life is consistent with a number of other contemporary developments in the conservation and environmental fields. Safeguards that were put in place to protect exploited wild life, such as placing a lower bound on exploitation – e.g. the US Marine Mammal Protection Act (MMPA 1972) and Endangered Species Act (ESA 1973) – are now dismissed in some quarters with the suggestion that exploitation of endangered species actually should be

Table 14.1 Causes of animal extinctions since approximately 1600[1]

Causes of extinction	Mammals (%)	All species (%)
Hunting (includes for food, skin, sport, live trade, feathers, and destruction of animals perceived as pests)	49	23
Habitat destruction	2	36
Introduced predators and competitors	49	39
Introduced disease, indirect effects, natural causes	0	2

[1] Approximately 488 species, of which 59 are mammals, are known to have become extinct since 1600. The data in the table concern only those animals for which a cause of extinction has been assigned (211 species, of which 18 are mammals). The cause of extinction for the majority of animals is not known (modified from Groombridge, 1992).

encouraged (e.g. S. Edwards, cited in Lavigne, 1991). Consistently, there are attempts afoot in the US to weaken the ESA, which is currently up for re-authorization (Noss and Murphy, 1995), and elements of Newt Gingrich's Contract with America, if passed, will effectively neutralize a large body of environmental and wild life legislation. Other examples include the recent attempt to water down the listing criteria for the Convention on International Trade in Endangered Species (CITES, 1994; also see Wold, undated) and the widespread view that environmental and wild life protection should not act as unnecessary obstacles to sustainable development or to free trade (UNCED, 1992; Lipsey *et al.*, 1994; Runge *et al.*, 1994).

The unanswered question is why individuals and organizations with an apparent interest in wild life conservation would promote the unsustainable marketing of wild life products. Some, perhaps, are simply unaware of the lessons of history. Others contend that historical considerations are irrelevant because (they believe) we have learned how to manage wild populations and have the means to regulate exploitation and trade. Still others have faith that traditional economic approaches offer a solution – perhaps the only solution – to current conservation problems.

Another possibility is that sustainable utilization, as currently defined, is actually part of the 'new world deception' that has enveloped sustainable development (Willers, 1994), especially since the publication of the Brundtland Report (WCED, 1987). After all, capitalist approaches – and placing value on wild life in the marketplace is clearly such an approach – rarely result in the transfer of the majority of benefits to working peoples. There is no compelling reason to believe that the result will be any different, for example, in Africa today.

One wonders if some of the recent developments alluded to above actually reflect a successful campaign by big business, multi-national organizations and the global banking community, among others, to maintain the status quo (e.g. Adams, undated) by ensuring the continued net flow of capital from the South to the North, while simultaneously feigning concern for the developing world through the rhetoric of Rio (UNCED, 1992; Johnson, 1993). While the debt crisis of the South might actually be mitigated through debt forgiveness by the North – an idea that was explicitly rejected by former US President, George Bush (Culpeper, 1992; Gorrie, 1992) – it surely will not be solved by encouraging the South to deplete its natural resources, including its mammal populations, even further to satisfy market demands and commercial interests in the North.

Critics may dismiss our views as being anti-exploitation and protectionist rather than conservationist, but this is not the case. Our point is that there are a number of ways in which wild life can be used (e.g. Eltringham, 1994); some of these have proved to be sustainable whereas others have not. Each case must be evaluated on its own ecological and socio-economic merits (MacNab, 1991) but, generally, local use to satisfy a local demand for food and fibre, and for recreation, has proved to be a sustainable activity (Geist, 1988; Eltringham, 1994). There are also numerous examples where tourism associated with wild life has proved to be biologically and economically sustainable. In both cases, the revenue generated comes largely from service-based activities and not from the wild life itself. In either case, some controls are likely to be required, but whether these arise from centralized or decentralized wild life management is probably a moot point.

Exploitation aimed at placing dead wild life in the marketplace to satisfy external demands is something quite different. Such exploitation, whether it involves market hunting, paid hunting, the advertising of hunting as sport or competition, game cropping (MacNab, 1991) or game ranching (Lavigne and Geist, 1993), has largely failed the test of sustainability (Geist, 1988; MacNab, 1991). In these instances, the situation will only be made worse if propositions to abandon regional, national or international wild life laws, in favour of decentralized wild life management (e.g. Child and Child, 1990), are implemented, especially when the products are intended for the international marketplace.

A major reason why the establishment of legal markets for valuable wild life products usually leads to uncontrollable depletion of populations is because they provide incentives for poaching. When the prices of wild life products are sufficiently high, they also attract criminal elements into poaching, making wild life protection not only increasingly difficult but also dangerous (Geist, 1985). And, as the markets for rhino horn, elephant ivory and tiger skins and penises attest, once markets are established, they cannot be readily legislated away (Geist, 1985).

Our review, therefore, reinforces the conclusion that placing value on dead wild life in the marketplace is rarely a sustainable activity. Indeed, as the North American experience suggests, it may only work in those few instances where the exploited species has a high reproductive rate and a relatively low economic value, and where there is also adequate legislation, enforcement and compliance (Caughley, 1993).

Our analysis is limited, however, because it has largely been concerned only with utilitarian values, neglecting for the most part the other objectives that society might have for wild life species, including those that are ecologically and ethically oriented (Lavigne and Geist, 1993). Space precludes such a discussion here, but one is reminded of Aldo Leopold's (1966) comment about the 'tedium of the merely economic attitude' toward wild life. If we only view wild life in economic terms, for example, how can we ever hope to conserve the numerous species that will never have economic value in the marketplace (Eltringham, 1994)?

Regardless, in 1921, C. Gordon Hewitt (p. 331) wrote that 'the sale of protected game must be prohibited if the disappearance of such game is to be prevented'. When he penned those words, the human population numbered about 1.8 billion people. In the intervening years, it has tripled to 5.5 billion and it continues to increase at some 1.7% per year (Johnson, 1994). The potential demand for wild life products is now enormous. For certain products, such as bear gall bladders, it is frequently said to exceed current world supply. If the advocates of sustainable utilization are successful in promoting the marketing of dead wild life under such circumstances, what hope is there, in the long term, for most of the species involved?

ACKNOWLEDGEMENTS

This manuscript is based on a paper given by the senior author at the UFAW Exploitation of Mammals Symposium, held in London on 25–26 November 1994. We particularly thank the conference organizers and the editors of this volume, Vicky Taylor and Nigel Dunstone, not only for the original invitation and the opportunity to contribute to the present volume, but also for their patience in dealing with tardy authors. Preparation of the manuscript and attendance at the meeting were funded by *BBC Wildlife* magazine (particular thanks to its editor, Rosamund Kidman Cox) and by the International Marine Mammal Association Inc., Guelph, Canada. We also thank John Twiss, US Marine Mammal Commission, for access to an unpublished report and Vernon Thomas and anonymous reviewers for comments on earlier drafts of the manuscript.

REFERENCES

Adams, A.L. (undated) *Sustainable Use of Wildlife and the International Development Donor Organizations*, Humane Society of the United States, Washington, DC.

Adams, J.S. and McShane, T.O. (1992) *The Myth of Wild Africa. Conservation Without Illusion*, W.W. Norton & Company, New York.

Anderson, S.B. (1987) Wild furbearer management in eastern Canada, in *Wild Furbearer Management in North America*, (eds M. Novak, J.A. Baker, M.E. Obbard and B. Malloch), Ontario Trappers Assoc., North Bay, pp. 1039–1048.

Anon (1990) *A wildlife policy for Canada*, Wildlife Ministers' Council of Canada, Canadian Wildlife Service, Environment Canada, Ottawa.

Anon (1994) The tragedy of the oceans. *The Economist*, March 19, 2–24.

Asibey, E.O.A. and Child, G. (1990) Wildlife management for rural development in sub-Saharan Africa. *Unasylva*, **41**, 3–10.

Bailey, T.N., Bangs, E.E., Portner, M.F., Malloy, J.C. and McAvinchey, R.J. (1986) An apparent overexploited lynx population on the Kenai Peninsula. *J. Wildl. Manage.*, **50**, 279–290.

Baskin, Y. (1994). There's a new wildlife policy in Kenya: Use it or lose it. *Science*, **265**, 733–734.

Bliss, M. (1987) *Northern Enterprise. Five Centuries of Canadian Business*, Mclelland and Stewart, Toronto.

Block, W. (ed.) (1990) *Economics and the Environment: A Reconciliation*, The Fraser Institute, Vancouver.

Bockstoce, J.R. and Botkin, D.B. (1983) The historical status and reduction of the western Arctic bowhead whale (*Balaena mysticetus*) population by the pelagic whaling industry, 1848–1914. *Rep. Int. Whal. Commn (Special Issue 5)*, pp. 107–141.

Bonner, R. (1993) *At the Hand of Man. Peril and Hope for Africa's Wildlife*, Alfred A. Knopf, New York.

Busch, B.C. (1985) *The War Against the Seals. A History of the North American Seal Fishery*, McGill-Queen's University Press, Montreal.

Caughley, G. (1993) Elephants and economics. *Conserv. Biol.*, **7**, 943–945.

Child, G. and Child, B. (1990) An historical perspective of sustainable wildlife utilisation. Paper prepared for Workshop No. 7, 18th IUCN General Assembly, Perth.

CITES (1994) *New Criteria for Amendment of Appendices I and II*. Doc. 9.41, CITES, Gland, Switzerland.

Clark, C.W. (1973a) The economics of overexploitation. *Science*, **181**, 630–634.

Clark, C.W. (1973b) Profit maximization and the extinction of animal species. *Journal of Political Economy*, **81**, 950–961.

Clark, C.W. (1989) Clear-cut economies. Should we harvest everything now? *The Sciences*, **29**, 16–19.

Culpeper, R. (1992) Who will save the Earth? *The Globe and Mail*, May 29, Toronto.

Cumming, D.H.M., Du Toit, R.F. and Stuart, S.N. (ed.) (1990) *African Elephants and Rhinos. Status Survey and Conservation Action Plan*, IUCN/SSC African Elephant and Rhino Specialist Group, Gland, Switzerland.

DeAlmeida, M.H. and Cook, L. (1987) Trapper education in North America, in *Wild Furbearer Management in North America*, (eds M. Novak, J.A. Baker, M.E. Obbard and B. Malloch), Ontario Trappers Assoc., North Bay, pp. 77–84.

Douglas, C.W. and Strickland, M.A. (1987) Fisher, in *Wild Furbearer Management in North America*, (eds M. Novak, J.A. Baker, M.E. Obbard and B. Malloch), Ontario Trappers Assoc., North Bay, pp. 510–529.

Ehrenfeld, D.W. (1970) *Biological Conservation*, Holt, Rinehart and Winston of Canada Ltd, Toronto.

Eltringham, S.K. (1994) Can wildlife pay its way? *Oryx*, **28**, 163–168.

Gaskin, D.E. (1982) *The Ecology of Whales and Dolphins*, Heinemann, London.

Geist, V. (1985) Game ranching, threat to wildlife conservation in North America. *Wildlife Society Bulletin*, **13**, 594–598.

Geist, V. (1988) How markets in wildlife meat and parts, and the sale of hunting privileges, jeopardize wildlife conservation. *Conserv. Biol.*, **2**, 1–12.

Geist, V. (1989) Legal trafficking and paid hunting threaten conservation. *Transactions of the North American Wildlife and Natural Resources Conference*, **54**, 172–178.

Geist, V. (1993) Great achievements, great expectations: successes of North American wildlife management, in *Commercialization and Wildlife Management. Dancing with the Devil*, (ed. A.W.L. Hawley), Krieger Publishing Company, Malabar, Florida, pp. 47–72.

Geist, V. (1994) Wildlife conservation as wealth. *Nature*, **368**, 491–492.

Giroux, A. (1987) The role of the trapper today, in *Wild Furbearer Management in North America*, (eds M. Novak, J.A. Baker, M.E. Obbard and B. Malloch), Ontario Trappers Assoc., North Bay, pp. 55–58.

Gorrie, P. (1992) Can global summit help heal the earth? *The Toronto Star*, May 30, Toronto.

Groombridge, B. (ed.) (1992) *Global Biodiversity. Status of the Earth's Living Resources*, Chapman & Hall, London.

Hash, H. (1987) Wolverine, in *Wild Furbearer Management in North America*, (eds M. Novak, J.A. Baker, M.E. Obbard and B. Malloch), Ontario Trappers Assoc., North Bay, pp. 575–585.

Hewitt, C.G. (1921) *The Conservation of the Wild Life of Canada*, Charles Scribner's Sons, New York.

Hoyt, J.A. (1994) *Animals in Peril: How 'Sustainable Use' is Wiping Out the World's Wildlife*, Avery Publishing Group, New York.

Innes, H.A. (1956) *The Fur Trade in Canada*, University of Toronto Press, Toronto.

IUCN (1992) *Criteria and Requirements for Sustainable Use of Wild Species. Proposed Policy, Second Draft, 15 July 1992*, IUCN/SSC Specialist Group on Sustainable Use of Wild Species and IUCN Sustainable Use of Wildlife Programme, Gland, Switzerland.

IUCN/UNEP/WWF (1980) *World Conservation Strategy. Living Resource Conservation for Sustainable Development*, Gland, Switzerland.

IUCN/UNEP/WWF (1991) *Caring for the Earth: A Strategy for Sustainable Living*, Gland, Switzerland.

Johnson, S.P. (1993) *The Earth Summit: The United Nations Conference on Environment and Development (UNCED)*, International Environmental Law and Policy Series, Graham & Trotman/Martinus Nijhoff, London.

Johnson, S.P. (1994) *World Population – Turning the Tide. Three Decades of Progress*, Graham & Trotman/Martinus Nijhoff, London.

Lavigne, D.M. (1991) Your money or your genotype. Special Report. *BBC Wildlife*, **9**, 204–205.

Lavigne, D.M. and Geist, V. (1993) Game ranching: A case study in sustainable utilization, in *Proceedings of the International Union of Game Biologists XXI Congress: Forests and Wildlife ... Towards the 21st Century*, (ed. I. Thompson), Petawawa National Forestry Institute, Petawawa, pp. 194–200.

Lavigne, D.M. and Kovacs, K.M. (1988) *Harps and Hoods. Ice-breeding Seals of the Northwest Atlantic*, University of Waterloo Press, Waterloo.

Leopold, A. (1966) *A Sand County Almanac*, Ballantine Books, New York.

Lipsey, R., Schwanen, D. and Wonnacott, R. (1994) *NAFTA: What's In, What's Out, What's Next?* C.D. Howe Institute, Toronto.

Ludwig, D., Hilborn, R. and Walters, C. (1993) Uncertainty, resource exploitation and conservation: Lessons from history. *Science*, **260**, 17 and 36.

MacNab, J. (1991) Does game cropping serve conservation? A reexamination of the African data. *Can J. Zool.*, **69**, 2283–2290.

McGee, H.F. (1987) The use of furbearers by native North Americans after 1500, in *Wild Furbearer Management in North America*, (eds M. Novak, J.A. Baker, M.E. Obbard and B. Malloch), Ontario Trappers Assoc., North Bay, pp. 13–20.

Melquist, W.E. and Dronkert, A.E. (1987) River otter, in *Wild Furbearer Management in North America*, (eds M. Novak, J.A. Baker, M.E. Obbard and B. Malloch), Ontario Trappers Assoc., North Bay, pp. 626–641.

Mitchell, E. and Reeves, R.R. (1983) Catch history, abundance, and present status of Northwest Atlantic humpback whales. *Rep. int. Whal. Commn, (Special Issue 5)*, pp. 153–209.

Monk, C.E. (1980) History and present status of fur management in Ontario, in *Worldwide Furbearer Conference Proceedings*, (eds J.A. Chapman and D. Pursley), Frostburg, pp. 1501–1523.

Mossman, A.S. (1975) International game ranching programs. *J. Anim. Sci.*, **40**, 993–999.

Mowat, F. (1984) *Sea of Slaughter*, McClelland and Stewart, Toronto.

Norse, E.A. (ed.) (1993) *Global Marine Biological Diversity*, Island Press, Washington, DC.

Noss, R.F. and Murphy, D.D. (1995) Endangered species left homeless in sweet home. *Conserv. Biol.*, **9**, 229–231.

Novak, M. (1987) Wild furbearer management in Ontario, in *Wild Furbearer Management in North America*, (eds M. Novak, J.A. Baker, M.E. Obbard and B. Malloch), Ontario Trappers Assoc., North Bay, pp. 1049–1061.

Obbard, M.E., Jones, J.G., Newman, R. *et al.* (1987) Furbearer harvests in North America, in *Wild Furbearer Management in North America*, (eds M. Novak, J.A. Baker, M.E. Obbard and B. Malloch), Ontario Trappers Assoc., North Bay, pp. 1007–1034.

Pagel, M. and Mace, R. (1991) Keeping the ivory trade banned. *Nature*, **351**, 265–266.

Pianka, E.R. (1970) On r and K selection. *American Naturalist*, **104**, 592–597.

Place, S. (1995) Ecotourism for sustainable development: Oxymoron or plausible strategy? *GeoJournal*, **35**, 161–173.

Rasker, R., Martin, M.V. and Johnson, R.L. (1992) Economics: Theory versus practice in wildlife management. *Conserv. Biol.*, **6**, 338–349.

Riedman, M.L. and Estes, J.A. (1988) A review of the history, distribution and foraging ecology of sea otters, in *The Community Ecology of Sea Otters*, (eds G.R. VanBlaricom and J.A. Estes), Springer-Verlag, New York, pp. 4–21.

Robinson, J.G. (1993) The limits to caring: sustainable living and the loss of biodiversity. *Conserv. Biol.*, **7**, 20–28.

Rorabacher, J.A. (1970) *The North American Buffalo in Transition: A Historical and Economic Survey of the Bison in America*, North Star Press, St Cloud.

Runge, C.F., Ortalo-Magné, F. and Vande Kamp, P. (1994) *Freer Trade, Protected Environment: Balancing Trade Liberalization and Environmental Interests*, Council on Foreign Relations Press, New York.

Shaw, J.H. (1985) *Introduction to Wildlife Management*, McGraw-Hill Book Company, New York.

Siemer, W.F., Batcheller, G.R., Glass, R.J. and Brown, T.L. (1994) Characteristics of trappers and trapping participation in New York. *Wildl. Soc. Bull.*, **22**, 100–111.

Strickland, M.A. and Douglas, C.W. (1987) Marten, in *Wild Furbearer Management in North America*, (eds M. Novak, J.A. Baker, M.E. Obbard and B. Malloch), Ontario Trappers Assoc., North Bay, pp. 530–547.

Swanson, T.M. (1992) Wildlife and wetlands, diversity and development, in *Economics for the Wilds*, (eds T.M. Swanson and E.B. Barbier), Earthscan Publications, London, pp. 1–14.

Talbot, L.M. (1995) *Living Resource Conservation: An International Overview*, Draft Report to The Marine Mammal Commission, Washington, DC.

Teer, J.G. (1993). Commercial utilization of wildlife: Has its time come? in *Commercialization and Wildlife Management – Dancing With the Devil*, (ed. A.W.L. Hawley), Krieger Publishing Company, Malabar, pp. 73–83.

Thomsen, J.B. (1992) Foreword, in *Economics for the Wilds*, (eds T.M. Swanson and E.B. Barbier), Earthscan Publications, London.

Todd, A.W. (1981) Ecological arguments for fur-trapping in boreal wilderness regions. *Wildl. Soc. Bull.*, **9**, 116–124.

Todd, A.W. and Boggess, E.K. (1987) Characteristics, activities, lifestyles and attitudes of trappers in North America, in *Wild Furbearer Management in North America*, (eds M. Novak, J.A. Baker, M.E. Obbard and B. Malloch), Ontario Trappers Assoc., North Bay, pp. 59–76.

UNCED (1992) *Agenda 21, Programme of Action for Sustainable Development*, United Nations Publications, New York.

Usher, P.J. (1971) *The Bankslanders: Economy and Ecology of a Frontier Trapping Community. Vol. 2: Community and Ecology*, North Sci. Res. Group, Dep. Indian Aff. and North. Dev., Ottawa.

Willers, B. (1994) Sustainable development: A new world deception. *Conserv. Biol.*, **8**, 1146–1148.

Wilson, E.O. (1992) *The Diversity of Life*, W.W. Norton & Company, New York.

Wold, C. (undated) *The New Listing Criteria Proposed by the Standing Committee Violate CITES and Are Inconsistent with the Theory of Sustainable Use*, The Humane Society of the United States, Washington, DC.

WCED [World Commission on Environment and Development] (1987) *Our Common Future*, Oxford University Press, Oxford.

Wright, J.V. (1987) Archaeological evidence for the use of furbearers in North America, in *Wild Furbearer Management in North America*, (eds M. Novak, J.A. Baker, M.E. Obbard and B. Malloch), Ontario Trappers Assoc., North Bay, pp. 3–12.

15

Wildlife trade – a conserver or exploiter?

David Bowles

SYNOPSIS

A number of factors affect the potential sustainability of the international trade in mammal species. This chapter proposes and examines six major factors: the numbers of the target species being traded; the breeding potential of the species; trade demands related to the sex and size of the individual animal; the economic aspects of the trade; the source and nature of demand; and the socio-cultural aspects of the end market. Case studies of species listed in the Appendices of CITES (Convention on the International Trade in Endangered Species of Wild Fauna and Flora) are used to examine the effect of these factors on the trade's sustainability, and broad effects of the trade on these species are extrapolated. The applicability of published guidelines and standards proposed to gauge sustainability of trade in mammal species are assessed, by applying them to the case studies and other species which have been impacted by trade. The potential benefits and detrimental effects of trade in mammal species are discussed and compared with the non-consumptive use of mammals for conservation, such as ecotourism. Tentative conclusions based on the case studies examined are drawn on the historic and potential role of wildlife trade in the conservation and exploitation of endangered and threatened mammal species. It is concluded that only small-scale international trade in valuable threatened wild mammals has any potential to be sustainable in the long term. Concomitantly this would have the most potential for maintaining good enforcement procedures and the implementation of high animal welfare standards.

15.1 INTRODUCTION

The global trade in wildlife is one of the world's most valuable trades. Although a precise evaluation of its annual worth is difficult, and estimates range from US$8 billion (Geist, 1994) to approaching US$20 billion (Swanson and Barbier, 1992), it is reportedly the third most valuable global trade following arms and drugs. Illegal trade in wildlife alone is estimated to be worth over US$3 billion per annum. Calls for an international treaty to regulate this trade were made as early as 1963 by IUCN, but it was another 10 years before such a treaty, the Convention on the International Trade in Endangered Species of Wild Fauna and Flora (CITES), was established (Wijnstekers, 1992). CITES came into force in 1975 and at present has 130 member countries.

CITES regulates international commercial trade in species by assigning three levels of regulation or protection to organisms that are or could possibly be traded (Table 15.1). The biennial Conference of Parties (CoP) to CITES makes any decisions required to alter the level of protection given to a species. In this chapter, CITES-listed species are used as case studies to evaluate the historical and potential role of international trade in wild species as a conservation tool. Over 500 mammal species are currently listed on the Appendices, so a wide range of species are covered. Additionally, all species listed on Appendix I or II are endangered or threatened: any unsustainable trade will play a crucial role in their future conservation. Information presented primarily covers the past two decades as more accurate data are available for this period. However, for certain species, data on pre-CITES trade also provides useful information on the effects of trade.

The potential benefits and detrimental effects of trade on CITES-listed species have been discussed at a number of CoPs. In 1992 a Resolution (Conf. 8.3) was adopted which recognized that commercial trade may be beneficial to the conservation of certain species if it is carried out at levels not detrimental to the long-term survival of that species; for many other species any trade would be detrimental (Anon, 1992a). This Resolution recognizes that the impact of international trade will vary according to the species and certain factors. This chapter proposes and examines six factors: the numbers of the target species being traded; the breeding potential of the species; trade demands related to the sex and size of the individual animal; the economic aspects of the trade; the source and nature of demand; and the socio-cultural aspects of the end market. The eventual effect on the target species, and thus the potential for the trade's sustainability, will be a combined result of all these factors and any changes in a category may have positive or negative feedback on other categories. Table 15.2 summarizes five principal uses of mammals in international trade, giving specific species examples for each use. Case studies are used to examine the historical role of international trade in the five uses and to assess its potential as a conservation tool, but the chapter concentrates on two uses which involve the larger numbers of traded mammals: luxury and medicinal purposes.

It should be noted that in 1995 CITES issued a revised taxonomic guide for some of its listed species, mainly concerning genera of the cat species. This chapter (like others in this book) retains the species name as

Table 15.1 Summary of the three levels of protection given to species in international trade under CITES

Levels of protection	Definition under CITES	Status of international trade	Species examples
Appendix I	Species threatened with extinction	No commercial trade; trade in sport trophies, certain captive-bred species and scientific purposes allowed	*Panthera tigris* *Diceros bicornis* *Balaenoptera musculus* *Melursus ursinus*
Appendix II	Species may become threatened with extinction unless trade is regulated	Regulated trade based on a non-detriment finding	*Felis bengalensis** *Tursiops truncatus* *Ursus americanus*
Appendix III	Species protected under national law	Trade allowed under permit or certificate	*Odobenus rosmarus* *Hystrix cristata*

* The Chinese population of *Felis bengalensis* (now *Prionailurus bengalensis*) is on Appendix II; populations of the subspecies *bengalensis* in certain countries are on Appendix I.

it was at the time of its original CITES listing. However, Appendix 15.A also gives revised names where appropriate.

15.2   FACTORS AFFECTING THE SUSTAINABILITY OF TRADE

15.2.1 Numbers of the target species being traded

The number of specimens in trade, particularly as a proportion of the wild population, is a key factor in evaluating potential sustainability. There is a wide range in numbers of specimens in trade: the hundreds of thousands of South American cat skins that were traded annually in the past three decades and the limited trade in sport-hunted elephant trophies are used as examples of the extremes of this range.

(a) High numbers: South American cat skins

In the mid 1960s, prior to the establishment of CITES, the majority of cat skins in trade were of the largest South American cat, the jaguar (*Panthera onca*): an estimated 15 000 jaguars were killed annually in the Amazonian basin (Smith, 1976), but numbers were not monitored accurately (Broad, 1987). International trade was prohibited in 1975, when CITES came into force. Although a number of exporting countries were not party to CITES at this time certain countries, such as Brazil, had

Table 15.2 A summary of the five principal markets for the international trade in CITES-listed mammal species and the factors affecting their sustainable use

Factor	Market				
	Luxury/fashion	Medicinal (e.g. TCM)	Laboratory animal trade	Zoos/captive breeding	Pet trade
Numbers in trade[1]	Medium–high	Medium–high	Low–medium	Low	Low–medium
Economics: value of specimen[2]	Medium–high	Medium–high	Low	From no assigned value to high	Low–high
Source of species	Wild-caught; captive	Wild-caught; captive	Wild-caught; captive	Wild-caught; captive	Wild-caught; captive
Marketed type	Dead	Dead/Deriv.	Live/Deriv.	Live	Live
Main location of principal end market	Europe USA/Japan	East Asia Chinese communities	Europe USA	Global	Europe USA/Japan
Examples of past and present wild-caught species	Ivory Cat skins Whale meat	Tiger bones Bear gall Rhino horn	Macaque spp.	Golden-headed lion tamarins	Orang utan Marmoset spp.

TCM = Traditional Chinese Medicine
Dead = Parts of and entire species
Deriv. = Derivatives (e.g. medicinal products)

[1] The categories are used only to give a broad indication of the numbers in trade; individual variation may occur. (Low = fewer than 50 per shipment; Medium = above 50 but fewer than 500; High = more than 500 per shipment)
[2] The categories are used to give a broad indication of the product's value when retailed in the end market.

already instituted national export bans. The over-exploitation and ensuing commercial extinction of the jaguar resulted in a shift to other species (Broad, 1987; Anon, 1992b). Those smaller species, the ocelot (*Felis pardalis*), margay (*Felis wiedi*), tiger cat (*Felis tigrina*) and Geoffroy's cat (*Felis geoffroyi*) were placed on Appendix II in 1977, allowing a 'controlled' trade (which as such is dependent on scientific findings to prove that any trade will not be detrimental to the species' wild populations).

The target species changed again in the late 1970s and 1980s from ocelots to tiger cats and eventually to Geoffroy's cat (Brautigam, 1989; Broad, 1991). Imports of ocelot skin into the USA increased from 100 000 in 1960 to 133 000 by the late 1960s (Broad, 1987), a trend which continued into the mid 1970s (Broad, 1991). This level of trade coincided with a decrease of 60–70% in the species' range (Foreman, 1989, cited in Brautigam, 1989). As exports of ocelot skins from range states dropped by the early 1980s (Anon, 1989a), trade in margay and tiger cat (oncilla) rose (Anon, 1989b; Anon, 1989c; Brautigam, 1989). Exports of oncilla skins peaked at over 86 000 in 1983. By 1990, when ocelot, oncilla and margay were placed on Appendix I, the best available information indicated that populations of oncilla and margay had also been reduced to 60–70% of their historical range (Brautigam, 1989). Trade then concentrated on the only South American spotted cat not on Appendix I: Geoffroy's cat. Brautigam (1992) and Anon (1992b) reported that hunting pressure increased on this species as a result of the listing of the oncilla and margay in 1989. Finally it, too, was listed on Appendix I in 1992.

A number of conclusions can be drawn from this example. Firstly, regulation of the legal trade proved to be ultimately impossible to enforce. To the traders, profit maximization proved to be a greater incentive than guaranteeing sustainability. The drive to maximize profits meant that trade patterns exploited the weakest routes of enforcement and altered as different range states passed national legislation prohibiting export. For instance, Paraguay was the major exporter of ocelot skins in the 1980s, but was only a range state for the one ocelot subspecies on Appendix I – the majority of skins would have been laundered from other range states (Broad *et al.*, 1988a). Similarly, although Bolivia is not a range state for the oncilla, it was one of the largest exporters of the species in 1984 and 1985 (Anon, 1989b). Figure 15.1 highlights the difficulties involved in identification of spotted cat skins, and exposes the problems of enforcing an illegal trade in one species whilst allowing a 'controlled' trade in a similar species. The legal trade in Geoffroy's cat between 1989 and 1992 probably acted as a cover for continuing illegal trade in margay and ocelot skins (Anon, 1992b).

Secondly, the trade concentrated on the largest target species, which gives the highest financial returns per animal, before shifting in decreasing size to the smaller species (McMahan, 1986). The same size coat that

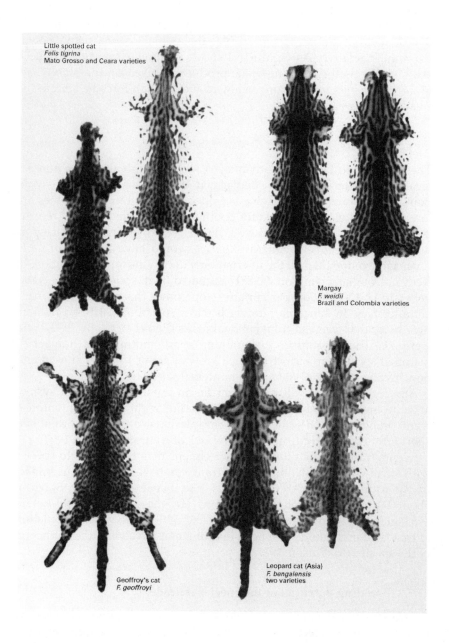

Little spotted cat
Felis tigrina
Mato Grosso and Ceara varieties

Margay
F. weidii
Brazil and Colombia varieties

Geoffroy's cat
F. geoffroyi

Leopard cat (Asia)
F. bengalensis
two varieties

Figure 15.1 The skins of four species of spotted cats showing the problems of identification.

requires eight jaguar skins needs the skins of 25 Geoffroy's cats (McMahan, 1986). The smaller the target species, the proportionally greater the effect such a shift will have on unsustainable trade, provided that demand and other factors such as the animal's reproductive potential remain relative. As Figure 15.2 shows, a similar trend occurred 30 years earlier as the trade in whale products focused on the blue whale (*Balaenoptera musculus*) before commercial extinction forced a shift to the smaller baleen whales.

(b) Low numbers: sport-hunted elephant trophies

There appear to be several advantages with a high value/low number legal international trade, particularly if the species has a large wild population. Numbers of mammals traded as sport-hunted trophies are low – the 1994 elephant quota for South Africa was 50. Financial returns per animal can be high: Leader-Williams (1990) estimated that sport hunting of 200–300 elephants in Zambia would have brought in more financial revenue than all the international donations for conservation. In Kenya, Douglas-Hamilton (1988) estimated that sport hunting could generate US$20 million per annum – four times that earned from park entrance fees. In Tanzania hunting licences contributed US$4.5 million, over twice that received from national parks (Poole, 1990). By 1990, there were 20 African countries encouraging sport hunting of a number of mammal species, an industry which grew at about 15% per annum and now involves about 10 000 tourist hunters (Makombe, 1994).

Any legal trade is still dependent on an enforceable control system and the high value/low number trade is not without its own enforcement problems. Both Kenya and Tanzania have previously prohibited sport hunting due to corruption and a lack of controls (Bonner, 1993).

Even a trade in low numbers of specimens may be a threat to species with small wild populations, as any trade can have an enormous impact on the future viability of that species. The illegal export from Brazil of about 60 golden-headed lion tamarins (*Leontopithecus chrysomelas*) in 1983 and 1984 represented between 25 and 50% of the world population at that time and could have been instrumental in causing the extinction of the species (Mallinson, 1987).

15.2.2 Breeding potential of the species in trade

If a trade is unsustainable, its effects will be greater on those species slow to reproduce (*K*-selected species) and therefore slow to recover from over-exploitation, than it will be on species which can reproduce rapidly (*r*-selected species) (Eltringham, 1994). In the mid 1980s an overall hunting mortality of 6.3% was calculated for African elephant populations

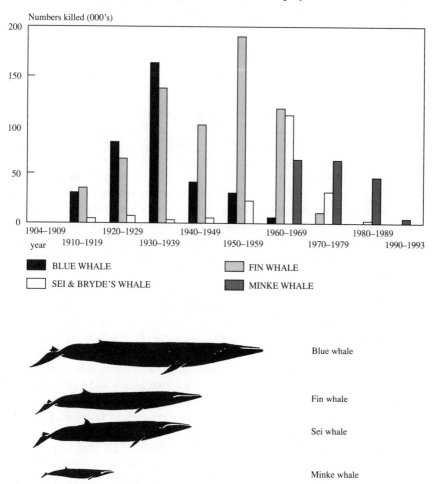

Figure 15.2 The relationship between size and numbers killed of four species of baleen whale in the Southern Hemisphere, 1904–1993. Source: IWC and the Bureau of Whaling Statistics.

when the maximum sustainable yield was calculated at 2–4% (Beddington *et al.*, 1989). There will be a long recovery period for this species as only one calf is produced on average every five years (Poole and Thomsen, 1989). Similarly, the blue whale population (*Balaenoptera musculus*), over-hunted in the 1940s and 1950s, remains at about 1% of pre-exploitation numbers and has yet to show any signs of a population recovery over much of its range, despite being totally protected from hunting since 1967 and from international trade since 1975 (Klinowska, 1991). Examples of non-mammal species where trade has been success- fully used as a conservation tool include butterflies (New, 1994) and crocodiles (Child, 1987) and have tended to be those species with a rapid breeding potential.

15.2.3 Different trade demands according to the sex and size of the animal

The potential for sustainable trade may decrease if a particular sex is targeted. Populations of musk deer (*Moschus* spp.) and saiga antelope (*Saiga tatarica*) have declined as a result of the trade in the products from these species as only the male is targeted (WCMC/IUCN, 1993; Anon, 1994a). In response to this, controls on the international trade in both species were increased in 1994. Similarly the African ivory trade, although not limited to male animals, was primarily targeted on males due to the sexual dimorphism in tusk weight. This resulted in an increase in the normal bias to females in a population, causing a decline in conception rates and increasing the difficulty of a population recovery (Poole and Thomsen, 1989).

In previous years, the specialized demands of the laboratory animal trade for wild-caught primates concentrated on animals below a certain size, often females and young of the larger wild-caught species. Any unwanted animals caught in the process were often killed in the source country as a result (Broad *et al.*, 1988b) so the number of animals in trade did not give a clear picture of the number of animals removed from the wild. Average exports of 2300 stumptail macaques (*Macaca arctoides*) from Thailand between 1964 and 1975 contributed towards a decline in their population and an eventual prohibition of exports by Thailand in 1976 (Treesucon, 1988). The numbers involved in the trade in live primates for laboratories are now very low (Table 15.2) and importation into some countries is decreasing (Jones and Jennings, 1994) but this example shows that even when the numbers involved in trade have been small, the effect it can have on the population in the country involved may still be substantial.

15.2.4 Economic aspects of the trade

Barbier *et al.* (1990) reported a number of economic trends common to the international animal trade:

- The profits are maximized at the middle-man stage, so the local people are not paid the real value of the product.
- A small number of traders control a large part of the trade.
- A large amount of revenue from the trade leaves the range state, benefiting only the traders and not the country.
- There is no long-term outlook by the traders.

Firstly, the majority of the rent of the resource (i.e. the difference between the returns gained from a species and the cost of supplying it) is captured at the middle-man stage, often a wholesaler or dealer outside the range state. Research into the illegal trade in tiger bone and rhino horn has found price increases of 50–300% at the middle-man stage for tigers (Mills and Jackson, 1994) and over 200% for Asian rhinos (Martin *et al.*, 1987). Similar results have been obtained from the legal trade in non-mammal species (Swanson, 1992).

Secondly, as a large percentage of the animal trade is controlled by a small number of traders, the largest profits from the trade leave the country in the form of foreign exchange and do not benefit the economy of the range state (Leader-Williams, 1990; Milner-Gulland and Leader-Williams, 1992). In the 1980s most of the profits from the ivory trade were paid into private bank accounts outside Africa or into the pockets of wealthy overseas entrepreneurs (Barzdo, 1989; Eltringham, 1994). Loss of revenue to the government will also occur from mis- or under-declaration of prices, profits or exported animals (Eltringham, 1994).

As a result, local people are paid a price for the commodity which does not reflect its true value (Anon, 1994b). The value per animal will be low and this invariably results in low welfare standards, with a corresponding rise in mortality rates in any live-traded mammal (Anon, 1994b). To compensate for mortalities caused by a drop in welfare standards, more animals will be taken from the wild, increasing unsustainability. Swanson (1992) reported that a relationship existed between welfare problems and potential sustainability in the trade in birds, although few data exist for the trade in mammal species.

Finally there is little incentive for the trader, once wealth is achieved, to promote a limited sustained exploitation of the resource (Ludwig *et al.*, 1993). This is due to a number of factors including low life expectancy, open access exploitation and the concept of 'getting rich quick', which would lead to increased social as well as financial power (Barbier *et al.*, 1990). In such cases the trader will aim for a short-term profit, not for a sustained long-term investment.

The quickest way to maximize those profits is by reducing costs. Evasion of controls is the cheapest option in achieving this (Barbier *et al.*, 1990), particularly as it is reasonably secure in the face of other market forces (Geist, 1994). If the price of the species starts to increase, avoidance of any control systems is also likely to increase. The legal ivory trade in the 1980s concealed an illegal trade which Douglas-Hamilton and Douglas-Hamilton (1992) estimated as 80% of the total trade. This coincided with the rising cost of ivory, which increased from an average of about US$60/kg (1979–1985) to US$120/kg (1987) and finally US$300/kg (1989), maximizing the incentives for short-term profits and illegal trade (Poole and Thomsen, 1989).

The short-term outlook of traders and the reluctance to wait for future returns from products encourages further over-exploitation (Barbier, 1992). Taken to its logical conclusion, provided that the market demand does not decline and controls are not enforced, the value of a species will continue to increase as its population declines, even to extinction (Barbier *et al.*, 1990). Table 15.3 summarizes certain species which are protected from international trade, where high prices have encouraged incidences of control avoidance which, in turn, increase pressure on the wild populations. A 10-tonne minke whale is worth about US$48 000 and so it is not surprising that illegal trade in whale meat still occurs (Mulvaney, 1993; Baker and Palumbi, 1994).

The trade in rhino horn is the principal cause of the extreme declines seen in rhinoceros populations over the past 17 years. The annual consumption of rhino horn in Taiwan and China, the two largest Asian consumers of horn, has been estimated at 400 kg (Milliken *et al.*, 1993) and 700 kg (Martin, 1990), respectively. There are estimated stockpiles of 9.8 tonnes in China (Leader-Williams, 1991) and 5–10 tonnes in Taiwan (Nowell *et al.*, 1992). At current rates of consumption these would take 12–14 years and 12–25 years, respectively, to clear (Martin, 1990; Milliken *et al.*, 1993). The continuing high levels of poaching in range states (Milliken *et al.*, 1993; TRAFFIC, 1994) suggest that traders are responding more to the possible extinction of the rhinoceros rather than to increased demand. A similar trend is occurring with the prices of tiger products (Martin, 1994).

15.2.5 Source and nature of demand

Any attempt to control or reduce demand depends on the type of use of the end product. Trade demands for products used for health reasons, such as Traditional Chinese Medicine (TCM), may be more difficult to control, as the product is viewed more as a necessity than a luxury. Products for CITES-listed mammals that are used in TCM, such as legally traded musk deer (Green, 1986) or illegally traded rhino horn (Milliken *et al.*, 1993), include some of the highest per unit values of any wildlife product (Table 15.3). Like rhinoceros populations, musk deer numbers have also declined – one species (*Moschus moschiferus*) by 50% in the three years from 1989 to 1992 alone (WCMC/IUCN, 1993).

15.2.6 Socio-cultural nature of end market

Consumer demand, and the enforcement of legal controls, can be influenced by the degree of public awareness and education, particularly from pressure by peers, media and non-governmental organizations (NGOs).

Attempts to enforce trade prohibitions in certain East Asian countries, which have little historical NGO presence, have been relatively unsuccessful until recently (Nowell *et al.*, 1992; Milliken *et al.*, 1993). However, retail habits have altered in markets such as the United States and western Europe, where NGO and media influences are stronger, making enforcement of legal controls less problematic. The trade in cat skins was centred on these regions in the 1960s and 1970s, particularly the USA and Western Europe (Broad, 1987). The UK alone imported over 76 000 ocelot skins in 1975 (Burton, 1976). Recent changes in consumer attitudes have resulted in a declining demand for spotted cat furs in these regions (Broad, 1987) and the market has moved to other areas, such as Greece and Japan, where media and NGOs are less influential (Walton, 1991; de Meulenaar and Gray, 1992; Fleming, 1994).

Similarly, a dramatic collapse in demand for ivory occurred in the USA and Europe (O'Connell and Sutton, 1990) as a result of reports that the ivory trade was responsible for a population decline in African elephants of about 50% from 1979 to 1989 (Poole and Thomsen, 1989). The market in the USA collapsed from its 1980s annual turnover of US$18–26 million and ivory imports fell by 95% in one year. As the demand and price fell, a complete loss of confidence occurred in the market, ensuring that the national and international prohibitions on the ivory trade, passed in 1989, were effective (O'Connell and Sutton, 1990). The low price for ivory and low poaching rate have generally continued in the following five years (Currey and Moore, 1994).

15.3 PROPOSED GUIDELINES FOR SUSTAINABLE TRADE

Over the past four years a number of principles and guidelines for sustainable use have been proposed (IUCN/UNEP/WWF, 1991; IUCN, 1993; IUCN, 1994). These are summarized in Table 15.4. Their applicability to international trade in CITES mammal species is examined here.

15.3.1 Principle 1: An assessment of the species' wild populations and its reproductive strategy

This principle is one of the most difficult to fulfil. Without accurate data, it is hard to assign a sustainable level of offtake. Yet accurate population data are rarely known until the species is close to extinction or confined to one area – a situation when any trade would be detrimental to the future viability of the species. For example, the population of the northern white rhino (*Ceratotherium s. cottoni*) is precisely known but only because the population is very low (32 specimens) and confined to one national park (Kemf and Jackson, 1994).

Population data are scarce for many species, particularly for pelagic

Table 15.3   Relationship between price of product (US$), population of mammal and degree of avoidance of controls in seven Appendix I mammal species where international trade has been prohibited for over 10 years

Species and product	Year of prohibition	Current wild population and trend	Mean retail price and trend in market price	Examples of control avoidance
Black rhino (horn) *Diceros bicornis*	1977 (CITES)	2480[1] Trend: 96% decline in 25 years[1]	$3250–10 000/kg[1] Trend: stable	Taiwan[1,4] China[1,3] Yemen[2]
Sumatran rhino (horn) *Rhinoceros sumatrensis*	1975 (CITES)	390–540[5] Trend: decline	$40 558/kg[6] No data on trend	Singapore[7]
Bengal tiger (bone) *Panthera t. tigris*	1975 (CITES)	3250–4700[8] Trend: decline[8]	$100/kg[8] Trend: 100% increase over 2 years	India[8] Cambodia[9] Republic of Korea[8]
Humpback whale (meat) *Megaptera novaengliae*	1975 (CITES) 1966 (IWC)	25 000[10] Pre-exploitation population: 150 000[10]	No recent data	Soviet misreporting in 1960s[11] Japan[12]
Minke whale (NE Atlantic) (meat) *Balaenoptera acutorostrata*	1986 (CITES) 1986 (IWC)*	Uncertain: 50 000–117 000[10,13] Trend: 70–80% decline from 1930s[10]	$350/kg[14] $48 000 for 10 tonne whale[15] Trend: stable in Japan	Japan[12] Norway[16]
Asiatic black bear (gall bladder) *Selenarctos thibetanus***	1979 (CITES)	50 000[17,18] Trend: declining[18]	$12.50–187.5/gm[23] Trend: increasing	East Asia[18]
Orang utan (live) *Pongo pygmaeus*	1975 (CITES)	30 000–50 000[19] No data on trend	$30 000[20] No data on trend	Taiwan[21,22] Thailand[21]

(continued)

1 Milliken *et al.* (1993)
2 Vigne and Martin (1993)
3 Loh and Loh (1994)
4 Nowell *et al.* (1992)
5 Kemf and Jackson (1994)
6 Martin and Martin (1989)
7 CITES (in preparation)
8 Mills and Jackson (1994)
9 Martin (1994)
10 Klinowska (1991)
11 Yablokov (1994)
12 Baker and Palumbi (1994)
13 IWC (1993a)
14 Anon (1994c)
15 Cawardine (1994)
16 Mulvaney (1993)
17 Servheen (1990)
18 Mills and Servheen (1991)
19 Anon (1992c)
20 S. McGreal, IPPL (personal communication)
21 Cater and McGreal (1991)
22 Martin and Martin (1991)
23 Mills (1995)

*   Norway and Japan hold reservations to the listing of the north-east Atlantic stock of minke whales on Appendix I of CITES. Norway additionally holds an objection to the 1986 IWC moratorium on commercial whaling. However, Norway currently has a national prohibition on exports of all whale products.

** Now *Ursus thibetanus*.

Table 15.4 Proposed principles and guidelines to assess the potential sustainability of international trade in mammal species (after IUCN/UNEP/WWF, 1991; IUCN, 1993; IUCN, 1994)

Categories	Principles and data or management required for sustainability
Population	1. An evaluation of wild population. An evaluation of the species' reproductive strategy.
Regulation and control	2. Ensure that the harvest does not exceed its capacity.
	3. Implement an adaptable management system.
	4. Establish a regulatory system.
Effects on other species	5. Ensure that where several species are being harvested the most vulnerable are protected. Reduce incidental take.
Local benefit	6. Fair sharing of costs and benefits to local people.

marine mammals. Miscalculation can also occur, even in well studied species. Until 1994, the best mean population estimate for the minke whale stock (*Balaenoptera acutorostrata*) in the north-east Atlantic Ocean was 86 736 (IWC, 1993a). New data presented by Cooke (1994) have shown that there were errors in the calculation and the actual figure may be as low as 53 000 (Chapter 8). This would cause a reduction in catch quotas granted under the sustainable management regime currently being developed by the IWC (IWC, 1993b) from approximately 330 animals to a negligible amount (Chapter 8). Once a management regime is put in place, it is difficult to rectify mistakes in numbers being taken. The two whaling management schemes used since the 1940s have failed to prevent over-exploitation of the target species (Figure 15.2).

15.3.2 Principle 2: Ensure that the harvest does not exceed its capacity

This principle relies on good enforcement and accurate data about other factors affecting the target species. These would include habitat destruction, the scale of domestic trade and welfare issues such as wastage from high mortality (e.g. from capture and, if the mammal traded is live, when the individual animal is in transit). The difficulty of obtaining and monitoring data on site and the problems of judging these variables mean that the optimum levels of exploitation are not detectable and are only gained through trial and error (Ludwig *et al.*, 1993).

15.3.3 Principle 3: Establish harvest levels that allow for uncertainty in population determination

Given the lack of data for the variables required under Principles 1 and 2, any offtake should be low if a sustainable level of trade is to be achieved. However, maximization of any benefits to the local community and the conservation of the species can only result from high-value trade, which may also increase the incentive to avoid controls. Such problems have already been mentioned for the trade in sport-hunted trophies.

15.3.4 Principle 4: Establish a regulatory and control system

Historically many of the regulatory control systems that have been established have not been adequately enforced. The trade controls for South American cat skins under Appendix II and for African elephant ivory, granted under the Ivory Control System, failed to address or rectify the unsustainable hunting that occurred during these years (Broad, 1987; ITRG, 1989). The implementation and enforcement of a control system are vital, particularly if the introduction of other specimens from outside the management programme is to be prevented. Beissinger and Bucher (1992) examined this issue when evaluating potential sustainable management of parrot species.

A possible solution to the problem of separation between control and ownership was suggested by ITRG (1989), in the form of centralization of control by the range state. Suriname is one of the few countries where the control of wildlife trade is done centrally, through the establishment of an export management system. A price is established in hard currency before any commercially traded mammal is exported. This is set by the government department that authorizes and manages export quotas. Once the sale occurs, the monies earned are paid in foreign exchange into a government-controlled bank. The local trader is paid by the bank in the local currency and so the country, rather than an elite group of traders, benefits directly from the trade.

Low quotas are set in Suriname. The advantage of low export quotas should be an increase in welfare standards for the species in trade (Anon, 1994b) and enforcement. The disadvantage lies in a reduced income, particularly if that income has to pay for the enforcement. Many countries which have the potential to export mammals have the least resources to monitor and control the trade.

15.3.5 Principle 5: Ensure that where several species are being harvested the most vulnerable are protected and incidental take is reduced

Traps used for lethal collection (e.g. fur trapping) and non-lethal collection (e.g. primate trapping) are generally not species-specific. This results in incidental take, which needs to be reduced for a sustainable use programme to be successful; otherwise any increased demand for the species may quickly cause the trade to become unsustainable. A number of sustainable use programmes using non-mammal species in trade, such as parrots (Beissinger and Bucher, 1992) and turtles (Dodd, 1979), have experienced similar problems.

15.3.6 Principle 6: Fair sharing of the costs and benefits to local people

Many socio-economic studies have concluded that local participation is mandatory if sustainable use programmes are to work. However, international trade is still largely controlled by a small number of traders, limiting the benefits to local people. Local participation in international trade in CITES-listed mammal species still tends to be limited to the low-volume trade such as sport hunting (Makombe, 1994; see also Chapter 13).

15.3.7 Other considerations

As wildlife trade is only one component of the potential total economic value (TEV) of a species, any decision to use international trade must be reconciled with other components. These other components have been summarized by Barbier (1992) as direct value of the species (e.g. from tourism, recreation, education), its indirect value (ecological role), its option value (any future use) and its existence value (cultural heritage, stewardship and bequest value).

It is difficult to place an economic value on some of these categories, such as bequest or cultural value. Certain species, particularly mammals, are seen as a pure public (or collective) good (Kuronuma and Tisdell, 1993), making them, in the opinion of some, off-limits to international trade, irrespective of any potential sustainability. Whales have been seen as a public good since the 1970s (Kuronuma and Tisdell, 1993) and a number of other mammals, such as the giant panda and tiger, would also probably fall into this category.

It is possible to generally compare income generated from international trade with that from non-consumptive use, such as tourism. The foreign exchange earned by Kenya from tourism in a year is its second highest income earner, attracting over half a million visitors in 1985

(Barnes *et al.*, 1992). The viewing potential of elephants alone in Kenya is worth US$25 million per year – about 10 times the value of the country's poached ivory exports (Barbier *et al.*, 1990) and a greater foreign exchange revenue earner than the loss of revenue from ivory exports (Barnes *et al.*, 1992).

Similarly, in 1991 direct income generated from whale watching was estimated to total US$75 million per annum, which was more than the income that could be generated from consumptive use of whales in international trade from many range states (Hoyt, 1992). In the late 1980s, Rwanda's income from tourists to view 600 mountain gorillas (*Gorilla g. beringei*) generated nearly US$1 million in park revenues and US$6–8 million total income to the local economy, making this one of the largest foreign exchange earners for Rwanda before the civil war started (Vedder and Weber, 1990).

Thus ecotourism has the potential to replace international trade as a revenue earner. However, if this is to occur, it is important that the problems outlined here with international trade, such as a lack of local involvement or poor enforcement, do not get repeated in any ecotourism projects.

15.4 CONCLUSIONS

Any effect of international trade on an individual species depends on a number of factors, including the numbers being traded, the nature and economics of the trade and the end market. The effects of these factors vary according to the species and the level of trade. However, historically much of the large-scale international commercial trade in CITES-listed wild-caught mammal species has been unsustainable, resulting in population depletions for many species and imminent extinction for some of them. A crude indication of this pattern can be seen by the additional protection given to mammal species over the last four CITES meetings (1987–1994) (Appendix 15.A). Sixteen species or subspecies, including the African elephant and Siberian tiger, have been listed in Appendix I, the majority qualifying as a result of unsustainable trade. In the same period, five species have been downlisted from Appendix I, although only one, the vicuna (*Vicugna vicugna*), would be traded internationally on a large scale.

Many of the problems that resulted in unsustainable trade in these species have been outlined in this chapter. Historically, a short-term approach has been adopted in trade in the species discussed here, increasing the incentive to avoid regulatory controls. The trade has also been dominated by a lack of local involvement and the maximization of profits to a small number of traders. For certain species, once the trade has created a market demand, enforcement of controls is difficult – even after the trade is prohibited. Many of the problems are also common to the trade in other taxa (Beissinger and Bucher, 1992).

Employing international trade as a use should only be seen as one of a number of options. In certain cases other options such as tourism may be more profitable and have fewer welfare problems than those which occur with traded mammals. In other cases, ethical, cultural or ecological factors may preclude the option of trade as a use at all.

Habitat destruction and an increasing human population mean that competition for space and the pressure on individual species will increase in the future. There is no doubt that long-term solutions guaranteeing the future conservation of mammals *in situ* are required. The potential for the international trade in CITES-listed species as a future sustainable use depends on solutions regarding the same factors that caused many species to be overused historically. Data on population sizes and threats on the species are usually scarce, which means that if the harvest size is not to exceed its capacity, low limits on numbers must be established and maintained. This should also increase the welfare standards and the potential for good enforcement, which suggests that it is unlikely that a large-scale (in terms of numbers) international trade in CITES-listed mammals could be successfully used as a conservation tool. Finally, if international trade on any scale is to play a role as a conservation tool for CITES-listed species, solutions are first required for the biological, socio-cultural and economic problems outlined here, which caused the historical overuse of many CITES-listed mammals.

ACKNOWLEDGEMENTS

The author prepared this chapter whilst at the Environmental Investigation Agency.

Some useful suggestions were provided by anonymous referees.

REFERENCES

Anon (1989a) Proposal to transfer *Felis pardalis* from Appendix II to Appendix I. CITES, Switzerland.
Anon (1989b) Proposal to transfer *Felis tigrina* from Appendix II to Appendix I. CITES, Switzerland.
Anon (1989c) Proposal to transfer *Felis wiedii* from Appendix II to Appendix I. CITES, Switzerland.
Anon (1992a) Resolution Conf. 8.3. Adopted by the 8th Conference of the Parties to CITES, Switzerland.
Anon (1992b) Proposal to transfer *Felis geoffroyi* from Appendix II to Appendix I. CITES, Switzerland.
Anon (1992c) Draft Red Data Sheet – Orang Utan. IUCN, Gland, Switzerland.
Anon (1994a) Proposal to include the Mongolian population of *Saiga tatarica* in Appendix I, and the other populations in Appendix II. CITES, Switzerland.

Anon (1994b) Paying the rent: a role for pricing and quotas in conservation. An issues paper prepared by TRAFFIC International to Animals Committee AC 10.12. CITES, Switzerland.

Anon (1994c) Whale meat – the red diamond. *Dagbladet* (Norway) April 29 1994.

Baker, C.S. and Palumbi, S.R. (1994) Which whales are hunted? A molecular genetic approach to monitoring whaling. *Science*, **265**, 1538–1539.

Barbier, E.B., Burgess, J.C., Swanson, T.M. and Pearce, D.W. (1990) *Elephants, Economics and Ivory*, Earthscan Publications, London.

Barbier, E.B. (1992) Economics for the wilds, in *Economics for the Wilds: wildlife, wildlands, diversity and development*, (eds T. M. Swanson and E. B. Barbier), Earthscan Publications, London, pp. 15–34.

Barnes, J., Burgess, J. and Pearce, D. (1992) Wildlife tourism, in *Economics for the Wilds: wildlife, wildlands, diversity and development*, (eds T.M. Swanson and E.B. Barbier), Earthscan Publications, London, pp. 136–152.

Barzdo, J. (1989) Elephant crisis. *Oryx*, **23**, 181–183.

Beddington, J., Mace, R., Basson, M. and Milner-Gulland, E. J. (1989) The impact of the ivory trade on the African elephant population, in *The Ivory Trade and the Future of the African Elephant: Interim report prepared for the Second Meeting of the CITES African Elephant Working Group, Gabarone, Botswana, 4–8 July 1989*, (Ivory Trade Review Group), CITES, Switzerland.

Beissinger, S.R. and Bucher, E.H. (1992) Sustainable harvesting of parrots for conservation, in *New World Parrots in Crisis*, (eds S.R. Beissinger and N.F. Snyder), Smithsonian Press, USA, pp. 72–115.

Bonner, R. (1993) *At the Hand of Man: Peril and Hope for Africa's Wildlife*. Simon & Schuster, London.

Brautigam, A. (ed.) (1989) *Analyses of proposals to amend the CITES Appendices*, IUCN/TRAFFIC/WCMC, Cambridge, UK.

Brautigam, A. (ed.) (1992) *Analyses of proposals to amend the CITES Appendices*, IUCN/TRAFFIC/WCMC, Cambridge, UK.

Broad, S. (1987) International trade in skins of Latin American spotted cats. *TRAFFIC Bulletin*, **9**, 56–63.

Broad, S. (1991) Contracting cats. *BBC Wildlife*, **9**(3), 201.

Broad, S., Luxmore, R. and Jenkins, M. (1988a) *Felis pardalis*, in *Significant Trade in Wildlife: a review of selected species in CITES Appendix II*, Vol. I, IUCN/WCMC, Cambridge, UK, pp. 43–58.

Broad, S., Luxmore, R. and Jenkins, M. (1988b) *Macaca fascicularis*, in *Significant Trade in Wildlife: a review of selected species in CITES Appendix II*, Vol I, IUCN/WCMC, Cambridge, UK, pp. 112–124.

Burton, J.A. (1976) Wildlife imports in Britain. *Oryx*, **13**(4), 330–331.

Cater, W. and McGreal, S. (1991) The case of the bartered babies. *BBC Wildlife*, **9**(4), 254–260.

Cawardine, M. (1994) *On the Trail of the Whale*, Thunderbay, UK.

Child, G. (1987) The management of crocodiles in Zimbabwe, in *Wildlife Management: Crocodiles and Alligators*, (eds G.J. Webb, S.C. Manolis and P.J. Whitehead), Surrey Beatty & Sons Ltd, NSW, Australia.

CITES (in preparation) Review of alleged infractions and other problems of implementation of the Convention. Proceedings of the 9th CITES Conference of the Parties. CITES, Switzerland.

Cooke, J. (1994) *Further analyses of Barents Sea minke whale CPUE in the Northeast Atlantic, with estimates of yield rates*. IWC/SC/45/NA7, IWC, Cambridge, UK.

Currey, D. and Moore, H. (1994) *Living Proof. African Elephants: the success of the CITES Appendix I ban*, Environmental Investigation Agency, London.

Dodd, C.K. (1979) Does sea turtle aquaculture benefit conservation?, in *Biology and Conservation of Sea Turtles*, (ed. K.A. Bjorndal), Smithsonian Press, USA, pp. 473–480.

Douglas-Hamilton, D. (1988) *Identification study for the conservation and sustainable use of the natural resources in the Kenya portion of the Mara-Serengeti ecosystem.* Final report December 1988, Nairobi, Kenya.

Douglas-Hamilton, I. and Douglas-Hamilton O. (1992) *Battle for the Elephants*, Doubleday, London.

Eltringham, S.K. (1994) Can wildlife pay its way? *Oryx*, **28**(3), 163–168.

Fleming, E. (1994) *The implementation of enforcement of CITES in the European Union*. TRAFFIC Europe, Brussels, Belgium.

Geist, V. (1994) Wildlife conservation as wealth. *Nature*, **368**, 491–492.

Green, M.J.B. (1986) The distribution, status and conservation of the Himalayan musk deer *Moschus chrysogaster*. *Biological Conservation*, **35**, 347–375.

Hoyt, E. (1992) *Whale Watching Around the World*. Whale and Dolphin Conservation Society, Bath, UK.

ITRG [Ivory Trade Review Group] (1989) *The Ivory Trade and the Future of the African Elephant. Vol 1, Summary and conclusions*. Report prepared for the 7th CITES Conference of the Parties, Switzerland.

IUCN (1993) *Guidelines for the Ecological Sustainability of Nonconsumptive and Consumptive Uses of Wild Species*, IUCN, Gland, Switzerland.

IUCN (1994) Note by the Director General on Guidelines for the ecological sustainability of non-consumptive and consumptive uses of wild species. Addendum 1 to GA/19/94/3. IUCN, Gland, Switzerland.

IUCN/UNEP/WWF (1991) *Caring for the Earth. A strategy for sustainable living*, Gland, Switzerland.

IWC (1993a) International Whaling Commission, Chairman's Report of the 44th Annual Meeting. *Rept. Int. Whal. Commn*, **43**, 64–65.

IWC (1993b) International Whaling Commission, Report of the Sub-Committee on Management Procedures (Annex 1). *Rept. Int. Whal. Commn*, **43**, 93–102.

Jones, B. and Jennings, M. (1994) *The Supply of Non-human Primates for Use in Research and Testing: welfare implications and opportunities for change*, RSPCA, Horsham, UK.

Kemf, E. and Jackson, P. (1994) *Rhinos in the Wild*, WWF Species Status Report 20, WWF International, Gland, Switzerland.

Klinowska, M. (1991) *Dolphins, Porpoises and Whales of the World. IUCN Red Data Book*, IUCN, Gland, Switzerland.

Kuronuma, Y. and Tisdell, C.A. (1993) Institutional management of an international mixed good. The IWC and socially optimal whale harvests. *Marine Policy*, July, pp. 235–250.

Leader-Williams, N. (1990) Black rhinos and African elephants: lessons for conservation funding. *Oryx*, **24**, 23–29.

Leader-Williams, N. (1991) *The World Trade in Rhino Horn: a review*, TRAFFIC International, Cambridge, UK.

Loh, J. and Loh, K. (1994) A spot check on the availability of rhino products in Guangzhou and Shanghai, China. *TRAFFIC Bulletin*, **14**(2), 79–80.

Ludwig, D., Walters, C. and Hilborn, R. (1993) Uncertainty, resource exploitation and conservation: lessons from history. *Science*, **260**, 17 and 36.

Makombe, E. (ed.) (1994) *Sharing the Land: Wildlife, People and Development in Africa*, IUCN/ROSA Environmental Issues No 1. IUCN/ROSA, Zimbabwe and IUCN/SUWP, USA.

Mallinson, J.J.C. (1987) International efforts to secure a viable population of the golden-headed lion tamarin. *Primate Conservation*, **8**, 124–125.

Martin, C.B. and Martin, E.B. (1991) Profligate spending exploits wildlife in Taiwan. *Oryx*, **25**, 18–20.

Martin, E.B. (1990) Medicines from Chinese treasures. *Pachyderm*, **13**, 12–13.

Martin, E. (1994) Prices up, tigers down. *BBC Wildlife*, **12**(10), 61.

Martin, E.B. and Martin, C.B. (1989) The Taiwanese connection: a new peril for rhinos. *Oryx*, **23**(2), 76–81.

Martin, E.B., Martin, C.B. and Vigne, L. (1987) Conservation crisis – the rhinoceros in India. *Oryx*, **21**, 212–218.

McMahan, L. (1986) The international cat trade, in *Cats of the World: Biology, Conservation and Management*, Proceedings of the 2nd International Symposium 1982, (eds S. Douglas Miller and D. Everett), Texas A & I University, Kingsville, pp. 461–488.

de Meulenaar, T. and Gray, J. (1992) *The Control of Wildlife Trade in Greece*, TRAFFIC International, Cambridge, UK.

Milliken, T., Nowell, K. and Thomsen, J.B. (1993) *The Decline of the Black Rhino in Zimbabwe: implications for future rhino conservation*, TRAFFIC International, Cambridge, UK.

Mills, J. (1994) *International Trade: a summary of the demand in consuming Asian countries*. Paper presented to the International Symposium on the Trade of Bear Parts for Medicinal Use, TRAFFIC International, Cambridge, UK.

Mills, J. (1995) Asian dedication to the use of bear bile as medicine, in *Proceedings of the International Symposium on the Trade of Bear Parts for Medicinal Use*, 9–11 September 1994, Seattle, Washington, TRAFFIC USA/WWF, Washington DC, USA, pp. 4–17.

Mills, J. and Jackson, P. (1994) *Killed for a Cure: a review of the worldwide trade in tiger bone*, TRAFFIC International, Cambridge, UK.

Mills, J. and Servheen, C. (1991) *The Asian Trade in Bears and Bear Parts*, WWF, USA.

Milner-Gulland, E.J. and Leader-Williams, N. (1992) Illegal exploitation of wildlife, in *Economics for the Wilds: Wildlife, Wildlands, Diversity and Development*, (eds T.M. Swanson and E.B. Barbier), Earthscan Publications, London, pp. 195–214.

Mulvaney, K. (1993) Norway caught flogging a minke. *BBC Wildlife*, **11**(12), 62.

New, T.R. (1994) Butterfly ranching: sustainable use of insects and sustainable benefit to habitats. *Oryx*, **28**, 169–172.

Nowell, K., Chiy, W-L. and Pei, C-J. (1992) *The Horns of a Dilemma: the market for rhino horn in Taiwan*, TRAFFIC International, Cambridge, UK.

O'Connell, M.A. and Sutton, M. (1990) *The Effects of Trade on International Commerce in African Elephant Ivory: a preliminary report*, WWF, Washington, USA.

Poole, J.H. (1990) *Elephant Conservation in Eastern Africa: a regional overview*, IUCN/EARO, Nairobi, Kenya.

Poole, J.H. and Thomsen, J.B. (1989) Elephants are not beetles. *Oryx*, **23**, 188–198.

Servheen, C. (1990) *The Status and Conservation of Bears in the World*, International Conference on Bear Research and Management 8, Monograph Series No. 2.

Smith, N.J.H. (1976) Spotted cats and the Amazon skin trade. *Oryx*, **13**, 362–371.

Swanson, T.M. (1992) Economics and animal welfare: the case of the live bird trade, in *Perceptions, Conservation and Management of Wild Birds in Trade*, (eds J.B. Thomsen, S.R. Edwards and T.A. Mulliken), TRAFFIC International, Cambridge, UK, pp. 43–57.

Swanson, T.M. and Barbier, E.B. (eds) (1992) *Economics for the Wilds: Wildlife, Wildlands, Diversity and Development*, Earthscan Publications, London.

TRAFFIC [Trade Records Analysis of Fauna and Flora in Trade] (1994). No respite for Asian rhinos. *TRAFFIC Bulletin*, **14**(3), 2.

Treesucon, U. (1988) A survey of stump-tailed macaques (*Macaca arctoides*) in Thailand. *Natural History Bulletin of Siam Society*, **36**, 61–70.

Vedder, A. and Webber, W. (1990) Rwanda: Mountain gorilla project, Volcanoes National Park, in *Living with Wildlife: wildlife resource management with local participation in Africa*, (ed. A. Kiss), World Bank Technical Paper no. 130, Washington, DC.

Vigne, L. and Martin, E.B. (1993) Yemen's rhino horn trade increases. *Oryx*, **27**(2), 91–96.

Walton, B. (1991) Catcalls. *BBC Wildlife*, **9**(3), 198–202.

Wijnstekers, W. (1992) *The Evolution of CITES. A reference guide to the Convention on International Trade in Endangered Species of Wild Fauna and Flora*, 3rd edn, CITES, Switzerland.

WCMC/IUCN [World Conservation Monitoring Centre/International Union for the Conservation of Nature] (1993) Musk deer, in *Significant Trade in Wildlife: A Review of Selected Animal Species in CITES Appendix II*, Draft report to the CITES Animals Committee, Cambridge, UK, pp. 59–63.

Yablokov, A.V. (1994) Validity of whaling data. *Nature*, **367**, 108.

APPENDIX 15.A. LIST OF APPENDIX I AND II MAMMAL SPECIES WHOSE PROTECTION HAS ALTERED IN THE FOUR CoPS FROM 1987 TO 1994

1. Species given increased protection from international trade due to unsustainable trade (*c.*14 spp.) and the associated look-alike problems (76 spp.)

Species	Change in Appendix	Reason for increased protection
American black bear *Ursus americanus*	II	Look-alike problem with trade in bear products
Sloth bear *Melursus ursinus*	II–I	Unsustainable trade in bear parts
Ocelot *Felis pardalis* (now *Leopardus pardalis*)	II–I	Unsustainable trade in skins – large illegal trade
Margay *Felis wiedii* (now *Leopardus wiedii*)	II–I	Unsustainable trade in skins – look-alike problems
Oncilla *Felis tigrina* (now *Leopardus tigrinus*)	II–I	Unsustainable trade in skins – look-alike problems
Geoffroy's cat *Felis geoffroyi* (now *Oncifelis geoffroyi*)	II–I	Unsustainable trade in skins – look-alike problems
Jaguarundi *Felis yagouarundi* (now *Herpailurus yaguarondi*)	II–I	
Siberian tiger *Panthera tigris altaica*	II–I	Increasing trade in bones – provided cover for illegal trade in other tigers
Red panda *Ailurus fulgens*	II–I	Declining population; concerns at trade for zoos
Crab-eating fox *Dusicyon thous*	II	Look-alike problem with species already in skin trade
Pampas fox *Dusicyon gymnocerus*	II–I	
African elephant *Loxodonta africana*	II–I	Unsustainable trade in ivory
Flying foxes (70 spp.) *Acerodon* and *Pteropus* spp. (2 species)	II–I	Look-alike problems for trade in meat
Jentinck's duiker *Cephalophus jentinki*	II–I	

African pangolin (3 spp.) *Manis gigantea* *Manis tetradactyla* *Manis tricuspis*	II	Concern at level of trade in scales for TCM and look-alike problems
Saiga antelope *Saiga tatarica*	II	Unsustainable trade in horn for TCM
Peccary (2 spp.) *Tayassu* spp.	II	Traded for skins
Hippopotamus *Hippopotamus amphibius*	III–II	Problems with trade in ivory
Chacoan peccary *Catagonus wagneri*	I	Newly discovered species; low population
Vietnamese ungulates (2 spp.) *Pseudoryx nghetinhensis* *Megamuntiacus vuquangensis*	I	Newly discovered species – not heavily in trade but low populations
Iberian lynx *Felis pardina* (now *Lynx pardinus*)	II–I	Declining small population

2. Species where protection from international trade has been relaxed and commercial international trade has resumed

Species	Change in Appendix	Reason
Vicuna (certain populations) *Vicugna vicugna*	I–II	Traded; population increasing
Southern white rhinoceros (South African population) *Ceratotherium s. simum*	I–II	To allow trade in live animals and hunting trophies only – population increasing

3. Species which are not in large-scale international trade and where protection from international trade has been relaxed

Species	Change in Appendix	Reason
Mexican bob cat *Felis rufa escuinapae*	I–II	10-year review
Northern elephant seal *Mirounga angustirostris*	Deletion from II	Not in trade

Brown hyaena *Hyaena brunnea* (now *Parahyaena brunnea*)	I–II	Pre-listing criteria
Roan antelope *Hippotragus equinus*	Deletion from II	Not in trade
Hopping mouse *Notomys* spp.	Deletion from II	Pre-listing criteria
Heath rat *Pseudomys shortridgei*	Deletion from II	Pre-listing criteria
Kangaroo rat *Dipodomus phillipsii phillippsis*	Deletion from II	Pre-listing criteria
Ground squirrel *Lariscus hosei*	Deletion from II	Pre-listing criteria
Sumatran rabbit *Nesolagus netscheri*	Deletion from II	10 year review
South African hedgehog *Erinaceus frontalis*	Deletion from II	10 year review
Aardvark *Orycteropus afer*	Deletion from II	Not in trade
Mato Grosso anteater *Tamandua tetradactyla*	Deletion from II	Pre-listing criteria
Cape pangolin *Manis temminckii*	I–II	Pre-listing criteria; put into Appendix II for look-alike reasons

Ten-year review and Pre-listing criteria: a change in status occurred as a result of nomenclature changes or because a species was placed on an Appendix before criteria listing was agreed.

16

The tiger – road to extinction

Valmik Thapar

SYNOPSIS

The tiger has been exploited throughout its range since at least the sixteenth century and persecuted for trophies and many other products apart from its skin. This chapter examines the historical context of human interaction with the tiger – from its revered position in Asian society to its present exploitation, almost to the point of extinction. Human population growth, habitat modification and fragmentation are contributory factors in the global decline of all subspecies of tiger but the most important factor in recent years has been the trade in tiger bone – a constituent of traditional medicines.

16.1 A BRIEF HISTORY

The tiger has been exploited throughout its range since at least the sixteenth century and has been hunted for products apart from its skin. In Chinese medicine, every part of the tiger's body played a role in the treatment of human illness. A Chinese materia medica dating from the sixteenth century recommends the use of tiger fat to combat vomiting and to treat dog-bite wounds; tiger blood for building up the constitution and strengthening the willpower; its testes can be used to cure scrofula; its bile to calm convulsions in children; even its eyes are used to treat epilepsy and malaria. The flesh of the tiger was supposed to prevent stomach and spleen disorders. The whiskers and claws provided great strength and courage, and the roasted skin, especially when mixed with water, was a cure for all ills. Tiger grease diluted with oil cured stomach ailments, while bones from the end of the tail and floating ribs destroyed evil and brought good fortune. The Chinese also believed that the tiger had great sexual powers, since it is able to copulate several times an hour and over a hundred times in the course of a few days. The tiger's penis was therefore considered a most powerful aphrodisiac.

In China, the image of the tiger acted as a charm against the influence of spirits and a likeness of the animal in any form was also considered a formidable protection against disease. A tiger's claw averted evil; small bones of the feet were potent charms and believed to prevent convulsions in children. Otherwise incurable fevers were calmed by reading verses about tigers! For further historical details, see Thapar (1992).

The question arises: how did the tiger survive in the wild when its skin and every part of its body were in such demand? It seems likely that China's forests were full of tigers before the advent of the gun. Tigers were certainly trapped and killed throughout the country but only on special ceremonial occasions, and there was not the large-scale destruction that came with commercial and sport hunting. The lives of men and the image of the tiger were interwoven in the ancient culture of the area. The disintegration of these elements has accelerated over the last hundred years as a result of various events in the lives of the forest communities, the forest and the tiger.

Chang Tao-Ling, the founder of Taoism, was dedicated to the search for a dragon–tiger elixir which would grant eternal life. Only in conjunction with the tiger could he fight evil and find the essence of eternal life. The tiger plays a comparable role in Hinduism as the vehicle of the goddess Durga. A variety of objects bearing the image of the tiger have survived from the sixth and seventh centuries BC. Chains, cases and boxes of gold decorated with tiger emblems evoke vast, dark forests where fierce struggles were a fact of life. Such battles between animals were reproduced on ornaments. It is obvious that an intense feeling of religious awe accompanied the sighting of a wild tiger.

In the mountains of Tibet, the people used tiger-skin rugs to ward off snakes, scorpions, insects and other creatures. Those in authority would sit on a tiger rug to pronounce verdicts or mete out punishments. A tiger rug could be part of the ritual adornment of a throne. The image of the tiger was sacred to all Tibetans. The belief in the tiger and its ability to cure while alive or dead was pervasive throughout the range of the tiger.

For the people who lived in the forest, the tiger was the most important and most powerful representation of nature. The tiger seemed to symbolize the force that could provide life, defeat evil and act as an 'elder brother' to humans, defending crops and driving out evil spirits. It was the protector, the guardian, the intermediary between heaven and earth. It was the symbol of fertility and regeneration. What is remarkable is that this belief prevailed despite the fact that tigers sometimes killed people. Forest communities accepted the tiger's right to intervene in their lives – that which gave life also had the right to take it away. On very special occasions people trapped tigers for rituals and ceremony.

There was an extraordinary balance throughout the tiger's range. This was primarily due to a way of life that was intrinsically bound up with and deeply dependent on nature, but as cultures changed and populations grew, the worship of the tiger faded, leaving in its wake the ever increasing exploitation.

16.2 NEW RELIGIONS, POLITICAL IDEOLOGIES AND TECHNOLOGY

H.R. Caldwell, a Christian missionary in China, documents the changes that occurred early in the twentieth century. Caldwell's principal reason for being in China was to advance the knowledge of Christianity. He spent 24 years in Fukien, a land full of tigers, where man and tiger had mutual respect. Caldwell's memoirs provide a remarkable account of how traditional beliefs were wiped out (Caldwell, 1925).

Caldwell thought he could change traditional beliefs with his gun. In an area where tigers were revered because the people believed their supernatural energies had to be propitiated, he slaughtered them at every opportunity. Caldwell records that an elder from one of the forest communities said to him, 'Teacher, I am afraid those people would not have heard of Christ until this day had you not killed that Tiger.' A new kind of exploitation of the tiger had started. Did a conflict of new religions spread by the Europeans tear into traditional beliefs? Did the tiger and the forest suffer with the arrival of guns? One thing is certain: a way of life in Asia was rocked at its foundations by people like Caldwell.

As the twentieth century progressed and communism held sway over China, the importance of cultivation was stressed so that the land would produce more. Like so many political systems, this ignored the fine balance that existed between humans and nature and the links between the past and the present. By the middle of the twentieth century, many wild creatures in China, including the tiger, had been declared pests. Hunters were encouraged to kill tigers and were paid a bounty for each one. For the forest communities a price had been put on the head of their god.

Many countries contributed to the demise of the tiger. The Dutch went to Sumatra and declared the tiger a pest. Rewards were given for shooting it – and this on an island where the tiger was an all-pervasive figure of worship. The French did the same in Vietnam and the British on the Indian subcontinent. The merciless slaughter started, both of tiger habitat and of the traditional beliefs of forests and tribal communities. Gone were the checks and balances; sport hunting turned forests into tiger graveyards. The technology of the gun became the modern tool of destruction and by the twentieth century was freely available for all to use.

By the 1970s the tiger had been squeezed to the point of extinction (Gee, 1969) and only some rapid conservation measures such as Project Tiger seemed for a moment to delay the inevitable. By the 1980s, concepts of free market economies and open trade in tiger products only fuelled the tiger's demise. The repercussions were fatal for the tiger. The deep effect of traditional beliefs had completely disintegrated under the pressure of population growth and new economic models. A natural balance had been toppled. It was a billion dollar trade in the use of endangered species and queues of people were waiting to enter it to make a quick profit. Added to this was the huge human population boom. Some 2.9 billion people or 59% of the world's population live in the tiger's range. This vast human population, through a series of rapid and new economic measures, were turning into voracious consumers. Deforestation in this area of Asia is 4.7 million hectares annually. The tiger and its forest were the first to suffer. Does the tiger have any chance of survival?

Human exploitation of the tiger had entered a new era as the population of the tiger plummeted. Three subspecies are extinct, two are virtually extinct, and from the 100 000–150 000 tigers that might have existed 150 years ago we are left with 5000–7000 animals today (Figure 16.1; CSG-IUCN, 1994).

16.3 INDIA – THE TIGER'S PLIGHT

At the turn of the century there were still at least 50 000 tigers left in India even though their population was declining rapidly due to sport hunting. The British and the Maharajas, both of whom ruled, competed with each other for the sake of record tiger kills. The Maharaja of Surguja won the race, with a personal tally of 1200 tigers. Undoubtedly countless thousands of tigers died from gunshot wounds, and many that were injured finally turned into man-eaters. Newly independent India fuelled this endless destruction as Shikar travel agencies sprung up everywhere, enticing the hunter into what humans had created as the greatest sport available: tiger hunting.

It was as recently as 1970 that tiger hunting was banned and a population estimate revealed only 1800 tigers left in India. In 1973 Project Tiger was launched; 10 years later tiger populations had increased to an estimated 4000. But by 1987/88 something started to go wrong. A huge demand from traditional medicine, especially for tiger derivatives, had drastically reduced populations of tiger in the wild over South-East Asia. The demand increased so rapidly that it now overflowed into India. Poachers in India could make extraordinary profits if they could fulfil the demands of gangs that now liaised between Taiwan, Hong Kong, China and India. It was noticed that tigers were missing first in Dudhwa on the Indo-Nepalese border, followed by Chitwan in Nepal, and the first marginal indications of this new problem were evident in the 1989 census of tigers held in India.

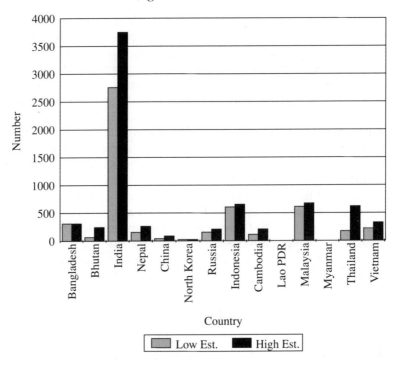

Figure 16.1 Estimates (high and low) of total world tiger population in 1994. Source: CSG-IUCN (1994).

16.4 THE RANTHAMBHORE EVENT

India's tiger population had been estimated as 1800 in 1972, 3000 in 1979, 4000 in 1984 and then only 4344 in 1989 (Anon, 1972, 1993, 1994). This drop in population growth occurred despite the fact that suitable tiger habitats (which make up 3% of India's land mass) were available. Given the prey density available, these areas could accommodate at least 6000–8000 tigers. It was only in 1992 that the entire world got a rude shock from a series of events that shook the small but world-renowned Ranthambhore Tiger Reserve.

An independent group of people led by the author monitored the census results. From a high of 45 tigers, Ranthambhore was left with no more than 15–20 tigers (Ranthambhore Foundation, 1992). A month later a poaching gang was captured and tiger and leopard skin and bones seized: they admitted to killing 20 tigers. This illegal trade had been going on for at least four years.

The Ranthambhore event reflected the state of affairs over the entire range of the tiger. By the census of 1993 it was clear that India was left with between 2750 and 3750 tigers; up to 1600 tigers had been butchered for the trade in derivatives. While the skins had entered the global market, the bones and other parts were smuggled into Nepal, China and other countries in South-East Asia. Figure 16.2 shows the number of tigers exported from various countries to South Korea and Japan and this is contrasted with the number of wild tigers in those countries. Business in the free market economy was booming, particularly in Taiwan and China. The Sumatran population of tigers had also fallen with both the Siberian and Indo-Chinese following. At least 2000 tigers had been massacred between 1989 and 1993 (Table 16.1).

By August 1993 the biggest ever haul of big-cat derivatives in India's history took place with nearly 400 kg of tiger bones, eight tiger skins and over 60 leopard skins being seized. The flesh on the bones was still rotting. The seizure took place in a Tibetan refugee community in Delhi and further investigations revealed that many Tibetans were involved in the smuggling of tiger bones through Nepal and Ladakh into China. This was a community that had once worshipped the tiger's image. The Dalai Lama stated that this activity was against every traditional norm of Tibetan belief and must be stopped. Appendix 16.A lists the recent initiatives to save the tiger.

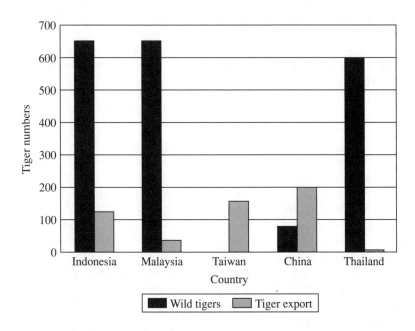

Figure 16.2 Number of tigers exported from various South-East Asian countries to South Korea and Japan and number of wild tigers in those countries. Source: Jackson and Mills (1994).

Table 16.1 Tiger bones
seized in India from 1989
to 1993 (kg)

Year	Weight (kg)
1989	0
1990	90
1991	0
1992	18
1993	667

16.5 THE EFFECT OF EXPLOITATION

In India, when the tiger disappears from a particular region there is little interest in sustaining the natural ecosystem it inhabited and large tracts of forest are made available for agriculture and industry. Even when they are not developed, the lack of natural predators can turn deer and antelope populations into pests and seriously upset the natural balance; then sport hunting may be allowed. Irrespective of this, few people realize that when India loses her tigers, and subsequently her forests, then she will also lose her vital water catchment areas. This will seriously affect water tables, causing soil erosion and desertification, and decreasing agricultural productivity. Such losses will have fatal repercussions on humankind.

16.6 CONCLUSION

One single action is suggested. A Global Tiger Forum, historic in its approach and principles, has been initiated by India and will soon come into legal being. Let every country, organization and lobby group participate in it. Let it be devoid of self-interest (Figure 16.3). Let it find strategies that are implemented without delay. Let it be empowered by the world community to act, with the resources to act, and the global political will to enforce its recommendations. Let this forum become an example of what humans can do for the tiger. If this can be achieved then we will find a ray of light, in what is today a hopeless situation. If we fail, then the tiger will be virtually extinct in five years – and its extinction will trigger the extinction of many other species. Even today as the tiger suffers, so do leopards, rhinos and elephants, to talk of only a few. Once gone, we will never be able to recreate them and it will be a horrifying example of the deep failure of modern civilization, which killed off one of the most remarkable creations of nature.

Figure 16.3 The tiger on the road to extinction.

16.7 POSTSCRIPT

By April 1995 it is clear that the tiger could be virtually extinct in India by 1999. Intensive poaching and an apathy in government, both central and state, to deal with field issues has been resulting in the demise of the tiger. The Global Tiger Forum is still not activated since there are insufficient members. Most new policies remain on paper rather than being translated into the field. The tiger has no time for endless meetings, rhetoric or lip service. It requires urgent action in the field. Tiger Link is the only new initiative in India outside of government that may help: a link between people and organizations in the interest of the tiger, that solely prioritizes actions. The final battle has started to save the tiger; if lost, as is likely, then in India it will signal the virtual devastation of our rich fauna and flora. After all, if we fail to save the tiger, what can we save?

REFERENCES

Anon (1972) *Project Tiger*. Report from Government of India, Ministry of Environment and Forests.

Anon (1993) *A Critical Review of Project Tiger*. Report from Government of India, Ministry of Environment and Forests.

Anon (1994) *All India Tiger Census Report*. Report from Government of India, Ministry of Environment and Forests, Steering Committee.

CSG-IUCN [Cat Specialist Group, IUCN] (1994) *The Status Report on the Tiger, 1994*, IUCN, Gland, Switzerland.

Caldwell, H.R. (1925) *Blue Tiger*, London.

Gee, E.P. (1969) *The Wildlife of India*, Fontana, London.

Ranthambhore Foundation (1992) *Census of Tigers in Ranthambhore*, Ranthambhore Foundation, New Delhi, India.

Thapar, V. (1992) *The Tiger's Destiny*, Kylie Cathie Ltd, London.

APPENDIX 16.A INITIATIVES TO SAVE THE TIGER

The death of tigers in Ranthambhore catalysed a series of events from 1992 to 1994:

May 1992: A census conducted in Ranthambhore reveals a decline in the tiger population.

June 1992: A poaching scandal erupts in Ranthambhore. Full media coverage starts all over India and the rest of the world.

September 1992: The Ministry of Environment and Forests reconstitutes the Steering Committee of Project Tiger and the first meeting is held after four years to assess the situation. A tiger census is announced in India in 1993.

November 1992: The Cat Specialist Group (IUCN) meets in Delhi and the chairman Peter Jackson assesses the information from experts and declares a serious problem, fearing the virtual extinction of the tiger within 10 years.

February 1993: An international symposium is held on the tiger sponsored by the Ministry of Environment and Forests. More information trickles in on the crisis. The meeting culminates in the Delhi Declaration, a set of national and international initiatives to save the tiger.

March 1993: WWF and other international organizations realize the gravity of the problem.

April 1993: Regional Cat Specialist Group comes into being in India.

June 1993: The first reports of the tiger census indicate a declining trend.

August 1993: The biggest ever seizure of tiger derivatives takes place in Delhi.

September 1993: China legally bans trade in tiger derivatives.

October 1993: The Ministry of Environment and Forests announces the following new initiatives.

(a) An anti-poaching Strike Force for all Project Tiger reserves.
(b) An intelligence gathering Task Force to monitor trade and traffic in endangered species, especially the tiger.
(c) A tiger allowance for all Forest Guards, as an incentive to protect the tiger.
(d) Projects concerning ecological development with local communities around Project Tiger reserves.
(e) International coordination by Tiger Range countries to initiate the Global Tiger Forum.

October 1993: At the Forestry Forum for Developing Countries, India holds the first meeting of Tiger Range countries to initiate the Global Tiger Forum.

December 1993: The tiger census reveals that there could be 2750–3750 tigers left. Degradation of habitat has also increased. The Minister of Environment and Forests declares a Tiger Crisis and creates a Tiger Crisis Cell to monitor developments.

March 1994: A meeting of Tiger Range countries is called in Delhi to constitute legally the Global Tiger Forums; national and international NGOs are present.

April 1994: For the first time ever the United States uses the Pelle amendment and imposes sanctions on Taiwan for trade in tiger derivatives and rhino horn.

May 1994: Taiwan and Hong Kong change their rules on wildlife trade in an attempt to make them stricter.

June 1994: Minister for Environment and Forests in India declares 'Operation Monsoon' as a special effort to check poaching. In India from 1992 to 1994, the Steering Committee of Project Tiger has met four times, the Tiger Crisis Cell four times, Cat Specialist Group six times. Two international meetings just on the tiger have taken place; WWF has initiated new strategies for tiger conservation, and there has been vast media coverage.

17

The exploitation of Asian elephants

Jacob V. Cheeran and Trevor B. Poole

SYNOPSIS

Asian elephants have a unique status as the only domesticated animal whose population depends on the recruitment of individuals caught in the wild. Elephants are used in their native countries such as India, Malaysia, Sri Lanka and Thailand for forestry work, religious festivals and in timber yards and are also housed in temples and logging camps. They are trained in many countries to perform in zoos and circuses. Their capture often, but not invariably, involves cruelty and rough treatment. Elephants are highly intelligent and, in the confined conditions of captivity, boredom is a common problem. Elephants are usually underworked in zoos but, at the other extreme, overwork in other situations can cause exhaustion, particularly in combination with inadequate opportunities for feeding.

Wild Asian elephants conflict with humans where the size of their reserve is inadequate and they have taken to crop raiding. Attempts to translocate the elephants are of limited value as, frequently, the animals will return from over 100 km away. In some areas which are heavily populated by humans, wild elephants have little chance of long-term survival, particularly where their translocation or the creation of habitat corridors is impracticable. In such situations, it should be possible to develop a policy of humane domestication of the wild elephants to meet the requirements of local people. There is also a poaching problem in reserves to acquire ivory, which diminishes the male elephant population.

There are few, if any, self-sustaining captive elephant groups; thus elephant populations are declining both in the wild and in captivity. There are real welfare problems for both wild and domesticated elephants. To maintain populations of Asian elephants, it is essential that they should be bred more efficiently in captivity.

17.1 WILD ELEPHANTS

Much attention is drawn in western countries to the endangered status of the African elephant, while the plight of the Asian species (*Elephas maximus*) tends to have been ignored. However, the population of the Asian elephant is more endangered and is estimated to be between

25 000 and 40 000 in the wild. Asian elephant habitat is fragmented and the animal is in competition with people whenever it strays into human settlements. The small size of undisturbed habitats and the high degree of mobility of these animals often leads to crop raiding.

Farmers try to frighten off the elephants by shouting and waving their arms and using firecrackers. In some cases the animals are attacked with firearms, or burned with lighted torches soaked in vegetable oils. These actions make the animals aggressive so that they are provoked to attack humans and sometimes kill them. Farmers can lose their livelihood through the actions of a single wild elephant, so the animal is very unpopular in some regions. Even where electric fences are used to protect crops, some male elephants learn that ivory is an insulator and break the wires with their tusks. Sukumar (1989) and Jayewardene (1994) give excellent accounts of the conflict between elephants and people and carefully consider possible ways of improving the situation.

In many instances, the real problem lies in the fact that reserves are too small to sustain a viable population of elephants, so that they invade agricultural land either to reach other groups or to raid crops. One technique to allow gene flow between small groups is by providing corridors between such areas. Alternatively, in India and Sri Lanka, elephants have sometimes been translocated to larger areas where they can contact other herds. Such animals may be moved as far as 100 km from their original site. However, translocation of elephants requires great expertise and a high level of organization. In some instances, one or more animals have returned to the original site (Jayewardene, 1994; Sukumar, 1994).

Asian elephants are also exploited by ivory poachers. This only affects the males but leads to a sex ratio of males to females which averages one adult male to five adult females in India, but in some areas it may be as low as one male to 24 females (Sukumar, 1989). The effect of poaching is not as serious as it would be for a monogamous species, but the low male:female sex ratio in some areas does give cause for concern. In Sri Lanka, elephants are strictly protected and only one in 20 male elephants are tuskers so that there is less incentive for ivory poaching and the adult sex ratio has been estimated to be of the order of one male to two females (McKay, 1973; Kurt, 1974).

17.2 CAPTIVE ELEPHANTS

The Asian elephant has a unique status as the only domesticated animal which is taken directly from the wild. It has been caught and domesticated for at least 3000 years in India, Malaysia, Sri Lanka and Thailand and has formed an integral part of the culture and religion of these countries. Sukumar (1989) believes that up to 50 000 elephants may have been captured for domestication over the past 100 years and that, at present, in India alone, there are between 2500 and 3000 elephants in captivity. Nowadays, domesticated elephants are mainly used for forestry (logging), in religious festivals, in timber yards and kept in zoos and circuses.

17.2.1 Capture

Even today, in spite of the endangered status of the Asian elephant, the great majority of captive elephants are taken from the wild. In some cases these animals may be calves abandoned by their mothers; they are thus saved from an early death. Typically, however, working elephants are caught in the wild between the ages of 15 and 20 when they are sufficiently large to be used for heavy work. Traditional methods of capture, taming and training may involve some cruelty. Details of elephant training are given by Fernando (1988) and Jayewardene (1994) who emphasize the importance of positive reinforcement and the building of a good relationship between the animal and its handler. Both capture and training can be made less stressful by taking more time and being patient, and utilizing tranquilliser drugs such as diazepam also has a taming effect.

17.2.2 Role of personal bond

In the past, the job of mahout was a family tradition and the profession was passed on from father to son. The children of mahouts also had intimate knowledge of the elephant which was looked after by their father. The relationship between humans and elephants was a lifelong one and bonded with affection and trust and the domesticated elephant would normally respond only to commands from its personal mahout. Unfortunately, nowadays, mahouts are employed on a temporary basis and often change their employer. As a result, when a new mahout is employed he has to use brute force to make his new elephant obey him. Sometimes the level of violence is such that the elephant suffers permanent injuries which in some cases have proved fatal. This treatment can be avoided if the period allowed for the new mahout to take control is lengthened and employers are educated to appreciate the need for adequate time to be allowed for the take-over period. Hart (1994) has provided an excellent assessment of mahout–elephant relationships in Nepal.

17.2.3 Housing and husbandry

In nearly all instances captive elephants are chained when not working, usually in open houses, and some temple elephants may spend nearly all their lives on a tether. Elephants require a varied diet but are often fed on only one type of fodder; also the quantity fed is often quite inadequate.

Gokula (1994) studied a sample of temple elephants in southern India and found that many of them relied entirely on begging for their food and receive only a quarter of the amount which they eat in the wild. In addition, the food is often unsuitable. Generally zoo elephants and those on logging camps where there is natural vegetation have a much better diet. Veterinary care in temples is usually inadequate or absent and reliance is often placed solely on traditional medicines. Elephants need proper veterinary care; they are also susceptible to parasitic infections and so need periodical de-worming with a broad-spectrum anthelminthic.

17.2.4 Activities in captivity

The best traditions for daily activity for elephants are found in those source countries where generations of mahouts have cared for these animals. Elephants are easily bored so working is beneficial for them and they can carry on piling logs in a timber yard, for example, without supervision, even expanding the base of the pile when necessary. They also enjoy participating in religious processions and contact with people. However, they can be overworked, as is common in timber logging, when the animal is not supervised by the owner. Captive elephants in some areas are also overworked during the temple festival season when they are used continuously in ceremonial parades, all day and night. The daytime procession progresses inch by inch on a hot, bitumenized road in the full heat of the sun, which can prove very tiring for the shade-loving elephant; the lack of food and water also makes the animals temperamental. At festivals in the Indian state of Kerala, the mahout is commonly plied with intoxicating liquor by the local people and this often leads to him maltreating the elephant and losing control over it. This may result in the killing of the mahout or the animal running amok in the crowd, creating panic. The first author and his veterinary colleagues in Kerala have been called out to control these situations nearly 350 times in the last 15 years, requiring the use of tranquillizer darting on approximately 250 of these occasions.

17.3 SPECIAL WELFARE PROBLEMS WITH ELEPHANTS

The exploitation of captive elephants can create welfare problems for the animal, the most common of which is prolonged chaining. Tethered elephants become bored and develop behavioural problems (Schmid, 1995). Where there is no enclosed housing for the animal, chaining overnight may be acceptable, provided that the tether is not too short and plenty of food is available. In India, chaining elephants in zoos is illegal, but many still do so. There is no doubt that elephants should be willing to be chained for short periods of an hour or two to allow such procedures as veterinary examinations. It is unsatisfactory, however, to tether them for periods longer than a few hours. On the other hand, confining them to small stalls, as in some zoos and circuses, is probably no less restrictive than a reasonably long tether.

Older male elephants (over 25 years) can present a serious problem because they are liable to come into annual musth. Musth does not occur in all males – some show it occasionally and others regularly – but it invariably causes the animal to become aggressive, intractable and difficult. Traditionally, male elephants have been tethered with heavy chains throughout musth, which usually lasts from one to three months. This is clearly inhumane and the animals may show serious stereotypy during this period. Musth creates a significant management problem and this is probably why so few male elephants have been kept in circuses and zoos. Attempts are being made in Kerala to solve the problem of musth by using anti-androgens and long-acting depot-sedatives. However, the lack of availability and the high cost of such drugs place constraints on their extensive use. Factors which do seem to minimize the severity of musth seem to be a good relation with the mahout, being kept in a social group with females (Poole *et al.*, in press) and having plenty of space to move around.

Another problem for the future of captive elephants is their poor record of breeding, which has resulted in the lack of a self-sustaining captive population (Frost, 1986; Haufellner *et al.*, 1993). In general, the best breeding records are from elephants in forest camps in source countries where they are kept in groups and females can either mate with a captive bull or a wild bull in the forest. However, wild bulls have been known to entice such cows when in heat, and take them into the forest for courting and mating. While this in itself does not create problems, there have been instances where a wild bull has challenged and killed a domestic bull kept in the forest. The record of breeding in zoos and circuses has been very poor probably because they keep very few bulls, often house the sexes separately and seldom have groups as large as six, a number which Roocroft and Zoll (1994) regard as the minimum size of social group for elephants. UFAW recently carried out research on breeding elephants in Sri Lanka (Taylor, 1995) and it is hoped that such work will eventually lead to improvements in breeding success.

17.4 CONCLUSIONS

The exploitation of Asian elephants creates both conservation and welfare problems. There can be little disagreement that it would be a tragedy if the Asian elephant was driven to extinction. Wild elephants face two major problems: poaching and conflict with farmers. Attempts are being made to address these problems in the countries where Asian elephants are indigenous (Sukumar, 1989; Jayewardene, 1994) but it is difficult to resolve conflicts between wild elephants and humans in areas where there are small islands of habitat surrounded by human habitation. In some areas which are heavily populated by humans, wild elephants have little chance of long-term survival, particularly where the translocation of the animals or the forming of habitat corridors is impracticable. In such situations, it should be possible to develop a policy of humane domestication of the wild elephants to meet the requirements of local people, through lumbering and so on. This would not only ensure that elephants and local people work together and are not in conflict, but will also save 'pocketed' herds which have no future.

To encourage reproduction in captivity, elephants should not be kept in barren enclosures lacking vegetation, or confined to indoor housing for long periods. An elephant needs opportunities to graze and wallow, or a regular daily routine of work. Staffing must be adequate to ensure that familiar, trusted keepers are employed to care for elephants, and staff should have received training to ensure that they are competent in both handling and training them. Management must be certain that the relations between the elephant and its handler are good and that both the motivation and ability of the staff member are adequate for the task.

Elephants are highly intelligent and, in domestication, should work with a well trained and trusted human companion; this situation will not only utilize the animal's great strength but also its skills and versatility. In some areas, elephants can be used instead of vehicles to carry tourists on safari into areas of natural habitat. Such a programme of elephant domestication should only be adopted if the highest standards of welfare can be applied. This will require an extension of the training programme for mahouts already initiated by the Elephant Welfare Association of Trichur in southern India. Every elephant should be provided with its own mahout or keeper, but also have the opportunity to mix freely with other elephants of both sexes in a familiar group to facilitate captive breeding. There should be strict welfare guidelines for the keeping and husbandry of elephants and these should be enforceable. If such a programme was adopted this would benefit both people and elephants and help to secure a future for this highly endangered species.

17.4.1 Recommendations

We should like to make the following suggestions for ameliorating the problems facing the existing populations of Asian elephants.

1. Efficient management of the few large reserves of natural habitat with effective anti-poacher strategies and an incentive to local people to conserve and maintain elephants in their immediate vicinity.

2. In pocketed areas of habitat, where translocation or the use of corridors is not practicable, elephants should be either:

   (a) domesticated for work to help local people; or
   (b) assigned to good zoos able to provide for the management of a breeding group of elephants, basic training by a specific keeper assigned to the animal permanently and adequate mental stimulation through either daily access to a large area of natural vegetation or full domestication with a daily work schedule.

3. Strict guidelines for the humane treatment and training of captive elephants should be produced and keepers and mahouts trained to act accordingly.

17.4.2 Caution

The recommendations and suggestions made in this article are all known to be practicable, but readers must appreciate that elephants should be handled with great care and sensitivity because they can be dangerous. The authors offer their advice in good faith but cannot accept any liability for problems or accidents arising from its application.

REFERENCES

Fernando, S.B.U. (1988) Training working elephants, in *Animal Training: a review and commentary on current practice*, Universities Federation for Animal Welfare, Potters Bar, UK, pp. 101–113.

Frost, J. (1986) *A Survey of Elephants in Captivity in the British Isles* (unpublished).

Gokula, V. (1994) The status and management of temple elephants with forest camp and zoo elephants: a comparative study. *ZOO ZEN*, **10**(4), 1–28.

Hart, L.A. (1994) The Asian elephants–driver partnership: the drivers' perspective. *Applied Animal Behaviour Science*, **40**, 297–312.

Haufellner, A., Kurt, F., Schilfarth, J. and Schweiger, G. (1993) *Elephanten in Zoo und Circus. Teil 1: Europa*, European Elephant Group.

Jayewardene, J. (1994) *The Elephant in Sri Lanka*, Mortlake Press Ltd, Colombo, Sri Lanka.

Kurt, F. (1974) Remarks on the social structure and ecology of the Ceylon elephant in the Yales National Park, in *The Behaviour of Ungulates and its Relationship to Management*, Vol. 2, (eds V. Geist and F. Walther), IUCN, Switzerland, pp. 618–634.

McKay, G.M. (1973) Behaviour and ecology of the Asiatic elephant in Southeastern Ceylon. *Smithsonian Contributions to Zoology*, **125**, 1–113.

Roocroft, A. and Zoll, D.A. (1994) *Managing Elephants*, Fever Tree Press, Ramona California.

Poole, T.B., Taylor, V.J., Fernando, S.B.U. *et al.* (in press) Social behaviour and breeding physiology of a group of captive Asian elephants. *International Zoo Yearbook*.

Schmid, J. (1995) Keeping circus elephants temporarily in paddocks – the effects on their behaviour. *Animal Welfare*, **4**(2), 87–101.

Sukumar, R. (1989) *The Asian Elephant: Ecology and Management*, Cambridge University Press, Cambridge, UK.

Sukumar, R. (1994) *Elephant Days and Nights: Ten Years with the Indian Elephant*, Oxford University Press, Delhi.

Taylor, V.J. (1995) The social behaviour and breeding physiology of a group of captive Asian elephants, in *Proceedings of the 8th Elephant Keepers' Workshop*, (eds N.G. Spooner and J.A. Whitear), North of England Zoological Society, Chester, England, pp. 58–68.

Part Five

Ecotourism: Making Mammal Populations Pay

18

The impact of ecotourism development on rainforest mammals

Nigel Dunstone and Jane N. O'Sullivan

SYNOPSIS

Nature-based tourism is one of the few economic uses of natural areas that could be compatible with protection of the environment and its wildlife. However, the demands placed on these ecosystems from tourism can destroy or disturb the very attractions the clients pay to enjoy. The impact of ecotourism on distribution, status, and activity of rainforest mammals, will be discussed with reference to the Manu National Park, Peru.

The distribution, status and behaviour of a variety of mammal species were assessed in the vicinity of a tourist lodge, along existing trails of varying usage and antiquity, along newly cut trails and in undisturbed forest. A comparison was made of the abundance and diversity of mammals in areas of differing disturbance. The progressive habituation of mammals along newly cut trails was also investigated. Further information on mammalian diversity was obtained by live-trapping within 200 m of the trail system and by radio-tracking selected species (ocelot and marsupials).

The results are related to the expectation of what mammals tourist groups might see during a typical sojourn in the jungle. Dawn and dusk forest walks were most profitable for mammal viewing, although many telemetered individuals rested within a few metres of busy tourist trails.

Regrowth of disused tourist trails can enhance habitat diversity, providing a patchwork of areas of varying succession and species composition. Since this secondary growth is often more dense than the surrounding forest it can provide refuges for wildlife. In most instances the land lost to trail creation is negligible and telemetered animals frequently used trails within hours of their being cut. Strategies for reducing the disturbance effects on sensitive species are discussed.

18.1 INTRODUCTION

Ecotourism (or nature tourism) is becoming increasingly popular both for international visitors who flock to areas of high biological diversity, and for the governments of impoverished nations who see this development as a means of generating considerable revenue for regional and local economies. Nature-based tourism is one of the few economic uses of natural areas that could be compatible with the protection of the environment and its wildlife (Boo, 1990), and could produce a revenue which may be utilized to implement conservation strategies (Groom *et al.*, 1991). However, the demands placed on these ecosystems from tourism can destroy or disturb the very attractions that clients pay to enjoy (Dunstone and Shoobridge, in press). Ecotourism has been slow to develop in South American rainforests partly because of difficulties of access to wilderness areas, political instability and weak marketing (Dunstone and Shoobridge, in press). Furthermore, the flagship mammalian species of interest to tourists are secretive in their habits and are less well known than their African counterparts.

The southern rainforests of Peru boast an astounding diversity of species with in excess of 200 species of mammals described, including 13 species of primate, giant otter and jaguar (Janson and Emmons, 1991). Manu Biosphere Reserve protects some of the world's most important wild-lands. To describe it one must inevitably resort to superlatives since few protected areas in the world are as large as Manu and probably none of them as pristine and rich in wildlife species. The park protects an entire watershed where there is no hunting and no logging. It includes an enormous variety of habitat types, from the Andean grasslands at 3900 m, through elfin and cloud forest on the eastern slopes of the Andes, to vast lowland Amazonian rainforests at 300 m.

The Manu Biosphere Reserve was established in 1977, and comprises a core area of 1 532 806 ha preserved in its natural state, an experimental or reserved zone of 257 000 ha serving as a buffer zone which is set aside for experimental research and ecotourism, and a cultural zone of 91 394 ha which provides an area of permanent human settlement where sustainable uses of land and forest are promoted.

The Department of Madre de Dios, within which the park is situated, is not well served with an infrastructure to allow for efficient handling of tourism. There is an international airport at Puerto Maldonado but until recently its services were very unreliable. Most of the clients visiting the area will have organized an inclusive tour to the jungle as an extension of their cultural visit to Cusco and Machu Picchu national monument. Of the 100 000 tourists that visit Cusco each year, perhaps only 10% will venture to the southern rainforests. The tourist potential of the Madre de Dios remains poorly advertised locally. At present there are five privately owned tourist lodges in the proximity of Puerto Maldonado and more are planned. Manu Biosphere Reserve has one lodge in the reserved zone and two in the cultural zone.

18.2 TOURISM DEVELOPMENT

From about 1980, the reserved zone began receiving a steady trickle of tourists, usually travelling individually or as small groups in privately chartered canoes. In 1984, organized tour parties started using the Zona Reservada. These groups were based in organized campsites along the Rio Manu, in forest clearings or alongside ox-bow lakes (cochas). One company still conducts tours of this type. In 1986 the first permanent tourist facility (Manu Lodge) was built close to Cocha Juarez in the reserved zone. In the first few years of these ventures some 500 tourists per year visited Manu, usually during the dry season from May to September (Figure 18.1a). However, this number is likely to increase as a result of enhanced marketing in the USA and Europe. The main effect has been to increase the boat traffic and hence level of disturbance on the Manu river, the only access route to the tourist facilities in the park.

Since 1980 the Park has developed a limited tourism potential. It was of interest to determine the effect of this disturbance on the populations and distribution of mammal species and to suggest how the tourist potential of the park can be realized in a way that will not jeopardize the wildlife. Few studies have investigated the impact of such developments on the fauna. This study investigates the effect of a low-key tourist development on the local mammal fauna in terms of changes in status, distribution and activity brought about by their interaction with humans. Since the lodge had been in operation for some 10 years before the present study began, it was not possible to know whether observed behaviour was contingent on ecotourism activity. Instead, wherever possible, comparisons are made with data from Cocha Cashu Field Station, which is some 50 km up-river of the Manu Lodge within the core area, and where no ecotourism activity occurs.

18.3 THE STUDY AREA AND OBJECTIVES

The bulk of the Biosphere reserve lies within the watershed of the 250 km long Manu river. The Manu is a meandering river approximately 150 m wide, but is subject to extensive flooding during the rainy season when it may rise 5–8 m above its dry season level. The currents that accompany these floods are responsible for eroding the banks and depositing extensive beaches. Because of these destructive forces it is unlikely that the forest in the vicinity of the lodge is of any great antiquity, except on ridges and other elevated ground untouched by the river currents (Terborgh, 1983). Most of the lowland regions of the park are at about 400 m above sea level and receive approximately 2000 mm of rainfall per year. Most of this rain falls from October/November to April/May (Figure 18.1b).

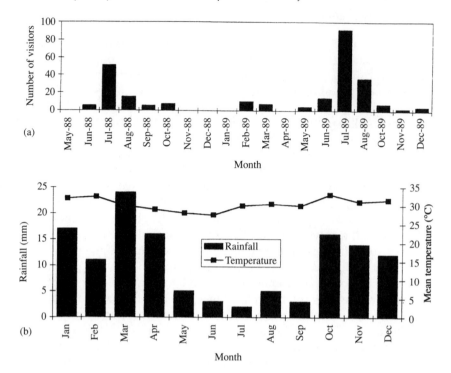

Figure 18.1 (a) Tourist statistics for visitors to the Manu Biosphere Reserve from May 1988 to December 1989; (b) temperature and precipitation recorded at Manu Lodge for the period January to December 1989.

Studies were conducted in the vicinity of the Manu Lodge tourist facility on Cocha Juarez, an ox-bow lake off the Rio Manu. The lodge is in the humid tropical life zone. The area immediately surrounding the lodge that is accessible by the trail system consists of several habitats: humid terra firma forest, humid low-lying forest, tropical rivers and their margins, and ox-bow lakes and their margins.

The objectives of the study were as follows.

1. To determine the status and distribution of mammal species in the vicinity of Manu Lodge (and to compare this with Cocha Cashu, approximately 50 km up-river).
2. To identify the main types of disturbance event caused by tourists, both directly and indirectly.
3. To investigate habitat utilization and range use by target species.
4. To determine activity and movement patterns of these species in order to assess whether there is spatial and/or temporal avoidance of humans.

The bulk of the data presented in this chapter were collected over two successive July–August dry season visits to Manu Lodge.

18.4   STATUS AND DISTRIBUTION OF MAMMAL SPECIES IN THE VICINITY OF THE TOURIST LODGE

The distribution, status and behaviour of mammal species were assessed in the vicinity of the tourist lodge, along existing trails of varying usage and antiquity, along newly cut trails and in undisturbed forest.

The vegetation in the vicinity of the lodge was relatively undisturbed and in many areas impenetrable, particularly where successional growth was evident. The trail system (Figure 18.2) was based on linking viewing points (elevated observation areas and lake-side hides) to encompass a wide range of habitats, and including fruiting trees (e.g. *Ceiba pentandra*) which provide viewing opportunities when a variety of species aggregate to feed there. A grid-based trail system, as exists at Cocha Cashu research station, was not used. Trails were mapped and marked at 50 m intervals, allowing sightings and movements of mammals to be recorded with that accuracy. Various new trails were cut to link up existing trails. In this way the trail system was extended from 15.8 to 21.8 km. The average length of trails walked by tourists was 5.54 km (± 0.44) and each trail took approximately three to four hours to walk, depending on sightings.

The methods employed involved standard walks, transect, circuit and focal sampling techniques. A comparison was made in terms of the abundance and diversity of mammals in areas of differing levels of disturbance. The progressive habituation of mammals to human activity along newly cut trails was also investigated. Further information on mammalian diversity and distribution was obtained by live-trapping in a broad band within 200 m of the trail system.

18.4.1   Vegetation survey and feeding resources

Vegetation classification was made according to Terborgh (1983) at 50 m intervals along the entire 28 km trail system. Eight major vegetation types were distinguished:

1. **High ground forest**. Mature tall forest on high ground (> 700 m) above the zone of annual inundation.
2. **Transitional forest**. Forest with closed canopy exposed to seasonal flooding. Subdivided into:

    (a) sub-canopy dominated by palm;
    (b) sub-canopy dominated by *Ficus insipida* and *Cedrela odorata*, with understorey dominated by *Heliconia* and *Costus*.

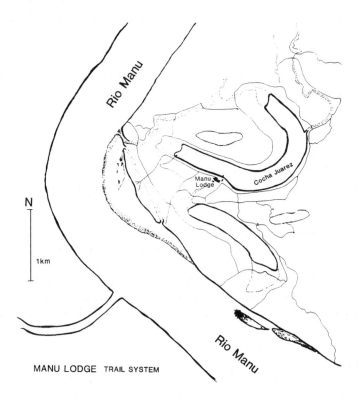

Figure 18.2 The trail system at Manu Lodge.

3. **Emergent**. Transitional forest with many tall emergent trees, typically *Ceiba pentandra*, *Dipteryx* and strangler figs (*Ficus* spp.).
4. **Swamp forest**. Fig swamp dominated by *Ficus trigona* and *Heliconia* spp.
5. **Forest openings and vine tangles**. Caused by tree falls.
6. **Quebrada margins**. Forest stream margins.
7. **Lake margins**. Shrubs, typically *Annona tessamannii* and thick vine growth.
8. **Zabolo**. Pioneer vegetation that represents the first stage of plant succession on recently deposited river banks. Dominated by *Gynerium* spp. reaching a height of 8–10 m.

The presence of fruiting and flowering trees (keystone resources for many mammalian species) above or adjacent to trails was also recorded to give an indication of the spatial and temporal distribution of food resources. Most trees along the trail system had previously been identified and tagged. Care has to be taken in estimating fruit availability because, to a certain extent, the meandering trails system had been developed to encompass as many of these keystone resources as possible in order to maximize the possibility of observing foraging animals.

18.4.2 Expectation of what tourist groups might see

The trail system was divided into five sections following, in general, the route taken by guide-led tourist groups. All trails originate and terminate at the lodge. Two of the sections were short, taking approximately one hour to complete; two were of medium length, requiring 2–3 hours to complete; whilst one, the ridge trail, required 6–8 hours of daylight to circumnavigate, and some participants preferred to camp overnight along the route.

Each trail was walked by groups of three or four participants (UK biology undergraduates) at either dawn, midday or dusk on each day. All observations of mammals were recorded; in addition, field signs (footprints, food remains and faecal material) were noted. The recording of mammal presence and absence was aided by the construction of print traps using soft sand or mud across trails.

The daily count of mammal species observed and the cumulative number of different species seen over a 32-day period in the Park are shown in Figure 18.3a, while Figure 18.3b gives similar data for mammal species observed from trails. Approximately 70% of the total number of species seen were viewed within two weeks in the park, and 90% by day 21. Thereafter a further 11 days of systematic trail walking only added six more species to the list. The daily count ranged from two to nine species and it was difficult to predict what these might be. There is a fairly good agreement between these two sets of data, indicating that standard trail walks are fairly representative of mammalian diversity and abundance. Appendix 18.A gives a complete list of mammal species observed, tracked or trapped during the study period.

It is pertinent to consider these results in terms of the typical time period that tourists stay in the Park. Terborgh *et al.* (1984) records the presence of 74 species of non-volant mammals at Cocha Cashu research station. Of this total number potentially available, participants in the survey located 35 (47%) over a six-week survey period. On the basis of the present observations, tourists may expect to see well over half of the common fauna during a one-week stay at the lodge. Dawn and dusk walks were most profitable for mammal viewing (Table 18.1). The number of mammal species observed during morning (commencing at dawn) and afternoon (ending at dusk) walks was broadly equivalent. These data can be compared with the numbers of species observed off-trail during the course of casual observations made during radio-tracking studies (see later).

Table 18.2 attempts a comparison between trails of varying distance

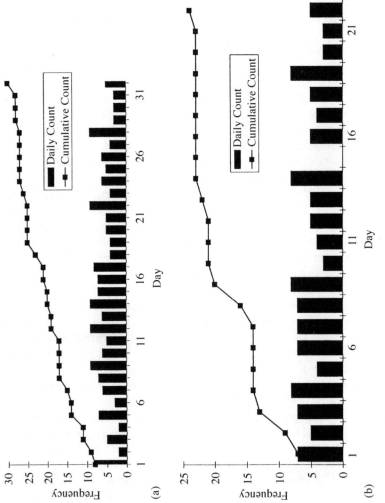

Figure 18.3 Cumulative number of 'new' mammal species encountered and daily count of mammal species observed: (a) in the Park; (b) along tourist trails.

Table 18.1 Temporal distribution of mammal encounters

(a) Mean no. species observed per 5 km

	Dawn	*Midday*	*Dusk*
Along trail	5.7 ± 0.47	2.3 ± 0.47	3.0 ± 0.8
Off trail	6.2 ± 0.8	4.7 ± 1.6	5.1 ± 0.7

(b) Mean no. individuals observed per 5 km

	Dawn	*Midday*	*Dusk*
Along trail	44.0 ± 27.0	17.7 ± 15.4	60.0 ± 24.1
Off trail	42.6 ± 28.0	24.8 ± 20.1	52.0 ± 27.9

from the lodge. Data are presented as the number of species of mammal observed per 10 km of trail walked with dawn, dusk and midday walks combined. The trails are ranked in terms of distance of their mid point from the lodge. It is interesting to note that the lodge trail, which is closest to the tourist base, provides the greatest frequency of encounters with mammals. However, care must be taken in such comparisons because the trails are not equivalent in the types of habitat traversed or in the opportunities for mammal observation (e.g. availability of keystone resources).

There were two occasions when it was found necessary to cut entirely new trails. The Cocha Antigua trail was cut in order to give access to the area used by a radio-collared ocelot during daytime activity (see later). This trail, amounting to some 5 km in length, encompassed a mixture of emergent forest, caña brava and seasonally flooded swamp land. The Quebrada trail was also approximately 5 km long, following the course of an old creek, and included stream-edge vegetation as well as transitional and emergent types. The more open aspect of the canopy along this trail may have aided the visibility of some species. These data were compared with sightings along a 5 km straight-line transect through broadly equivalent habitat types. The frequency of sightings of species of mammal along these trails are shown in Figure 18.4.

18.4.3 Trapping studies

Thirty 'Tomahawk' live traps constructed of galvanized wire with single swing doors were available. Twenty of these traps measured 120 cm × 65 cm × 50 cm and the remainder measured 80 cm × 30 cm × 25 cm. Various trapping strategies were employed: traps were either set along a trap line, along or in the vicinity of trails or, occasionally, set opportunistically where field signs indicated the frequent presence of a target species.

Table 18.2 Comparison of encounters with mammal species along four tourist trails (trails ranked in increasing distance from lodge; walks were conducted over three months and each trail surveyed at dawn, midday and dusk)

Trail	Trail length (m)	No. walks	No. species per 10 km	Mean no. individuals	SD
Lodge	2800	9	13.2	38.2	28.6
Lake	5750	9	6.43	40.5	35.9
Kapok	5950	9	5.04	42.8	42.6
Quebrada	5200	9	6.54	36.4	32.1

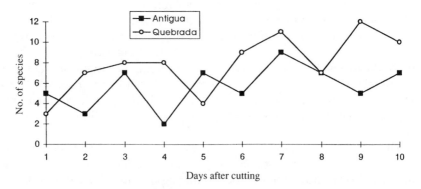

Figure 18.4 Observations of mammal species from newly cut trails.

Most mammalian captures were of marsupials, principally the common opossum *Didelphis virginanus* and the brown four-eyed opossum *Philander opossum*; of the rodents trapped, *Proechimys* spp. predominated.

Figure 18.5a–c shows trapping success expressed as captures per 100 trap-nights at various distances from the lodge, in different habitats and at varying distances from the trails. These data are somewhat difficult to compare because of a number of intercorrelated variables. For example, there appears to be a greater trapping success at about 1 km from the lodge, but this simply represents the fact that the most suitable habitat, zabalo, was located at that distance. Perhaps the most interesting finding is that trapping success was marginally better at a short distance into the forest from trails. This might reflect the secondary vegetation present there.

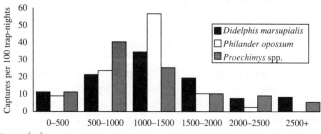

(a) Distance (m)

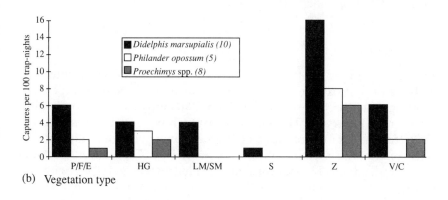

(b) Vegetation type

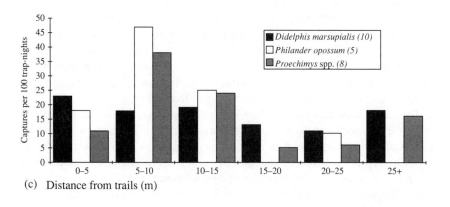

(c) Distance from trails (m)

Figure 18.5 Trapping success: (a) at varying distance from the lodge; (b) in various vegetation types; (c) at varying distances from tourist trails. (P/F/E, transitional forest; HG, high ground forest; LM/SM, lake and stream margin; S, swamp; Z, zabalo; V/C, vine tangles and clearings.)

18.4.4 Detailed study of selected animal groups

(a) Baseline surveys of primates

Primates are one of the main tourist attractions in Manu. Twelve species are found in the park (Terborgh, 1983), of which nine had been regularly recorded within the trail system of Manu Lodge. Neotropical rainforest primates are known to be sensitive to hunting and logging practices, but little is known of the effect of benign human disturbance. The potential effects of disturbance on primates involved research, conducted by undergraduate students acting as 'tourist' observers, which has not been fully reported elsewhere (Daniels *et al.*, 1989).

(b) Standard walks

In order to survey for primates, the trail system was divided into four sections of approximately equal length. In order to reduce duplicated sightings, each section was walked simultaneously by a group of either three or six observers. This ensured that sightings were of distinct primate troupes, and not of the same troupe moving between different sections of the trail system. Walks commenced at 0700 with all sections being walked on 12 occasions (six walks were conducted with six observers and six walks with three observers). In order to reduce the temporal variation in disturbance, the order in which the sections were walked was reversed daily. Observers walked at an approximate speed of 1.5 km/hr; in total, 228 km of trail were walked.

There was no difference in the numbers of primates seen by groups of three or six observers, indicating that their powers of observation and the behaviour of the primates were unaffected by the size of 'tourist' group. Nine species of primate were encountered during the three-month study period with a varying frequency ranging from rare (one occurrence) in the case of the monk saki *Pithecia monachus*, to occasional in the case of the howler monkey *Alouatta seniculus*, to regular in the case of squirrel monkeys *Saimiri sciureus* and the two species of capuchins *Cebus albifrons* and *C. apella*. The most frequently encountered species was the squirrel monkey, troupes of which were seen on 23 occasions over 12 days. In terms of number of monkeys sighted, they were also the most frequently seen (25.7 per 10 km walked) with a mean group size of 32.6 individuals. Large troupes of squirrel monkeys comprising 20–50 individuals were often seen low in the canopy, where they travel more slowly than other primate species. Furthermore they were relatively tolerant of human observers, often coming within several metres of 'tourists'.

The census was carried out during the dry season when food resources were in relatively short supply. This might have led to increased movements associated with the location of foraging sites, and hence greater opportunity for observation. It is likely that the different primate species were not encountered in proportion to their actual population levels. Emmons (1984) provides estimates of primate population densities from Cocha Cashu research station, approximately 50 km up-river from this study site. In the present study, squirrel monkeys were encountered more frequently than would be expected if present at the density recorded by Emmons. This may have resulted from their enhanced visibility when using the lower canopy or from their greater tolerance of human observers. With the exception of the two species of capuchins, all other primate species tended to move away from observers.

An attempt was made to determine to what extent primates use flowering and fruiting trees adjacent to the trail system to obtain food supplies and therefore potentially came into contact with ecotourists. The temporal abundance of feeding resources for primates was determined along the standard walk trails. This is undoubtedly a biased sample since the trail system was designed to link up these resources to provide viewing opportunities for the tourists. Nevertheless, Table 18.3 shows that during the dry season a large number of primate species rely on a relatively small number of food resources.

Temporal distribution of keystone resources for primates
For much of the dry season many of the frugivorous species are limited in their ranges by their often widely dispersed food resources. Furthermore these resources may become available in flushes (Figure 18.6). There was no discernible event which triggered the flowering and fruiting event that happened in weeks 7–10. For long periods there is relatively little available, and then a temporary super-abundance. Outside of the tourist season the animals become quite dependent on resources that are later visited by tourists, creating a conflict of interests and potential disturbance of feeding behaviour. An example of this concerns a platform constructed in the canopy of a *Ceiba pentandra* tree which serves as a major food source for a wide variety of animals during the dry season.

(c) Radio-tracking studies

Radio-tracking studies were carried out to determine whether there was spatial or temporal avoidance of humans by target species when in the vicinity of the lodge and its trail system. Usage of tourist trails, avoidance of tourist groups and reaction to inducements (e.g. artificial feeding sites) were monitored by radio-tracking two species of opossums, *Didelphis marsupialis* and *Philander opossum*, and a felid – the ocelot, *Felis pardalis*.

Table 18.3 Arboreal feeding resources for primates

Plant species	Description	Eaten by	Part eaten
Combretum fruticosum	vine	SM, BC, WFC, BSM, ET, ST	Nectar
Celtis iguanea	vine	ET, ST, DT	Fruit
Astrocaryum spp.	palm	BC, WFC	Nut
Scheela cephalotes	palm	SM, BC, WFC	Pith
Igna spp.	tree	WFC	Seeds, nut
Ficus perforata	tree	BSM	Fruit
Ficus trigonata	tree	NM, BC	Fruit
Apeiba echinata	tree	BSM, WFC	Fruit
Erythrina spp.	tree	BSM, RHM	Fruit
Matissia cordata	tree	BSM, SM, BC, WFC	Flowers/ nectar
Ceiba pentandra	tree	SM, BSM, BC, WFC, ET, ST	Flowers/ nectar

Data from Daniels *et al.* (1991).

Key to primate species:

BC = Brown capuchin
BSM = Black spider monkey
DT = Dusky titi
ET = Emperor tamarin
NM = Night monkey
RHM = Red howler monkey
SM = Squirrel monkey
ST = Saddle-back tamarin
WFC = White-fronted capuchin

Telemetry of opossums

During the course of two dry season studies a total of four *D. marsupialis* and two *P. opossum* were tracked in a variety of habitats and at varying distances from Manu Lodge. The opossums were fitted with small radio-collars amounting to less than 5% of body-weight and released at the point of capture. Observations were made opportunistically during the course of other studies, but in general at least two locations were obtained during each day and night over a six-week period. Telemetered individuals frequently rested within metres of busy tourist trails during the day, particularly favouring the dense secondary growth associated with trail creation. Regrowth of disused tourist trails can enhance habitat diversity, providing a patchwork of areas of varying antiquity and species composition.

Telemetry of ocelot

Two ocelots (one male and one female) were captured in successive dry seasons in cage-traps situated in the zabalo vegetation where tracks indicated frequent activity. Radio transmitters weighing less than 0.5% of

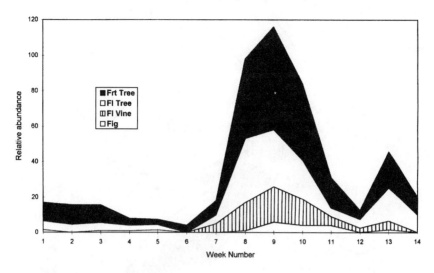

Figure 18.6 Sporadic availability of feeding resources along the trails during 14-week study period in June, July and August. (Frt, fruit; Fl, flower). Data from Daniels *et al.* (1991).

body-weight, set on leather collars, were attached under light anaesthesia. The ocelots were released near their point of capture. The signal from one animal, a male, failed after four days. Restricted signal reception in the dense forest habitat necessitated continual monitoring of the other cat's movements until an adequate estimation of the home range emerged. Thereafter, in addition to tracking on foot, monitoring was accomplished by location from a vantage point located 30 m up in the canopy of a *Ceiba pentandra*. Approximate location in this manner allowed a ground-based observer to be guided to the telemetered animal by VHF radio to facilitate closer observations. In this way radio-tracking observations were conducted almost continuously for approximately 30 days and nights. It was frequently possible to observe the subject, which appeared to become habituated to the authors' presence. Her total dry-season range included most of the habitat types available, but she particularly used emergent and swamp forest (Figure 18.7). The area encompassed by 75% of the day-time and night-time radio fixes obtained from this animal is also shown, and indicates a substantial shift from use of the tourist trail system during the night to more secluded areas during the day. A relatively small number of favoured resting points, including vine tangles and old tree-trunks, were utilized. Most of her activity was ground-based with very little taking place in the trees. The activity pattern (Figure 18.8) was not substantially different from that recorded by Emmons (1988) for female ocelots at Cocha Cashu field station further into the park. There was, however, a slight preponderance of late

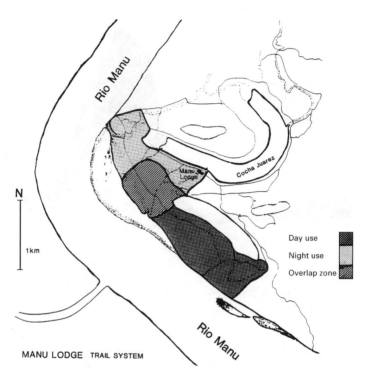

Figure 18.7 Range of a female ocelot in relation to the trail system.

afternoon and night-time activity with rather less occurring during the period when tourists might have been using the trails. This was also reflected in her utilization of tourist trails, which was predominantly overnight (Figure 18.9). On occasions her behaviour could be observed in the vicinity of tourists: typically she lay obscured in dense vegetation as the groups walked by. Emmons (1988) also recorded ocelots to make use of a grid of trails cut to facilitate scientific observations. Even though the Manu Lodge trail system was not cut for this purpose the ocelot made extensive use of it, occasionally within hours of their being cut. Since the secondary growth surrounding trails is often more dense than that in the adjacent forest, it provided a refuge for wildlife. Presumably the wide, man-made trails allowed a more silent approach to prey, many species of which were also seen to use the trails and a number of prey captures were observed.

18.4.5 Canoe-based observations

Canoe-based observation should be the least destructive and potentially minimally disturbing form of wildlife-watching, since paddle-powered

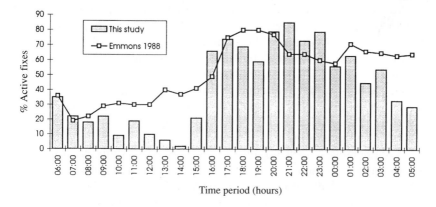

Figure 18.8 Daily activity of a female ocelot. Histogram, this study; data from Emmons (1988) given for comparison.

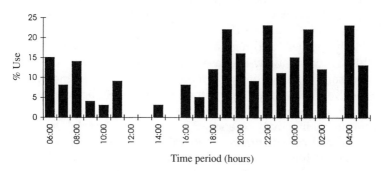

Figure 18.9 Use of the tourist trail system by a female ocelot during the day and night.

canoes approach animals quietly, are non-polluting and slow to follow retreating subjects. One of the flagship species for the Madre-de-Dios region is the giant otter *Pteronura brasiliensis*. Giant otters are now largely confined to ox-bow lakes (cochas) along the Madre-de-Dios, Tambopata and, particularly, the Manu rivers.

The species is endangered throughout its range; probably the best populations exist in Manu. A recent survey by Schenck and Staib (personal communication) gave a total population of 40 otters in the Madre de Dios, distributed over 24 cochas. Sixteen of these otters were resident on the three cochas most visited by tourists. Very few tourists out of hundreds questioned by Schenck and Staib expected to see these rare and endangered animals. Many more gave as their motive for visiting the protected area the remote possibility of seeing other flagship species, such as jaguar. Of those tourists who did view the otters, almost all thought the experience was the highlight of their trip.

Groups of giant otters have a preferred lake but will travel by river to others, seeking new reservoirs of fish as these become depleted during the dry season. River disturbance can therefore cause problems. Tourism in Manu is largely based along the river, since this is the only route of access into the Park. The river gives access to the ox-bow lakes, the preferred habitats of the giant otters. Three of these lakes are open to large-scale tourism and probably all of the 500 or more tourists visiting Manu annually will spend time at one or more of the lakes.

The otters are gregarious, noisy and diurnal – a feature that has made them vulnerable to pelt hunters in the past and to tourists now. They approach tourist canoes out of curiosity and noisily threaten the boats. Their normal behaviour is to approach, periscope and snort at canoe-borne tourists. This behaviour is often misinterpreted by tourist groups who, if inadequately guided, have been known to pursue the otters around the lake for enhanced photographic opportunities. On one of the lakes where this has frequently happened the otters have modified their behaviour by not approaching canoes (Schenck and Staib, personal communication).

There is some evidence that otters will habituate to people if given sufficiently frequent and regular exposure without detrimental contact, and this has been largely the case at Cocha Juarez. However, during the rainy season otters go for many months without experiencing humans, who then reach very high concentrations over a period of a month or so during the dry season. This coincides with the period when the otters concentrate their activity on a small number of ox-bow lakes.

Otters can also be disturbed if the trail system comes close to the shore of the lake. At Juarez the cocha is encircled by a trail. Here it is difficult for the otters to bring food to the shore, which they occasionally require to do with large fish. Whereas otters can avoid trail-based tourists appearing at a fixed place, they rarely habituate to tourists in a canoe. If at all practicable, no more lodges should be built on cochas containing giant otter populations. Coordination of tourist visits should occur to reduce over-use of particular areas, and the number of canoes using the lakes at one time should be regulated. Consideration should be given to the use of alternative methods of observation, e.g. canopy platforms overlooking the lakes.

18.5   DISCUSSION

Over-ambitious tourism development (e.g. poorly designed trail systems and exceeding the carrying capacity in terms of number of clients) can lead to the fauna moving from the vicinity of a lodge or to restriction of the animals' range and disturbance of feeding sites and other maintenance activities. On the other hand, sympathetic tourism development will allow for enhanced viewing prospects while at the same time causing minimal disturbance to the fauna. A preliminary analysis (Table 18.4) shows the distances at which a variety of unhabituated species react to the presence of human observers. Further studies of this type are needed to identify potential attractions and determine their resilience to visitor observation in order that sustainable use levels may be established.

Trail systems vary considerably in extent between lodges. In many cases their development and scope reflects the necessity for encompassing a variety of habitats, or to relieve tourist pressure within habitats. Occasionally new trails have to be cut around tree falls or landslides. Regrowth of vegetation cleared during trail development can be spectacularly quick and may occur during the off-season when tourist pressure is low. Given the greater density of secondary growth, it is frequently simpler to clear new forest for use by the tourists. There thus arises a situation similar to that produced by shifting agriculture, providing a patchwork of areas of differing antiquity which can offer refuges for wildlife. In most instances the extent of land loss caused by trail creation is negligible. It is usually found that animals begin to use the new trails very soon after they have been cut and our observations suggest that this is a general phenomenon affecting herbivores and carnivores.

In Manu, whilst it has been valuable to funnel all park visitors past the guard station, the increased river traffic has undoubtedly been one of the major factors in the decline of sightings of riverine fauna including giant otters, capybara, etc. While in general the tourist pressure in the Manu Biosphere Reserve is low, cautious consideration should be given to expansion of the number of tourists allowed to visit the area in future. Any national ecotourism policy should incorporate visitor management and dispersion strategies to minimize tourist pressure on a small number of species, particularly giant otter. There is an urgent need to survey the

Table 18.4 Reaction distances for unhabituated mammals to human observers

Species	Number of observations	Mean distance (m)	SD
Cebus apella	58	15	11
Saimiri sciureus	63	10	8
Felis pardalis	9	28	33
Nasua nasua	3	45	–
Pteronura brasiliensis	18	11	9
Pteronura brasiliensis (with cubs)	18	150	56
Sciureus spadiceus	8	10	5
Tayassu tajacu	13	33	13

mammalian diversity and to change visitor biases and preferences through education with the aim of making fuller use of the facilities and spectacular wildlife that Manu has to offer.

ACKNOWLEDGEMENTS

We are grateful to the Direccion General de Forestal for permission to work in the Manu National Park. We offer our profound thanks to Gillian Hinchcliffe and Robert Strachan for their field assistance and help in developing the concepts expressed in this paper. Particular thanks are due to successive groups of students from the University of Edinbugh (Mike Daniels, Chris Fairgrieve, Andrew Lamb and Jeanette Yates) and the University of Durham (Paul Jenkins, Matthew Porter, Samantha Jackson, Nicola White, Peter Cranswick and Gillie Sargent) who assisted with these projects and for allowing us to quote their unpublished data. We are particularly grateful to Tony Luscombe of ECCO for logistical support and to Boris Gomez and the staff of Manu Lodge for ensuring our comfort in the field.

REFERENCES

Boo, E. (1990) *Ecotourism: the potentials and the pitfalls*, Vol. 1, WWF, Baltimore.

Daniels, M., Fairgrieve, C., Lamb, A. and Yates, J. (1991) Baseline survey of the primates of the Manu National Park, Peru, July–August 1989 (unpublished expedition report).

Dunstone, N. and Shoobridge, D. (in press) Ecotourism in Amazonia: expectation and realisations, in *Tourism, Ecotourism, and Protected Areas*, (ed. Ceballos-Lascurçin), IUCN.

Emmons, L. (1984) Geographic variation in densities and diversies of non-flying mammals in Amazonia. *Biotropica*, **16**, 210–222.

Emmons, L. (1988) A field study of ocelots in Peru. *Rev. Ecol. (Terre et Vie)*, **43**, 133–157.

Groom, M.J., Podolsky, R.D. and Munn, C.A. (1991) Tourism as a sustainable use of wildlife: a case study of Madre de Dios, southeastern Peru, in *Neotropical Wildlife Use and Conservation*, (eds J.G. Robinson and K.H. Redford), University of Chicago Press.

Janson, C.H. and Emmons, L (1991) The non-flying mammal community at Cocha Casu Biological station, Manu National Park, Peru, in *Four Neotropical Rainforests*, (ed. A.H. Gentry), Yale University Press.

Terborgh, J. (1983) *Five New World Primates: a study of comparative ecology*, Princeton University Press.

Terborgh, J., Fitzpatrick, J.W. and Emmons, L. (1984) Annotated check-list of birds and mammal species of Cocha Cashu Biological Station, Manu N.P. Fideliana (Zoology) 21.

APPENDIX 18.1 LIST OF NON-VOLANT MAMMAL SPECIES OBSERVED, TRAPPED OR TRACKED

Common name	Scientific name	Observation
Southern opossum	*Didelphis marsupialis*	O T
Brown four-eyed opossum	*Metachirus nudiciudatus*	O
Red mouse opossum	*Marmosa rubra*	O
Southern tamandua	*Tamandua tetradactyla*	O
Giant anteater	*Myrmecophaga tridactyla*	O
Squirrel monkey	*Saimiri sciureus*	O
Saddle-back tamarin	*Sanguinus fuscicollis*	O
Emperor tamarin	*Sanguinus imperator*	O
Night monkey	*Aotus trivirgatus*	O
Dusky titi	*Callicebus moloch*	O
Brown capuchin	*Cebus apella*	O
White-fronted capuchin	*Cebus albifrons*	O
Red howler monkey	*Alouatta seniculus*	O
Black spider monkey	*Ateles paniscus*	O
Monk saki	*Pithecia monachus*	O
Kinkajou	*Potos flavus*	O
South American coati	*Nasua nasua*	O
Southern river otter	*Lutra longicaudis*	O
Giant river otter	*Pteronura brasiliensis*	O
Ocelot	*Felis pardalis**	O T
Margay	*Felis weidi**	S
Jaguaroundi	*Felis yaguaroundi**	S
Jaguar	*Panthera onca*	O
Tayra	*Eira barbara*	S
Puma	*Felis concolor*	S
Red brocket deer	*Mazama americana*	O
Grey brocket deer	*Mazama gouazabira*	O
Capybara	*Hydrochaeris hydrochaeris*	O
Collared peccary	*Tayassu tajacu*	O
Brazilian tapir	*Tapirus terrestris*	O
Southern Amazon red squirrel	*Sciureus spadicus*	O
Sanborns squirrel	*Sciureus sanborni*	O
Bolivian squirrel	*Sciureus ignitis*	O
Amazon dwarf squirrel	*Microsciureus flaviventer*	O
Spiny rat	*Proechimys steeri*	O T
Brazilian porcupine	*Coendou prehensilis*	O
South American tree porcupine	*Coendou bicolor*	O
Paca	*Agouti paca*	O T
Agouti	*Dayprocta agouti*	O T
Forest rabbit	*Sylvilagus brasiliensis*	O T
Nine-banded armadillo	*Dasypus novemcinctus*	O

Key: O: observed; T: trapped; S: field signs.
* Species names have since been altered: see Appendix 15.A for revisions notified by CITES in early 1995.

19

Ecotourism and mountain gorillas in the Virunga Volcanoes

Alastair McNeilage

SYNOPSIS

Ecotourism has played an important role, along with education and anti-poaching patrols, in conservation projects for mountain gorillas. Small groups of tourists are taken by trained guides to visit habituated groups of gorillas. This chapter examines the available evidence on the effects of tourism on the gorillas, both advantageous and deleterious. Tourism has provided a valuable source of income for the national parks in both Rwanda and Zaire, as well as benefiting the national economies and giving the local governments a strong incentive to protect the gorilla population. In addition, the gorilla groups visited by tourists and the areas of the forest which they inhabit are monitored daily and probably receive improved protection. The levels of illegal human disturbance in areas visited and not visited by tourists are compared. Over half the population is now in groups monitored for research or tourism. However, these visits could cause disturbance and stress to the gorillas and greatly increase the risk of human diseases being introduced into the population. Census results show that the proportion of immature animals in monitored groups is higher than in other groups, suggesting that any adverse effects of human visits are outweighed by the protection received. Civil war in Rwanda has demonstrated how political stability is crucial for successful ecotourism programmes.

19.1 INTRODUCTION

The mountain gorilla (*Gorilla gorilla beringei*) exists in just two small isolated populations in the Virunga Volcanoes on the borders of Rwanda, Zaire and Uganda and in the Impenetrable forest in Uganda. The total population in the wild is around 600, approximately 300 of which live in the Virungas. The forested area of the Virungas is protected as continuous national parks in each of the three countries but is completely isolated by human habitation and cultivation.

The Virunga gorilla population was first studied in 1960 by George

Schaller, who estimated that there were 400–500 individuals at that time (Schaller, 1963). During the 1960s and 1970s the population declined dramatically to just 250 by 1981. This decline can be attributed to a combination of factors (Weber and Vedder, 1983). Some 40% of the Rwandan section was converted to agricultural use in 1968/69 as part of a European funded scheme to grow pyrethrum as a cash crop. However, the majority of the decline occurred in the region of Mount Mikeno in Zaire, where direct hunting was likely to have been prevalent during the civil war in Zaire in the mid 1960s. During the 1970s, markets developed for gorilla trophies, primarily skull and hands, and for live young. It is not clear how many gorillas were killed to satisfy this market: there were at least 13 from known groups and possibly many more (Harcourt and Fossey, 1981).

It was in response to this macabre trade and the resulting publicity that the Mountain Gorilla Project was set in up in 1979 in Rwanda by a consortium of international organizations, African Wildlife Foundation, World Wide Fund for Nature, Fauna and Flora Preservation Society and People's Trust for Endangered Species (Harcourt, 1986). The first approach of this three-pronged project was to improve park security. Guard numbers were increased and their training and equipment improved. Secondly, a conservation awareness programme was initiated, targeting all levels of the human population particularly in the area around the park. Thirdly, a controlled ecotourism programme based on gorilla viewing was started. A similar project was launched in the Zairian section of the forest in 1984 by Frankfurt Zoological Society (Aveling and Aveling, 1989).

Ecotourism can constitute a valuable form of sustainable use of natural areas and wildlife, allowing income to be generated in a non-consumptive way. However, some studies have shown adverse effects of disturbance caused by tourists on large mammals such as increased vigilance at the expense of feeding time, flight, increased range size, and the avoidance of heavily disturbed areas (Jeppsen, 1987; Humphries *et al.*, 1989; Yalden, 1990). This chapter briefly describes the ecotourism programme and reviews the positive and negative effects of ecotourism on the gorillas, the protected area and the two countries involved. Evidence from various sources on the relative importance of these effects is presented.

Tourism in Rwanda has come to an abrupt halt following the recent civil war and massacres. Most of the arguments concerning tourism cease to be relevant under such circumstances, but they are discussed here in the hope that the situation in Rwanda will soon allow conservation programmes to continue there, as well as because of their relevance in Zaire and other countries.

19.2   THE ECOTOURISM PROGRAMME

Rwanda is one of the poorest, most densely populated countries in Africa. The area around the Virungas is the most densely populated part of Rwanda with over 400 people/km^2. There is therefore enormous pressure on the land around the park and the situation is little better in Zaire and Uganda. If conservation projects are to continue indefinitely and become independent of foreign intervention in such a poor country, it is a great advantage, if not essential, for them to be financially self-sufficient. Tourism has been a valuable source of much needed foreign currency in Kenya and other African countries for some time and seemed to provide the only obvious means by which the park could become financially viable. This would not only pay for better protection, but also emphasize the importance of the park and its wildlife in the national economy to the Rwandan government and local people. Revenue would be brought into the economy by tourists staying in hotels, buying souvenirs and so on, as well as the direct income in the form of park fees.

The late Dian Fossey, in her research work at Karisoke Research Centre in Rwanda, had shown how gorillas could be successfully habituated to the presence of humans and approached to within a few metres. Several groups of gorillas have been habituated for visits by tourists on both the Rwandan and Zairian side of the forest. A system was devised on the principle that the welfare of the gorillas is always put first and the disturbance to them kept to a minimum. One small group of tourists, accompanied by trained guides, visits each group each day for a limited period and the visit is terminated if there is an adverse reaction from the gorillas. The small numbers of tourists also maintain a high quality of visit, justifying relatively high fees. The park is small and would be susceptible to damage by large numbers of visitors, and there are a limited number of gorilla groups which are suitable for habituation for tourism. Such a high-value, low visitor numbers system is therefore an appropriate strategy.

19.3   POSITIVE EFFECTS OF ECOTOURISM

19.3.1 Economics

The most obvious positive effect of gorilla tourism has been the financial benefits brought to the national parks and each country as a whole. During the first five years of the programme in Rwanda the number of tourists visiting the park annually increased more than four-fold to almost 6000. With the higher fees charged for gorilla visits, revenue increased far more in the same period – by over 3000% (d'Huart *et al.*,

1985). In Zaire, park receipts rose from zero to £23 000 per month at peak times within three years of the start of the programme (Aveling and Aveling, 1989). The increased income for the parks has the potential to facilitate greatly their effective management and to improve the protection they receive. Gorilla ecotourism thus pays for the protection of not only a habitat containing many other species of plants and animals, but also an important watershed which maintains local water supplies and the fertility of the surrounding agricultural land. In addition, the national parks departments can use the income to subsidize the protection of other parks and reserves.

If economic arguments are to be used for the conservation of natural areas, then ideally it should be demonstrated that tourism is more profitable than other forms of land use. In 1984, gorilla tourism could be calculated to earn about $200/ha, whereas cattle grazing would earn only $15/ha. A cattle grazing scheme proposed for the park by the Ministry of Agriculture in Rwanda in 1979 could thus be rejected on the basis of economics alone (Harcourt, 1986).

In the country as a whole, foreign tourists who visit Rwanda primarily to see the gorillas bring in far more income than just the price of the park fees, spending on hotels and restaurants, buying souvenirs and so on. At its peak, tourism in Rwanda was one of the largest sources of foreign income, competing with the traditional cash crops, tea and coffee.

19.3.2 Surveillance

The daily presence of park personnel in areas used for tourism can improve surveillance in two ways. Gorillas can get caught in snares set for antelope and generally break free, leaving a noose around the hand or foot. This can cause severe injuries and deaths, especially if the noose is made from wire. A veterinary centre has been established by the Morris Animal Foundation which provides medical care as needed, and successful techniques have been developed to immobilize gorillas, remove such snares and treat the wounds (Foster, 1992). In 1991 alone, nine cases were reported of gorillas in monitored groups caught in snares, of which seven required human intervention. Of these, six were successfully released while one died from severe injuries, probably inflicted by other members of the group trying to pull the individual free (Macfie, 1992). Such interventions would be difficult in unhabituated groups and rely on regular monitoring of the groups to detect animals caught in snares. Tourist guides, visiting the groups each day, are ideally placed to undertake this monitoring.

In addition, such snares are more likely to be found before a gorilla or any other animal is caught in them in areas visited regularly by park personnel. The presence of guides and guards might also be expected to act as a deterrent against poachers entering those areas of the forest in the first place. A sizeable portion of the forest (about 30%) is used for tourism or research and so monitored in this way (Figure 19.1).

The levels of human disturbance were surveyed in various areas of the parks, as part of a larger study on the ecology of the gorilla population in the Virungas. Study sites were selected on the Rwandan side of the forest: one in an area used for gorilla tourism, another around the Karisoke Research Centre and a third in an area on the south side of the park used for neither (Figure 19.1). The research centre, run by the Dian Fossey Gorilla Fund, operates its own full-time anti-poaching patrols so that the area around it is particularly well protected. Each site, covering 12–20 km^2, was divided into strips 500 m wide and a randomly positioned line transect was walked across the site within each strip. Seven to 10 transects were walked in each site. Snares were noted, whether set or old, along with the perpendicular distance of each from the transect line. The numbers of snares found in the research and tourism sites were insufficient for normal distance sampling calculation procedures (Burnham *et al.*, 1980) but plotting the number of snares detected against distance from the line indicated that snares were reliably found within 5 m either side of the line. Each transect was therefore treated as a strip 10 m wide and snares seen outside that distance were not used in calculations. The number of poachers' tracks crossed by the transect was also noted, to give two indices of the level of human disturbance in each area, the density of snares/km^2 and the number of tracks crossed/100 km of transect.

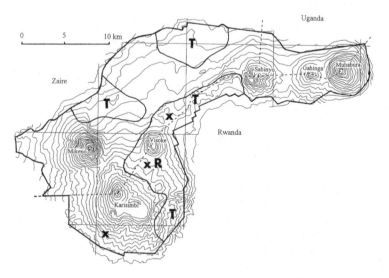

Figure 19.1 The Virunga Volcanoes: areas of forest used regularly for tourism (T) or research (R) and positions of sites (x) surveyed for signs of illegal human disturbance.

The density of snares and the number of poachers' tracks crossed were lower in both the research and tourism sites than in the less well monitored area (Figure 19.2). Using individual transects as the sampling unit, these differences were statistically significant (two-tailed Mann-Whitney U tests, $P < 0.05$ for snares and tracks in the tourism site; $P < 0.001$ and $P < 0.05$ for snares and tracks, respectively, in the research site; $n = 7$ in the tourism site and $n = 10$ in the other two sites). The density of snares appears higher in the tourism site than the research site, but most of the snares there were found in the most remote part of that site – an area visited relatively infrequently by tourists and guides. The differences in snare and track density between the research and tourism areas were not significant. These results indicate that the area used for tourism benefits considerably from the associated monitoring, such that levels of illegal disturbance are comparable with the well protected region around the research centre.

19.4 NEGATIVE EFFECTS OF ECOTOURISM

19.4.1 Disturbance to the gorillas

Habituated gorillas do not generally seem to react to the presence of humans, provided that the people are in small numbers and behave in an appropriately subdued way. However, the demand for gorilla visits has often exceeded the number of places available, putting considerable pressure on the controlled tourism system. The official quota of tourists for some groups has been increased from six to eight and illegal visits by

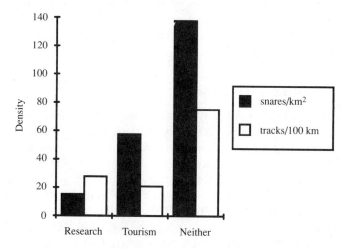

Figure 19.2 Density of snares and poachers' tracks crossed on transects in the three areas surveyed for human disturbance.

larger groups of tourists, more than once per day and to unauthorized groups have been made. The system has at times been in danger of breaking down, particularly in Zaire (Wrangham, 1992). Tour operators and tourists are often the instigators of these problems, pressurizing and bribing guides and officials to allow extra visits when the normal quota is fully booked. Much effort has been made by the local authorities and international organizations to keep the tourist system properly controlled, but clearly there is a risk that rules will be broken. It is much harder for guides to control the movement and behaviour of large groups of tourists, and considerable stress could be caused to the gorillas.

Also, repeated visits by tourists, even in an apparently controlled way, could cause slight but cumulative stressful effects on the gorillas, which might not be obvious to the human observer at the time, but which might still have a negative effect on the gorillas (d'Huart *et al.*, 1985). If stress caused by tourism has a serious negative effect on the gorillas, reproductive output of those groups would be expected to be reduced. The proportion of immatures within groups can be used as an index of reproductive health. Gorilla census techniques allow the age–sex composition to be determined from night nest counts and dung bolus size (Harcourt and Fossey, 1981; Weber and Vedder, 1983). Dung from infants of less than six months old is not generally found. Those individuals are only detected in habituated groups where composition can be determined by observation and so were excluded from the analyses described here. The proportion of immatures in groups monitored regularly because of research or tourism was found to be higher than those in non-protected groups during censuses in 1981, 1986 and 1989 (Figure 19.3). The high proportion of immatures suggests that these groups are reproductively healthy and that the benefits of added protection outweigh any negative effects of human visits.

19.4.2 Risk of disease transmission

Gorillas, being closely related to humans, might be expected to be susceptible to human diseases. Close proximity of tourists almost certainly increases the risk of disease transmission. An infection causing relatively minor illness in humans could potentially cause much more serious problems in a population without acquired immunity, such as the gorillas. It is quite difficult to measure this risk or to show conclusively that a particular disease was transmitted to the gorillas by humans. Little is known about the diseases which would be found in the population in the absence of humans, and people have entered and lived in the gorillas' habitat for hundreds of years so that many diseases have probably been shared for a long time. Gorillas in an orphanage have been

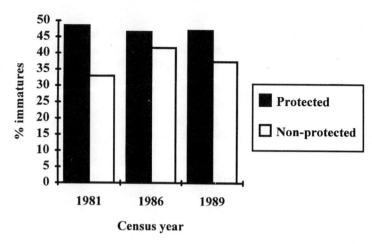

Figure 19.3 Proportions of immatures in 'protected' groups monitored regularly through tourism and research compared with non-monitored groups found in past censuses. Data for 1981 from Aveling and Harcourt (1984); for 1986 from Aveling and Aveling (1989); and for 1989 from Sholley (1991).

observed to catch colds readily from humans, particularly when stressed (Hudson, 1992). During an outbreak of respiratory illness in 1988, six gorillas died in monitored groups and in one case measles was suspected (Ramsey, 1988). In a second outbreak in one of the tourist habituated groups in 1990, one adult male died from bronchopneumonia which was unlikely to have been caught from humans, but which could have been secondary to a viral infection which was (Macfie, 1991). Mountain gorillas, living at high altitude in a cold damp climate, seem to be particularly susceptible to respiratory infections and it is unfortunate that these are perhaps the most likely diseases to be transmitted by tourists.

The fact that monitored groups have a high proportion of immatures and are at least as reproductively healthy as non-monitored groups suggests that this risk of disease transmission has not caused serious harm to these groups until now. That is not to say that a serious risk does not exist and every precaution should be taken to minimize it. A rule was established that a distance of at least 5 m should be maintained between visitors and gorillas, based on an estimate of the maximum distance that droplet-transmitted respiratory diseases might be passed, for example if a visitor sneezed. In addition tourists who are ill are not permitted to visit the gorillas, and children under 15, who are regarded as being particularly likely to carry infections, are also prohibited.

For the protection of the gorillas it is important that these precautions be carefully followed, but this has not always been the case. Gorillas, particularly curious juveniles, will sometimes approach tourists and guides are not always able, or willing, to maintain the 5 m limit. The possibility of large tips from tourists who are thrilled to have gorillas actually touch or even climb on them is presumably a strong incentive to break the rules. I have personally observed a young gorilla lick the nose of a tourist in Zaire; there could hardly be a more effective way to transmit respiratory infections.

19.4.3 Other disturbance in the park

As well as the risk of disturbance to the gorillas themselves, increased numbers of tourists could cause erosion of paths, litter in the forest and disturbance to other wildlife, including perhaps other, unhabituated groups of gorillas. With proper control of the numbers of tourists and their behaviour, it should be possible to keep all of these forms of disturbance to a minimum.

19.5 CONCLUSIONS

Ecotourism has played an important role in mountain gorilla conservation in the Virungas. Without the incentives and income generated by tourism it is doubtful whether the gorilla population would have survived, yet with improved protection the population decline seen through the 1960s and 1970s was halted. By 1989, when the most recent census was undertaken, the population had risen to an estimated 324 individuals (Sholley, 1991).

A note of caution must be sounded in this apparent success story. Tourism undoubtedly could cause stress and disturbance to the gorillas and increases the risk of transmission of human diseases into the population. With proper control of the numbers of tourists and their interactions with the gorillas these risks can be kept to a minimum and arguably acceptable level. For the most part, the system of visits has worked well, but despite the efforts of many dedicated people in the national parks departments and international organizations, the control has not always remained effective. For the welfare and protection of the gorillas, and therefore for the future of the ecotourism programme itself, it is essential that proper control be maintained.

On balance, the available evidence indicates that the potential risks of disturbance and disease transmission inherent in gorilla ecotourism have not had any serious deleterious effects to date. Both the human population of each country and the whole ecosystem of the Virungas seem to have benefited. However, there is as yet little information available to allow a full assessment of the effects of tourism on the gorillas. It is important that veterinary and demographic monitoring of the population be maintained. Ideally behavioural studies of the reactions of gorillas to tourists should be established, including observations on the effects of tourists on the movement patterns of gorilla groups and on the frequency of aggressive behaviour towards tourists, however mild.

The success of gorilla ecotourism has prompted similar projects in various other gorilla and chimpanzee populations, including the only other mountain gorilla population in the Impenetrable forest in Uganda. Many of the lessons learned in the Virungas can and should be applied elsewhere. While some problems such as the risk of disease transmission apply particularly to the great apes, other considerations might apply to many ecotourism programmes based on wildlife viewing. For instance, in many cases restricting the numbers of tourists might prevent disturbance to the wildlife and the area, while increasing the quality of the experience and so the fee that can be charged. Each case obviously presents its own circumstances and the costs and benefits of tourism always need to be carefully assessed.

Finally, the civil war in Rwanda has clearly demonstrated how vulnerable ecotourism is to political instability. At the start of hostilities in October 1990, tourism halted abruptly causing severe difficulties for the parks department who would have had to lay off many guards without emergency foreign support. Tourists did return in small numbers intermittently, but again stopped completely when renewed fighting and horrific massacres broke out in April 1994. The international media attention which these conflicts received means that most tourists will be discouraged from visiting Rwanda for some time to come. However, if the situation becomes stable, ecotourism will resume and continue to play an important role in gorilla conservation, as well as in rebuilding the Rwandan economy.

ACKNOWLEDGEMENTS

I am grateful to the governments and national parks departments of Rwanda and Zaire for permission to carry out research in the Virungas, and to Professor Stephen Harris and the Dian Fossey Gorilla Fund for support throughout my work there. The project was funded by a Study Abroad Studentship from the Leverhulme Trust. Ian Redmond provided helpful comments on the manuscript.

REFERENCES

Aveling, C. and Aveling, R. (1989) Gorilla conservation in Zaire. *Oryx*, **23**, 64–70.
Aveling, C. and Harcourt, A.H. (1984) A census of the Virunga gorillas. *Oryx*, **18**, 8–13.
Burnham, K.P., Anderson, D.R. and Laake, J.L. (1980) Estimation of density from line transect sampling of biological populations. *Wildlife Monographs*, **72**, 1–202.

Foster, J.W. (1992) Mountain gorilla conservation: a study in human values. *Journal of the American Veterinary Medical Association*, **200**, 629–633.

Harcourt, A.H. (1986) Gorilla conservation: anatomy of a campaign, in *Primates: the Road to Self-sustaining Populations*, (ed. K. Benirschke), Springer-Verlag, New York.

Harcourt, A.H. and Fossey, D. (1981) The Virunga gorillas: decline of an island population. *African Journal of Ecology*, **19**, 83–97.

d'Huart, J.P., von der Becke, J.P. and Wilson, R. (1985) *Parc National des Volcans: Plan de Gestion*, Office Rwandais de Tourism et Parc Nationaux, Kigali.

Hudson, H.R. (1992) The relationship between stress and disease in orphan gorillas and its significance for gorilla tourism. *Gorilla Conservation News*, **6**, 8–10.

Humphries, R.E., Smith, R.H. and Sibly, R.M. (1989) Effects of human disturbance on the welfare of park fallow deer. *Deer*, **7**, 458–463.

Jeppsen, J.L. (1987) Impact of human disturbance on home range movements and activity of red deer (*Cervus elaphus*) in a Danish environment. *Danish Review of Game Biology*, **13**, 1–38.

Macfie, E. (1991) The Volcano Veterinary Centre, Rwanda. *Gorilla Conservation News*, **5**, 21.

Macfie, E. (1992) The Volcano Veterinary Centre. *Gorilla Conservation News*, **6**, 22–24.

Ramsey, C. (1988) Six mountain gorillas die. *Primate Conservation*, **9**, 37.

Schaller, G.B. (1963) *The Mountain Gorilla: Ecology and Behaviour*, University of Chicago Press, Chicago and London.

Sholley, C.R. (1991) Conserving gorillas in the midst of guerillas. *Annual Conference Proceedings, American Association of Zoological Parks and Aquariums*, 30–37.

Weber, A.W. and Vedder, A. (1983) Population dynamics of the Virunga gorillas, 1959–1978. *Biological Conservation*, **26**, 341–366.

Wrangham, R.W. (1992) Letter to IZCN, Zaire. *Gorilla Conservation News*, **6**, 17–18.

Yalden, D.W. (1990) Recreational disturbance of large mammals in the Peak District. *Journal of Zoology, London*, **221**, 293–326.

20

Use, misuse and abuse of the orang utan – exploitation as a threat or the only real salvation?

Ashley Leiman and Nilofer Ghaffar

SYNOPSIS

The current plight of orang utans highlights many important welfare and conservation issues. This chapter explores human exploitation of the orang utan, covering past mis-use, present abuse and the future potential use of these animals. In particular, it concentrates on the situation today and confronts some of the questions, and controversies, posed by our continuing exploitation of the orang utan.

The earliest evidence of orang utan use by humans was as a food source – charred remains of these animals have been found dating as far back as 35 000 years ago. Modern exploitation is of a qualitatively different nature. This includes the keeping of up to 600 individuals as 'pets' in Taiwan; use in the television and film industries; use as advertising gimmicks to increase the sales of car tyres, tea, and fizzy drinks; and use as a symbol to promote tourism and as a valuable commodity to be visited by over 200 000 people a year. Future exploitation in the form of encouraging high-paying eco-tourists to visit the tropical forests of Malaysia and Indonesia may be the only way to ensure the survival of this magnificent red ape.

20.1 INTRODUCTION AND HISTORICAL PERSPECTIVES

20.1.1 Introduction

This chapter has three distinct parts. It begins by briefly documenting the manner in which humans have used the orang utan in the past, high-lighting why in this particular case it constitutes something of a misuse. Then it covers selected aspects of the present exploitation of this highly vulnerable ape, much of which may more reasonably be described as abuse. Finally, it considers ways in which future exploitation might be to the advantage of the conservation and welfare of the orang utan.

The aims in this chapter are relatively modest. They are to summarize what seem to be some of the more significant aspects of human exploitation of the orang utan. By its very nature, this treatment is both selective and subjective, but there is a need to draw greater attention to a number of points raised by the present situation and this is simply what is attempted here.

20.1.2 General ecology

Orang utans are the largest extant primates found in Asia where, currently, two subspecies are recognized: *Pongo pygmaeus pygmaeus* in Borneo, and *P. p. abelii* in Sumatra (Courtenay *et al.*, 1988; but see Röhrer-Ertl, 1988). These predominantly arboreal, frugivorous and highly sexually dimorphic apes have had a long and chequered history of exploitation, which appears to have originated in prehistoric times and continues right through to the present.

20.1.3 Prehistorical range and exploitation

Fossil evidence suggests that, during the Pleistocene, orang utan distribution extended from Java in the south, across mainland Asia, and reached up as far north as China (von Koenigswald, 1982). In more recent times populations have been restricted to pockets of forest on the islands of Borneo and Sumatra (for example, see Rijksen, 1978).

Fossils found at the Niah Caves in Sarawak show that anatomically modern humans were present in South-East Asia some 40 000 years ago, and also charred remains of the orang utan have been found and dated at about 35 000 years (von Koenigswald, 1982). At Niah and in the Padang Highlands of West Sumatra, orang utans comprise a large proportion of the remains found, suggesting that they were extensively used as a food resource. Additionally, it has been suggested that at least some individuals were kept as pets by these early cave dwellers (Harrison, 1962).

20.2   PRESENT DISTRIBUTION AND SITUATION

20.2.1 Causes of former range contraction

The present geographical range of these animals may best be described as something of a 'relic' pattern, and clearly represents just a fragment of their former extensive distribution area (Rodman, 1988). For example, even within a single area such as Borneo, the distribution is highly patchy and the animals are absent from large areas (Payne, 1986; Mackinnon, 1990).

Humans seem to have been largely responsible for this dramatic decline and range contraction (Sugardjito and van Schaik, 1991). Early use of orang utans as a source of meat is not altogether surprising: what is significant in this case is the unique vulnerability of these animals to such exploitation. Much of this may be attributed to extremely long interbirth intervals, typically in the region of eight years, making orang utans the slowest breeding species within the primate order (Galdikas and Wood, 1990). Traditional hunting by indigenous people has almost definitely been responsible for a number of local extinctions (Mackinnon, 1971); it has even been suggested that this may have replaced the practice of human head-hunting and cannibalism in the region (Rijksen, 1978).

Additionally, a variety of other factors have been proposed to explain the current dispersion of orang utans: for example, the distribution of minerals and the availability of suitable foods have often been suggested (e.g. Payne, 1986; Djojosudharmo and van Schaik, 1992). It is likely to be a combination of such influences that has led to the present fragmented range.

20.2.2 Current population status

Considering the current status within these restricted areas, the best estimates for population sizes suggest maximum totals of 5000 and 22 000 animals in Malaysia and Indonesia, respectively (Sugardjito and van Schaik, 1991). Table 20.1 gives the minimum and maximum figures available, together with the land areas of the various regions concerned (Payne and Andau, 1989).

20.2.3 Present-day direct threats

The most obvious threat now facing orang utan populations is the loss of habitat due to logging operations. In Kalimantan and Sumatra, human encroachment in the form of transmigration, which then leads to shifting agriculture, combined with a major logging industry has undoubtedly had the greatest impact (Collins *et al.*, 1991). In Sabah and Sarawak, logging is mainly selective and some secondary forest is eventually converted to plantation (Marsh and Greer, 1992). The ultimate conversion of primary forested habitats for agricultural use results in a permanent and irretrievable loss of habitat. However, even the practice of selective logging is thought to reduce orang utan densities from a maximum of three individuals/km^2 in primary forest to one individual/km^2 in logged/secondary forest (van Schaik and Azwar, 1991).

The direct threat posed by logging, coupled with the low reproductive rates, is a real cause for concern. Laws against the owning, killing,

Table 20.1 Current population estimates of orang utans

Region (country)	Estimated numbers	Land area (km²)
Sabah and Sarawak (Malaysia)	3000–5000	197 605
Kalimantan (Indonesia)	12 000–15 000	539 460
Sumatra (Indonesia)	5000–7000	473 606

capture or harming of orang utans were passed in Indonesia in 1931, in Sabah in 1963 and in Sarawak in 1958 (Payne and Andau, 1989). In spite of this formal protection, the habitat destruction leaves the orang utan particularly vulnerable to further exploitation.

20.3   CURRENT EXPLOITATION

Nowadays the exploitation of orang utans has taken on a qualitatively different nature: there are currently four major ways in which human beings continue to use orang utans. Of these, two – the pet trade (largely restricted to Taiwan) and the media industry – may more accurately be termed as an abuse of this species. The third, a recent resumption of hunting, is perhaps most appropriately considered as further misuse of these animals. The fourth, rehabilitation, originally seen as a means of law enforcement to prevent private ownership, now uses the orang utan as a commodity to be exploited for the purposes of ecotourism, education and local awareness. The following section highlights some of the more significant issues.

20.3.1 Pet trade in Taiwan

An archetypal example of exploitation is, without doubt, provided by the pet trade in Taiwan. During the period 1985 to 1990, political, economic and social factors combined to allow the smuggling of up to 1000 orang utans (Phipps, 1993). These individuals were poached from Kalimantan and taken into the Republic of China on Taiwan for sale as exotic pets. The trade continued unrecognized until the Orangutan Foundation raised the issue internationally. Although enforcement of Taiwan's 1989 Wildlife Conservation Law has now almost halted the trade in rare primates, the problem remains of what to do with the animals already on the island.

Estimates vary but for each orang utan reaching Taiwan as many as three to five individuals are thought to have died in the process (Cater, 1991). This represents a potential loss of more than 3000 animals. Even at its most conservative, such a figure suggests at least a 10% decline in the existing wild population. During 1990/91 the Orangutan Foundation

Taiwan conducted a detailed survey to monitor various aspects of the Taiwanese situation (Lee *et al.*, 1991). Some of the more significant findings of this study are reported below.

(a) The motivation for keeping orang utans

The demand was attributed to a popular 1986 television programme, 'The Naughty Family', which featured an orang utan named Hsiao Li, portrayed as an ideal companion and pet. Figure 20.1 documents the reasons people gave for keeping the animals as pets; the stated motivation of the overwhelming majority (35%) was that people liked orang utans (n = 48 owners). The original orang utan on television was replaced by a toy one at the beginning of 1990 but, as Figure 20.2 shows, the majority of acquisitions had already been made between 1987 and 1989 (n = 75 orang utans).

Taiwan, not being a member of CITES, passed their own Wildlife Conservation Law, which became effective in June 1990. As a result, owners were asked to register their pet orang utans. The total numbers in Taiwan are difficult to estimate: 283 were registered by November 1990 but, based on their enquiries, the Orangutan Foundation Taiwan believes that only one in three owners registered.

(b) The source of orang utans from within Taiwan

Of 85 individual animals whose sources were identified, 80% were obtained from newspaper advertisements, pet shops and weekend markets; the remaining 20% were 'given' to their then current owners. Today, any visitor to Taiwan may still easily find orang utans on show in the infamous Snake Alley.

(c) The major problems

A principal concern is that of illness: orang utans are generally susceptible to many human diseases and infections (Kaplan and Rogers, 1994). During two health checks conducted by the Orangutan Foundation Taiwan, 11 out of 30 animals examined had a positive reaction to hepatitis B; four tested positive to tuberculosis, and most had parasites. By law, all domestic animals are to be destroyed if they contract tuberculosis, but this regulation does not apply to protected wildlife species.

The second major (and growing) problem is what happens when individuals grow older. In the 1990/91 survey, it was found that some 64% of pet orang utans were aged between one and five years. These animals are now approaching sexual maturity, and this inevitably leads to serious

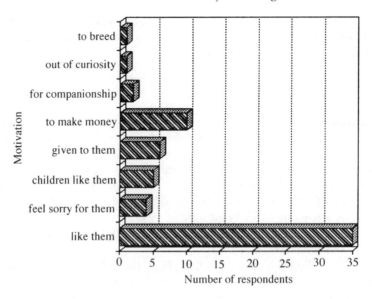

Figure 20.1 The pet trade in Taiwan: motivations for keeping orang utans.

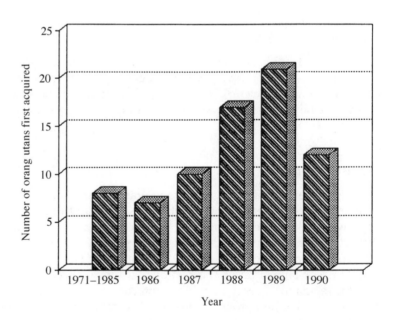

Figure 20.2 The pet trade in Taiwan: acquisitions of orang utans (1971–1990).

problems such as aggression and escape which often result in abandonment by the owners (Orangutan Foundation Taiwan, personal communication).

(d) The main solutions

Largely, the situation has been perceived as Taiwan's problem – governments, zoos, the Conservation (formerly Captive) Breeding Specialist Group and even animal welfare organizations have been reluctant to acknowledge this issue, let alone confront it. At present, the only solution has been repatriation to Wanariset, a reintroduction centre in East Kalimantan. To date three groups of orang utans have been sent back:

- The infamous 'Taiwan Ten' that were kept in Jakarta from November 1990 to October 1991. Seven of these were then sent to Wanariset and three to a medical research centre in Bogor; eventually two of these were also returned to Wanariset.
- Another 10 individuals were dispatched to Wanariset in January 1994.
- The most recent repatriation to Wanariset has been of 12 animals in September 1994. Of these, five had been abandoned, five were willingly given up and the cost of the repatriation was borne by the owners, and two were confiscated by police in a drug raid (Orangutan Foundation Taiwan, personal communication).

During the 1990/91 survey, 82% of owners claimed they would keep their animals when they grew older. The above figures contradict such an assertion. The fact that animals are being abandoned or voluntarily relinquished appears to be directly related to the problem of maturing animals.

An additional solution has been the establishment of a shelter at Pintung in southern Taiwan, attached to a polytechnic with a veterinary school. The original idea was that this would be upgraded to a permanent shelter for confiscated wildlife but so far this objective has not been realized as there is no adequate caging or husbandry expertise (Orangutan Foundation Taiwan, 1991).

20.3.2 Entertainment and advertising

The use of orang utans in the entertainment and advertising industries is another candidate for the category of classic exploitation. Here there appears to be something of a sudden upsurge in the numbers of orang utans involved; indeed, it would be fair to say that these are quite the 'flavour of the month' in this context. Some notable examples of products of recent advertising campaigns include a particular brand of tea (more commonly associated with their use of chimpanzees), a tyre company and two brands of orange drinks. With respect to the film industry, orang utans have featured in the world of Hollywood as stars of several films.

All of these orang utans, with the exception of the tea brand example, are based in the United States of America where the advertisements and films have been shot; with the more recent films, a disclaimer was run at the end specifically for broadcasting in the Asia–Pacific region, stating: 'Orang Utans are endangered and protected under International Law. Do not buy them' (Orangutan Foundation International, 1994).

Currently, there are approximately 62 privately owned orang utans in the United States. Most of them are hybrids of the two subspecies; the vast majority of these are or have been used in show business. This includes such diverse roles as cabaret artists, tourist attractions at Universal Studios, and as part of the entourage of travelling circuses. As a comparison, there are approximately 200 chimpanzees in private ownership, of which 25% are used for entertainment compared with some 60% of orang utans (P. Regan, personal communication).

20.3.3 Recent hunting

There have been some recent reports of a resumption of a form of exploitation that most people have long dismissed as a direct threat. Two independent film-makers (M. Linfield and S. Cunliffe, personal communications) visiting Kalimantan have reported an increase in local hunting pressure. This appears to be a result of recent fires, the increased intensity and frequency of which (compared with previous years) is likely to be due to a combination of factors. A particularly severe dry season together with shifting cultivation, the creation of plantations and a general opening up of the forest canopy following logging operations are all thought to be responsible. The presence of dry debris makes it possible for fires to sweep through, even reaching areas of primary forest, including peat swamps, that are not normally susceptible to such damage.

For example, within the area of West Kalimantan, particularly high densities of orang utans have been seen compressed into 50 m wide strips of riverine forest; satellite imagery has identified some 2000 sources of the fire in this area (M. Linfield, personal communication). The scale of this problem has led to indigenous people suffering a severe food shortage as they have lost their normal source of meat; furthermore, these people do not have cultural taboos against eating orang utans. The large size and slow movements of orang utans make them easy targets for modern hunters and has led to renewed exploitation reminiscent of that by early humans in this area. The female animals are hunted most often; inevitably these are accompanied by dependent young. Where families can afford to feed an extra mouth, these infants are often retained as pets. During August to September 1994, over 20 animals, including a six-week-old infant with its umbilical cord still attached, were confiscated in a period of 10 days. There has been a concomitant increase in the number of orang utan skulls now on display in the local towns – in one shop six skulls, both old and new, were available for sale (S. Cunliffe, personal communication).

20.3.4 Rehabilitation

Rehabilitation is included here both as a solution to the problems posed by the previously discussed 'use' of orang utans and as yet another form of exploitation of the species. The first rehabilitation centre was set up in 1961 (Harrison, 1962); nearly 35 years on, we are still faced with the problem of what to do with confiscated orang utans.

Currently, there are five main rehabilitation centres: one each in Sabah and Sarawak (the Malaysian States of Borneo) and three in Indonesia, of which one is in Sumatra and two in Kalimantan (Lardeux-Gilloux, 1994). The very fact that rehabilitation has been attempted in all parts of the orang utan's existing range emphasizes the important role it has assumed as a part of conservation strategies. Due to their solitary nature, and therefore lack of territorial disputes and problems of being accepted by a group (problems that are likely to apply to other great ape species), orang utans have been considered prime candidates for re-introduction to the wild. To date, according to communications from personnel at the various centres involved, at least 500 individual orang utans have passed through some sort of rehabilitation process.

A number of issues are raised by the question of rehabilitation. Originally seen as a means of law enforcement, and subsequently as a form of conservation, the centres are now considered absolutely vital in terms of dealing with the welfare of the ever increasing problem of confiscated animals (Sugardjito and van Schaik, 1991). However, in at least some of the existing rehabilitation centres, orang utans may most accurately be viewed as a commodity being exploited for the purposes of ecotourism. In the final analysis, orang utan rehabilitation has to be seen as a symptom of our failure to get to the underlying cause of the problem (Galdikas, 1991).

This notwithstanding, there are a number of possible benefits to be reaped from the process. The centres:

- enforce laws;
- remove orang utans from illegal trade;
- fulfil welfare considerations with a potential improvement in the quality of life for ex-captive orang utans;

- protect and save particular areas of forest (for example, the Centre Sepilok is located within one of the last remaining areas of primary forest in Sabah and, because of the orang utan releases at Wanariset, the Indonesian government may be persuaded to protect the surrounding forest);
- lead to educational benefits and raise conservation awareness, particularly among local people;
- are a source of revenue from visitors, potentially good for local business;
- provide easy access to orang utans for behavioural research purposes.

On the other hand, there are a number of obvious disadvantages associated with rehabilitation centres:

- they provide governments with a good image which deflects from the real threats of habitat destruction, hunting and poaching;
- they distract attention from the need to conserve more protected areas, and in most cases there is little or no follow-up of the released animals;
- there is a risk of transmitting diseases to the wild population and a danger of infection during quarantine (most confiscated orang utans have diseases due to their previous contact with humans).

20.4   OVERVIEW OF THE SITUATION

There is little doubt that past exploitation in the form of hunting pressure has shaped the destiny of the orang utan – it is particularly disturbing that this is now being resumed, as this species is unlikely to sustain such exploitation for long. More integrated conservation strategies and better management practices are clearly needed to prevent such misuse of orang utans from carrying on into the future.

With respect to the pet trade, the present predicament is perhaps best summed up by Phipps (1993):

It is a mistake common to the whole of mankind, thinking that we can meddle in things like the natural order, things that we don't completely understand, and then, if it doesn't work out, we can put it right. Well, who is going to put it right for the countless misplaced orang utans?

Few, if any, other species listed as CITES Appendix I are abused precisely for their human qualities. More significant is the problem of what the future holds for those orang utans still in Taiwan. Further repatriation to Indonesia does not seem like a feasible option. Rehabilitation centres and their surrounding areas are close to saturation point. In the case of severely diseased orang utans, there is probably little hope of them ever being released. We are then forced to ask ourselves the difficult question of what to do with these individuals. The frightening answer that confronts us is that, with such a diminished quality of life, euthanasia might be a fairer alternative. Although seemingly unacceptable, perhaps this is what it will take to galvanize the world and to force local governments and the international community to recognize the problem.

In light of the other abuse of these animals – using them as show business stars – this is the primary channel by which most people seem to recognize orang utans. At least Clyde, famous through Clint Eastwood films, raises public awareness that orang utans do exist. Surely we should be capitalizing on this and tying it into conservation efforts?

As to the process of rehabilitation, the overriding questions are: 'Does it work?' and 'How do we measure its success?' Ultimately, this remains the only humane, working solution to the problem of confiscated orang utans. It is suggested here that rehabilitation programmes are best regarded as yet another form of exploitation. This then represents a positive use of orang utans that needs developing further. Ecotourism, whilst not a viable prospect for wild populations, could be expanded to enable visitors to view semi-wild, rehabilitated individuals in a natural environment. As the orang utan has already been used as a national mascot to promote tourism for Malaysia, it would seem logical to move on and emphasize its role as a flagship species for conservation.

The challenge facing us is to come full-circle and use the public recognition of the orang utan through the media industry to highlight the need to save them in the wild. The contention put forward here is that we should use the ecotourism potential afforded by rehabilitation centres (which are necessary anyway) to help to achieve this end.

Not surprisingly perhaps, orang utans have long held a peculiar fascination for human beings – even the earliest European explorers to the region could not help but make mention of their close affinity to us. For example, Captain Beekman writes in 1714:

> The Monkeys, Apes, and Baboons are of many different Sorts and Shapes; but the most remarkable are those they call Oran-ootans, which in their Language signifies Men of the Woods: these grow up to be six foot high; they walk upright, have longer arms than men, tolerable good faces (handsomer I am sure than some Hottentots I have seen), large Teeth, no Tails nor Hair, but on those Parts where it grows on humane Bodies; they are very nimble footed and mighty strong; they throw great stones, sticks, and Billets at those Persons that offend them. The Natives do really believe that these were formerly Men, but Meta-morphosed into Beasts for their Blasphemy.

It would be tragic if this very humanness were to be responsible for their downfall.

ACKNOWLEDGEMENTS

The authors would like to acknowledge the help of Sarah Cunliffe, Michael Leach, Mark Linfield, the Orangutan Foundation International, Patti Regan and Ian Singleton. They also wish to thank Vicky Taylor for editorial assistance, and an anonymous referee for comments on an earlier draft of the manuscript.

REFERENCES

Beekman, D. (1714) *Voyage to and from the Islands of Borneo*, Republished (1973), Dawsons of Pall Mall, Folkestone and London.
Cater, B. (1991) The case of the bartered babies. *BBC Wildlife*, 9(4), 254–260.
Collins, N.M., Sayer, J.A. and Whitmore, T.C. (1991) *The Conservation Atlas of Tropical Forests: Asia and the Pacific (IUCN)*, Macmillan Press, London and Basingstoke.
Courtenay, J., Groves, C. and Andrews, P. (1988) Inter- or intra-island variation? An assessment of the differences between Bornean and Sumatran orang utans, in *Orangutan Biology*, (ed. J. Schwartz), Oxford University Press, Oxford, pp. 20–29.
Djojosudharmo, S. and van Schaik, C.P. (1992) Why are orang utans so rare in the highlands? Altitudinal changes in a Sumatran forest. *Tropical Biodiversity*, 1 (1), 11–22.
Galdikas, B.M.F. (1991) Protection of wild orang utans and habitat in Kalimantan *vis à vis* rehabilitation, in *The Great Apes Conference Proceedings, December 15–22 1991, Bohorok, Jakarta, Pagkalan Bun and Tanjung Puting, Indonesia*, Ministry of Forestry and Ministry of Tourism, Post and Telecommunication, Republic of Indonesia.
Galdikas, B.M.F. and Wood, J.W. (1990) Birth spacing patterns in humans and apes. *Am. J. Phys. Anthrop.*, 83, 185–191.
Harrison, B. (1962) *Orang-Utan*, Oxford University Press, Oxford.
Kaplan, G. and Rogers, L. (1994) *Orang Utans in Borneo*, University of New England Press, New South Wales, Australia.
von Koenigswald, G.H.R. (1982) Distribution and evolution of the orang utan, *Pongo pygmaeus* (Hoppius), in *The Orangutan. Its Biology and Conservation*, (ed. L.E.M. de Boer), D.W. Junk, The Hague, pp. 1–15.
Lardeux-Gilloux, I. (1994) Rehabilitation Centres: their struggle and their future. Paper presented at International Conference on Orang-utans: The Neglected Ape, 5–7 March 1994, Fullerton. California State University, Fullerton.
Lee, L.L., Phipps, M. and Chen, P.C. (1991) The Orang Utan in Taiwan, in *The Great Apes Conference Proceedings, December 15–22 1991, Bohorok, Jakarta, Pagkalan Bun and Tanjung Puting, Indonesia*, Ministry of Forestry and Ministry of Tourism, Post and Telecommunication, Republic of Indonesia.
Mackinnon, J.R. (1971) The Orang utan in Sabah today. *Oryx*, 11(2–3), 141–191.
Mackinnon, J. (1990) Species survival plan for orang utan in Borneo, in *Proceedings of the International Conference on Forest Biology and Conservation in Borneo*, (eds G. Ismail, M. Mohamed and S. Omar), Yayasan Sabah, Kota Kinabalu, Sabah, pp. 209–219.
Marsh, C.W. and Greer, A.G. (1992) Forest land-use in Sabah, Malaysia: an introduction to Danum Valley. *Phil. Trans. R. Soc. Lond. B*, 335, 331–339.

Orangutan Foundation International (1994) Universal promises to add disclaimer. *Pongo Quest*, **6**(1), 4.

Orangutan Foundation Taiwan (1991) *The Orang Utan Issue in Taiwan: Points for Consideration*. Unpublished Management Plan.

Payne, J. (1986) *Orang Utan Conservation in Sabah*, WWF Malaysia Project No. 96/86, WWF Malaysia, Kuala Lumpur.

Payne, J. and Andau, M. (1989) *Orang-Utan: Malaysia's mascot*, Berita Publishing Sdn, Bhd., Kuala Lumpur.

Phipps, M. (1993) *Report from the Orang Utan Foundation Taiwan*. Unpublished report.

Rijksen, H.D. (1978) *A Field Study on Sumatran Orangutans* (Pongo pygmaeus abelii *Lesson 1827): Ecology, Behaviour and Conservation*, H. Veenman and Zonen B.V., Wageningen, The Netherlands.

Rodman, P.S. (1988) Diversity and consistency in ecology and behaviour, in *Orangutan Biology*, (ed. J. Schwartz), Oxford University Press, Oxford, pp. 31–51.

Röhrer-Ertl, O. (1988) Research history, nomenclature and taxonomy of the orang utan, in *Orangutan Biology*, (ed. J. Schwartz), Oxford University Press, Oxford, pp. 7–18.

van Schaik, C.P. and Azwar (1991) *Orang utan densities in different forest types in the Gunung Leuser National Park (Sumatra), as determined by nest counts*. Report to L.S.B. Leakey Foundation.

Sugardjito, J. and van Schaik, C.P. (1991) Orangutans: current population status, threats, and conservation measures, in *The Great Apes Conference Proceedings, December 15–22 1991, Bohorok, Jakarta, Pagkalan Bun and Tanjung Puting, Indonesia*, Ministry of Forestry and Ministry of Tourism, Post and Telecommunication, Republic of Indonesia.

21

Elephant family values

Ian Redmond

SYNOPSIS

Elephants present a special set of problems when the question of exploitation arises. These fall loosely into the areas of ecology, economics and ethics. By drawing comparisons with the rules that different cultures follow for exploiting other species, this chapter looks at the different values attached to a family of elephants, alive or dead. It addresses the questions of whether the international trade in ivory and other elephant products should be reopened; whether elephant-viewing tourism is reaching its full potential; and whether those who decide on the fate of the elephant are taking into account the economic benefits of other species that are ecologically dependent on elephants. The debate is set against the wider picture of the decline of the Proboscidea and the concomitant rise of *Homo sapiens*.

The conclusions reached are that elephants are undervalued for their role as keystone species in African and Asian ecosystems and underused as tourist attractions, and that, wherever possible, they should be allowed to dwell in naturally regulated numbers in those ecosystems which have evolved to depend on their presence.

21.1   INTRODUCTION

Consider the consequences of successful ape conservation. If the example set by the Mountain Gorilla Project in Rwanda during the 1980s is widely followed (Chapter 19), apes will survive in well protected national parks, with revenues from controlled tourism paying the conservation bills, improving the quality of life for neighbouring communities and enriching the national treasury of the range state. This was the optimistic scenario which the civil war in Rwanda shattered, but if a lasting peace returns to the region, it could still be the pattern for the Virunga Volcanoes conservation area. If poaching were brought under control, however, the population of large herbivores would rise. There are few, if any, leopards and hyaenas left in the Virungas to curb this growth, so resources are likely to be the limiting factor. Eventually the time would come when there were too many mountain gorillas for the available land set aside for them, and the problem of overpopulation would then need to be addressed.

Advocates of the principles of sustainable utilization would presumably seek to introduce a system of culling and commercial exploitation along the following lines:

> Each year a proportion of the population equal to or less than the annual recruitment would be harvested. Adult apes would be shot (humanely, of course) and their meat and skins marketed; by-products might include novelty items made of ape leather, tourist curios or anatomical specimens made from hands and skulls, and the sale of baby apes to zoos, research establishments, the entertainment industry and for re-stocking former ranges. Whole family groups or, in the case of chimpanzees, communities, would be culled together, so as not to cause unnecessary distress by leaving any bereaved survivors; sport hunters would pay handsomely for the opportunity to display their courage by killing a silverback; and all the money raised would go towards the conservation of the species and its habitat.

Such a scheme is unlikely to receive much support from the international conservation community or the public at large, any more than a scheme to cull the inhabitants of overcrowded cities. Yet a virtually identical scheme is considered appropriate for elephants in a few southern African countries. Read the above paragraph substituting 'elephant' for 'ape' and the level of outrage subsides considerably in many people's minds. This chapter seeks to address some of the issues arising from this difference in attitude towards apes and elephants, and asks which animals it is acceptable to kill for commercial gain and why.

Perhaps the majority of people would feel uncomfortable eating chimpanzee steaks or wearing gorilla leather because of our evolutionary relationship with the great apes. When you share 98.4% of your DNA with a chimp, 97.7% with a gorilla, or even 96.4% with an orang utan (Sibley and Ahlquist, 1990, cited in Diamond, 1991) do you really feel comfortable cutting up and eating its body? For most people in most cultures the answer is an emphatic no, but it would be a resounding yes if you happen to have grown up in a Baka, Fang or Dayak family, respectively. In these and similar tribal communities, where ape meat is a normal part of the diet, no stigma is attached to its consumption, and indeed it may even be an item of choice or special significance (e.g. Sabater Pi, 1980; Galdikas, 1995). The belief that some of the strength or cunning of an ape can be attained by eating its flesh is not uncommon among those who regard ape meat as a delicacy. And in protein-hungry forest villages, where the alternative to ape meat might simply be no meat, who can blame parents for wanting to feed their children? (see Redmond, 1995).

Are those who object to ape meat just sentimental westerners who have lost all contact with the natural process of predation? No. As well as western objectors, there are many African tribes, perhaps even a majority, with similar qualms about eating apes. In 1989, for example, while investigating the illegal trade in gorilla and chimpanzee meat, body-parts and babies in Congo–Brazzaville, I asked many people in villages and on public transport, in a non-judgemental way, how they felt about eating ape meat. Some were enthusiastic about the merits of eating any ape; some would eat gorillas, but not chimpanzees because the latter are too human-like; others would eat neither for similar reasons, despite their daily consumption of other kinds of 'bush-meat' (Redmond, 1989). Clearly, the question of where the line is drawn between acceptable and unacceptable is a complex mix of cultural influences, religious doctrine and personal experience, with no easy distinction along geographical lines.

What happens if we substitute elephant for gorilla or chimpanzee in this discussion? To decide whether or not to eat an elephant steak is less of a problem for reasons of taxonomy and terminology. Eating an ape's leg smacks of cannibalism; eating the leg of a large quadruped that is commonly referred to by using cattle terms seems only a slight step away from the Sunday joint of roast beef. The inappropriate use of cattle terminology is an important factor in explaining our inconsistent attitude to ape and elephant meat, at least in the minds of Anglophones. By referring to elephants as bulls, cows and calves, we demean them. With a brain four times the size of ours, a complex society and the ability to plan, use tools, and communicate over several kilometres, elephants are much more than overgrown cattle. Big-game hunters used to refer to adult male gorillas as 'bull gorillas'; perhaps it originally served to exaggerate both the dangers of the hunt and the differences between hunter and hunted. Since the nature of gorilla family life has been revealed by Dian Fossey and her colleagues, however, 'bull gorilla' seems a disrespectful anachronism. Similarly, our growing knowledge of elephant society may lead to the abandoning of cattle terms in favour of something which more accurately reflects the nature of the animal described.

Elephants present a special set of problems when the question of exploitation arises. These fall loosely into three interrelated areas: economics, ecology and ethics. But the value placed on a family of elephants in each of these areas differs greatly between cultures, both human and elephant, and according to whether the elephant is alive or dead. Different cultures follow different rules for exploiting other species, and these rules may be changed by the whim of fashion or a swing in public opinion.

During the successful campaigns to ban the international ivory trade, the value of elephants was often portrayed in the media as a simple dichotomy: tourism or ivory. However, even with these two apparently straightforward values there are complications. Tourism is generally perceived as benign and 'non-consumptive', and the ivory trade as being destructive to elephants. But uncontrolled or badly planned tourism can harass elephants, destroy elephant habitat and lead to conflicts with local people; conversely, ivory can be gathered benignly from carcasses resulting from natural mortality. Clearly it is not the bare concept of elephant-based tourism or trading in ivory which is good or bad; it is the way these concepts are put into practice. The ban on international commercial ivory trading was only necessary because all attempts to control that trade during the 1970s and 1980s had failed, and as a result the continental estimate of African elephant numbers had declined from 1.3 million in 1979 to 609 000 in 1989 (ITRG, 1989). The ban was the last resort, but its widely hailed success in halting the African elephant population crash was the result of its combined effect with two other interrelated factors: a change in consumer habits due to 'elefriendly' campaigns to dissuade the public from buying ivory, and improvements in anti-poaching and law-enforcement efforts in range states, entrepôts and consumer countries (also influenced by public pressure and the ban).

One unfortunate side-effect of the campaign to ban the ivory trade was that it focused the attention of most people on the cash value of ivory to the exclusion of all other elephant values, except for the alternative cash value of tourism. What, then, are these other, unsung elephant values? Depending on your point of view, a dead elephant can be many things:

- To an economist, it can yield commercial products of low, medium and high value (meat, hides and ivory, respectively).
- To an ecologist, it can be seen purely as biomass – five or six tons of nutrients and minerals to be recycled into the ecosystem that produced it.
- To an elephant (and any human whose ethics include respect for non-human societies) it can be seen as a dead family member, to be mourned, visited and (if human interpretations of the behaviour are correct) to be remembered by the rest of its family and wider social circle.

A live elephant has the potential for all the above values when it expires, but its value while it walks the earth is immeasurably greater. Moreover, the above-mentioned values placed on dead elephants are greater if the animals are allowed to reach old age. This is because of the unusual growth curve which elephants display. Instead of growing rapidly during their youth and then remaining the same size, as do most mammals, elephants grow rapidly during their youth, but then continue to grow slowly all their lives. This is why older elephants tower above young adults, although the rate of growth is very slight in the last decade or two (Laws, 1966).

Let us examine the range of values attached to elephants in the light of the above observations.

21.2 IVORY

Such was the high price and high demand for ivory during the 1980s that the market value of African ivory to Asian carvers was estimated to be US$50 million per annum, of which only US$10–20 million remained in Africa. Most of the profits accrued to the traders and carvers (ITRG, 1989). Good quality ivory was fetching up to US$200/kg or more, which meant that a big tusker with, say, 50 kg of ivory on either tusk could be valued at US$10 000 for his ivory alone. The fact that tusk growth in male elephants increases exponentially year by year, however, means that this same elephant's tusks will be worth more every year it is allowed to live. Not only is the absolute weight of ivory greater, but also the value per kilogram is greater for large tusks than for small ones, because they offer more opportunities for the skilled craftsman. Computer modelling of elephant management strategies shows that the best way to manage a population of elephants for ivory is to let them die of old age. Any offtake above natural mortality eventually leads to extinction (Pilgram and Western, 1984). Culling whole family herds is an uneconomic method of generating revenue from ivory, yielding as it does a range of tusk size from babies to matriarchs.

The increasing value of tusks as elephants grow old has led to suggestions that a trade in 'ivory futures' could be used to raise money for conservation today, by offering investors the chance to invest in future ivory sales (J. Beddington, in Tudge, 1991). The practical drawback to this is that the cost of protecting this elephant from poachers (estimated at US$215/km^2 per year at 1981 levels: Leader-Williams, 1993) for an elephant lifetime of up to 60 years far outweighs any likely value of its tusks. Experience over the few years the ban has been in effect has shown that whenever a proposal is made to CITES to re-open the trade, there is an increase in speculative poaching. The consequent costs to elephant range states are measured in both the financial burden of increased anti-poaching patrols and the loss of life in shoot-outs between armed poachers and rangers (Currey and Moore, 1994). There are persuasive economic arguments for re-opening the legal trade in ivory (Barbier *et al.*, 1990) but they underestimate the skills and resources of those who control the illegal trade (in some cases, the same individuals who controlled the former legal trade). Logical though an ivory trade based on natural mortality might seem, in practice, if elephant front teeth are viewed as white gold, natural mortality will seldom occur. Mines can be fenced and patrolled until their treasure is in a bank vault, but if ivory again becomes a 'treasure', elephants will again become a walking invitation to organized crime. The criminal networks that made millionaires of certain 'ivory barons' have not gone away. Any move to re-open a legal ivory trade while the international smuggling and poaching syndicates still exist would result in a return to uncontrolled elephant slaughter in all but the most intensively protected areas.

21.3 MEAT AND SKINS

The culling of elephants to control their numbers, as practised in Kruger National Park, South Africa (as well as in Zimbabwe and Botswana), yields a valuable quantity of meat and skins as well as ivory. Effectively, it becomes an annual harvest (even if the argument for the cull is based on ecological grounds) and is often held up as an example of sustainable utilization. Whole family groups are killed on humane grounds, so as not to leave distraught survivors who have lost their social group (although their infrasonic distress calls are probably detected by members of the family's wider clan within a radius of several kilometres). This results in the slaughter and butchering of young and old elephants. Meat is sold fresh, dried and in cans of 'corned elephant'; hides are tanned; feet are hollowed and dried for sale as bins, stools or table-legs; and ivory is stockpiled for the anticipated future trade. It is claimed that all these products are incidental by-products of the perceived need to prevent the Kruger elephant population rising beyond 7000–7500, but this is clearly not the best system if sustainable utilization is an aim. Young elephants have less skin and meat than old ones. As with ivory, it would be more logical to allow each elephant to attain the maximum body size before cropping. Body-weight increases throughout an elephant's life, slightly in females and markedly in males (Laws, 1966). Thus, if killing elephants and marketing their meat for the maximum profit is the goal, ethical considerations aside, some form of euthanasia of elephants in their last few years of life would improve the yield. This would also mean that each individual would have realized its reproductive potential, and been able to pass on accumulated cultural knowledge and personal experience to the next generation. The drawback, of course, is the practical difficulty of killing elephant grandparents surrounded by their family, to say nothing of the morality of the exercise.

Is the marketing of elephant meat morally acceptable? Many of those who would object to eating ape meat on the grounds of near-cannibalism might also object to elephant meat. Are they all western 'bunny-huggers' with an urban outlook? Again, the answer is no. My Rwandan colleagues at the Karisoke Research Centre, for example, tell me that most Rwandans do not eat elephant meat because female elephants have breasts like a woman's and care for their babies. In the Nyungwe Forest of south-west Rwanda, Bahutu ivory hunters gave the meat to the pygmoid Batwa because tribal lore held that it was bad for the health (although not apparently for the Batwa's health: Lloyd, 1992). Other tribes, such as the Bambuti pygmies of Zaire and the Wata, or

Waliangulu, of Kenya, specialize in elephant hunting and love elephant meat. In other words, some tribes do, some tribes don't, just as some Caucasian tribes eat pork and others don't. For those whose traditions revolve around elephant hunting, there would be a strong case for allowing the equivalent of 'aboriginal whaling' for local, domestic use if kept within sustainable limits.

Elephant hides have both local and, prior to the CITES Appendix I listing, international markets – mainly for exotic leather cowboy boots in Texas, and other such products in Europe. Trade in elephant hides was derived almost entirely from culling operations in Zimbabwe, South Africa and Botswana (in that order), with the highest revenues reported by Zimbabwe as US$2 million per year prior to the CITES Appendix I listing (Thomsen, 1988). There have been no reported instances of poaching for elephant hides in Africa, although it could happen if demand grew and it became profitable enough. Asian elephant hide products have been reported on sale to tourists in Thailand, allegedly a result of poaching in neighbouring countries.

21.4 LIVE ELEPHANTS

The practice of capturing young elephants and training them is one which dates back more than 2000 years, but in all that time the elephant was never truly domesticated. Every new generation was captured from the wild, and where females did breed in captivity, the baby was often sired by a wild bull. Both species of elephant have been valued in the past as work animals or living tanks for military operations, but it is the Asian elephant which is most frequently associated with this role. The reverence which is shown to elephants in much of Asia stems partly from the worship of Ganesh, the elephant-headed Hindu deity, but this reverence sometimes manifests itself in ways which Ganesh would presumably deplore. Some temple elephants spend most of their time chained to the ground, rocking from side to side like the worst stereotypic zoo or circus elephant. For all the tender loving care of their devoted captors, being worshipped as the earthly manifestation of the god of good fortune and the remover of obstacles has its negative side.

Asian elephant numbers have dwindled to between 34 000 and 54 000, with a further 15 000 domesticated elephants (Santiapillai and Jackson, 1990). Thus the capture of elephants from the wild is banned in most Asian range states, and greater efforts are being made to encourage breeding in the captive population (Chapter 17). In southern Africa, however, young elephant survivors of the cull are regarded as just another marketable commodity. There has been an influx of such cull-orphans into the UK in recent years for zoos and circuses. It is true that to the owners of captive elephants, whether Indian mahout or British cir-

cus trainer, these animals are highly valued as the source of their livelihood, and probably loved as an intelligent companion animal. Sadly,
they and the public who watch their meaningless empty movements see
nothing of the greater elephant values. To see an elephant in a domestic
context is rather like watching a cog slowly turning on an industrial
museum wall, but without the connecting gears and crank shafts which
make its function in the machine so beautiful. An elephant without its
ecosystem is like a cog without its machine (Plate 7).

21.5 SPORT HUNTING

The greatest cash return on a single elephant is usually to sell a licence to
a sport hunter to take its life. The amount paid varies, but is frequently
measured in thousands of American dollars. In addition, sport hunters
must pay for the whole safari, and this increases the cash value of the
elephant life they take. In 1985, for example, safari hunting brought
Zimbabwe US$3.8 million in foreign exchange, and although this
included shooting many other species, it is estimated that each elephant
bull (for it is usually male tuskers that are sought for trophies) attracted
business worth US$26,000 (Child and White, 1988). If the proceeds of
sport hunting accrue to the local communities and are used wisely, as in
the best of Zimbabwe's CAMPFIRE programmes, one can argue that killing elephants can benefit the species. Earnings from sport hunting, however, could be greatly exceeded by the potential earnings from elephant
viewing, but the most profitable, close-quarters elephant viewing is
incompatible with practices which lead every elephant to suspect every
human or vehicle to be a harbinger of death.

21.6 TOURISM

The importance of elephants to Africa's tourist industry is evident from
the number of safari brochures and advertisements featuring elephants.
Can this attraction be given a cash value? What proportion of tourist revenues is directly attributable to elephants? One attempt to estimate this
elusive figure suggested that Kenya's elephants contributed US$25 mil-
lion per year to the country's economy (Brown and Henry, 1989). This
was on the basis of questionnaire responses from some of Kenya's
annual influx of 300 000 adult visitors, who collectively spent some
US$200 million per annum in Kenya in the late 1980s. As Barbier *et al.*
(1990) observe: 'This may be as much as ten times the value of its
poached ivory exports ... a powerful financial case for keeping elephants
alive for their non-consumptive value rather than harvesting them for
their ivory.' Moreover, the revenues from elephant viewing could readily
be increased. These figures are primarily from conventional viewing

methods, i.e. viewing from a safari-vehicle or lodge. Little effort has yet been made to diversify the viewing experience and market a range of elephant experiences with prices varying according to the quality of the view. Wildlife tourism can be categorized according to proximity achieved and excitement afforded (Table 21.1).

For most visitors, quality of experience is understood to be correlated with proximity and clarity of view or photo-opportunity; another factor for some is the excitement of a novel and apparently (to the unfamiliar) slightly dangerous encounter. For these visitors, the ability to recount tales of their 'risky' exploits afterwards is valued at least as highly as the experience itself. Thus it follows that such tourists will pay much higher prices for the privilege of closer than average encounters. Why is it, then, that medium to distant views from lodges and vehicles account for most of the present safari revenues? The main advantage of distant viewing is that large numbers of people can experience wildlife-watching at a relatively low cost per head. The main drawback is that to bus large numbers of people through a habitat, there must be a substantial investment in infrastructure (roads, lodges, etc.) if damage to the environment is to be avoided and client satisfaction maintained.

The example set by mountain gorilla tourism offers an alternative: an outstanding experience for a small number of people at a relatively high per capita cost (Chapter 19). This can be achieved with a minimum of investment in infrastructure, but requires a limited financial investment to develop and maintain a level of mutual trust between groups of wild animals and the guides who oversee the tourist encounters with them.

Given the high prices visitors are prepared to pay for a quality encounter, on foot with trusting animals in the wild, the potential returns on investment are enormous. Equally obvious is the fact that such a system must not be jeopardized by any consumptive uses or invasive management practices, such as culling, sport hunting or translocation. As yet, this system of viewing has not been developed for elephants. Hand-reared orphan elephants can be visited in several countries, e.g. Kenya, Rwanda, Zimbabwe, etc., but this has not been commercialized. Vehicular close encounters with habituated elephants are possible where long-term research programs exist, but again this has not been commercialized, except perhaps with the 'Presidential Herd' in Zimbabwe; and

Table 21.1 Categories of wildlife viewing

Cost	Degree of excitement	Proximity	Vantage from
Inexpensive	Distant view	> 50 m	Lodge
	Medium view	5–50 m	Hide
	Close view	1–5 m	Vehicle
Expensive	Interactive	< 1 m	On foot

in Garamba National Park, Zaire, and Botswana it is now possible to view elephants from the backs of domesticated African elephants along similar lines to those familiar in Asian parks. But nowhere is there an elephant equivalent of gorilla or chimpanzee watching, where visitors can thrill to a quiet approach on foot to within a few feet of trusting, curious animals. This is surprising, given that during the same period that Jane Goodall and Dian Fossey developed techniques to win the trust of chimps and gorillas, respectively, Iain and Oria Douglas-Hamilton (1975) demonstrated that trusting relationships with free-ranging elephants were also possible. Some would argue that elephants are too dangerous for such a system of viewing, but 30 years ago the same would have been said of a proposal to take tourists within a few feet of wild gorillas (Plate 8).

The success of wildlife tourism in Kenya has shown that wildlife – particularly elephants – can be profitable for both the government and the people who benefit from jobs created or profit-sharing schemes with local communities. The growth in private conservancies is also evidence that wildlife can be a more profitable land-use than farming. Some elephant range-states, however, will never be able to compete with Kenya or South Africa in terms of tourism. They may lack the infrastructure; they may be subject to wars or civil unrest; and, if their elephants are forest dwellers, the viewing opportunities may be very limited even if there is a good hotel and a stable government. These are the countries which could benefit most from the development of novel wildlife encounters, and where the ecological value of elephants must be appreciated for its economic benefit to the country.

21.7 ECOLOGY

Elephants play such an important role in whichever habitat they frequent that they are described as a keystone species (Western, 1989). Just as an arch will collapse if the keystone is removed, the disappearance of elephants is likely to presage the collapse of their ecosystem. Their role is so wide-reaching that they have been dubbed 'super-keystone species' (Shoshani, 1992). But when this subject is raised in discussions about the value of elephants, it is frequently dismissed as merely of academic interest. No one has yet attempted to put a cash value on the many species which humans utilize but which depend on elephants, directly or indirectly, for some part of their life-cycle. These elephant-dependent species may in fact have a greater combined economic value than all the other values put together, but it is difficult to quantify them and they receive only passing mention as 'indirect-use value' in economic analyses (e.g. Barbier *et al.*, 1990).

Elephants in both Africa and Asia eat many species of fruit and grain, and so act as seed dispersal agents for numerous plant species, some of which are of direct economic, cultural or medicinal value to people living in the same area. For example, the oil-rich nut of *Panda oleosa* is ground into a buttery paste by forest-dwelling pygmies, but the seeds are dispersed by elephants (M. Fay, in Chadwick, 1992). In Congo-Guinean rainforests, about one third of all tree species whose seed dispersal mechanism is known have their seeds dispersed by elephants (Alexandre, 1978). Some species are so adapted to elephant seed dispersal that their seeds will not germinate unless they have passed through an elephant's gut. Many elephant-dependent tree species may also provide food for monkeys or bush-pigs, which are then hunted for the bush-meat trade. Elephants also push over trees and break branches, bringing browse down to ground level and enriching the habitat for animals such as duiker, which also form an important item in the diet of people dependent on bush-meat for their protein.

There is little doubt that forest-dwelling people have some understanding of the elephant's role in the ecology of their habitat, but commercial hunters from outside the forest show little sensitivity. Driven by the profit motive, they will deplete the elephant numbers in one area, then move on, unaware or unconcerned that they have removed the major cog in the forest's natural machinery. The problem is that the effects of this loss may not be seen for decades. The knock-on effects of local elephant extinctions will only become apparent when older trees die, but are not replaced because their seeds have not been scattered in elephant dung. Gradually, there will be a loss of biodiversity, leading to an impoverished ecosystem. And it will be the local people who suffer.

In Rwanda, for example, a preliminary study of local attitudes to the almost extinct elephants of Nyungwe Forest revealed that the loss of the Nyungwe elephant population was much regretted. As recently as 1973 an estimated 2000 elephants lived in Nyungwe's swamps and forests, but ivory poachers from outside had begun to make a noticeable impact by 1976. Fewer than eight now survive and the local hunters who participated in their slaughter for a pittance now lament their loss to the ivory dealers in neighbouring countries, 'The power of money slaughtered the elephant. Now there are as good as none. We would never let it happen again' (Lloyd, 1992).

While forest elephants are credited with maintaining the African forests, bush elephants paradoxically are responsible for keeping the savannah open. Cattle ranchers have found it expensive to control the growth of scrub when elephants have been eliminated from range-land, and some are now reintroducing elephants to restore the balance. The characteristic acacia trees of the African savannah are also elephant-dependent species; elephants (and impala) eat the pods, thereby protecting the seeds from insects and other predators and depositing them in a ball of dung several miles from the parent plant. Many species then benefit from the shade trees thus planted.

The list of elephant ecological effects is long and complex, ranging from the digging of dry-season waterholes in arid areas to the excavation of subterranean salt-licks in the elephant-caves of Mount Elgon, Kenya (Redmond and Shoshani, 1987; Redmond, 1991, 1992). Even their apparently destructive utilization of *Colophospermum mopane* trees in southern Africa has been shown to have a positive effect, although 'hedged' trees were formerly considered an eyesore. The improved foliage growth which results from elephant utilization is important for browsing ungulates and the annual crop of 'mopane worms' – the larva of the mopane emperor moth *Imbrasia belina*. This is a large, nutritious caterpillar which is harvested for consumption by local people as a tasty snack. Although they are only present for a seven-week period, a villager can earn more than £400 by gathering and selling the caterpillars, which are eaten fresh or are canned for export to neighbouring countries. The mopane worm industry in Botswana is said to be worth about £4.42 million per annum (Styles, 1995) and provides just one example of the kind of economic value which stems from, but is seldom attributed to, the ecological role of elephants.

21.8 ETHICS

The ecological and economic arguments for elephants are (or should be) based on objective facts, but perhaps the most difficult issue is the moral question. Is it right to kill and consume animals whose intelligence and social organization approach our own in terms of complexity? The Great Ape Project (Cavalieri and Singer, 1993) demands that we should include gorillas, chimpanzees and orang utans in the 'community of equals', based largely on genetic grounds as our closest living relatives. But if it is the apes' intelligence and social awareness that we respect, should we not accord this same respect to proboscideans and cetaceans?

Commenting on the Great Ape Project, Robert Baron (1993) observed that 'what makes human beings morally relevant is their possession of consciousness; in particular their consciousness of pain and ... of themselves as individuals with present and future desires that they wish to fulfil. The degree to which chimpanzees (or any form of life) share these features is the degree to which they command ethical consideration' (Baron, 1993). The existence of consciousness, or self-awareness, is a difficult characteristic to prove conclusively in other species (or in non-communicating humans, such as infants and the severely disabled). Mirror self-recognition (MSR) has been demonstrated in the four species of great apes (although not every individual tested) and possibly in dolphins (see papers in Parker *et al.*, 1994), but MSR is only relevant for animals whose main sense is vision. It is difficult to conceive an equivalent test for animals such as elephants, whose primary sensory input is olfactory. Perhaps it is more useful to ask what the circumstances are in which self-awareness might evolve.

The ability to view oneself and one's actions within the context of the surrounding social matrix might be expected to arise in any situation where it would confer a selective advantage, i.e. where individuals live in a complex society in which reproductive success is related to social ability. The prerequisites would appear to be:

- a large complex brain with a good memory and an ability to reason;
- a long life-span with an extended period of development during which cultural transmission of behaviour patterns and accumulated knowledge occurs;
- a complex society in which relationships last for decades rather than weeks or months, so that the consequences of today's social interactions can affect events (and reproductive success) at a much later date.

This combination of circumstances has evolved independently in three orders of mammals: primates, cetaceans and proboscideans. Comparisons between their social behaviour are, therefore, particularly illuminating in our attempts to understand our own. Similarities between ape and human societies may stem from our common ancestor, but similarities between the three orders must represent convergent evolution and so should cast light on the selective pressures which acted in the formation of human nature.

Recognition of the consciousness of elephants or cetaceans certainly poses a moral dilemma for those who wish to kill them, whether for profit or other reasons. As with most moral decisions, it comes down to individual choice guided by cultural or religious edicts and personal experience. It is a question of where to draw the line between the acceptable and unacceptable; a follower of Ahimsa, the non-injury path of Jainism will not countenance the killing of any creature, whereas a member of a tribe that practises cannibalism will, for equally valid cultural reasons, happily consume human flesh. The vast majority of humans would condemn the latter practice (unless in a survival situation such as a plane crash or shipwreck) while ridiculing the Jain ascetic for wearing a face-mask to avoid accidentally ingesting flying insects and sweeping the path with peacock feathers for fear of squashing terrestrial invertebrates (Smart, 1989). However, the recognition of social bonds between companion animals and their keepers prevents many humans from eating dogs, cats and horses, so it is not unreasonable to respect the social bonds of elephants. I suspect that few people with any first-hand experience of a cross-species friendship with an elephant would feel comfortable about eating one, but the opportunities for such friendships are rare. Thus, it is up to those who have experienced the depths of elephant cognition to share that knowledge with those who are unlikely ever to do so.

One frequently heard objection to such philosophical arguments is that they are fine for Western armchair academics, but that they carry little weight among subsistence farmers in developing countries, who require compensation for crops destroyed by elephants. If the commercial killing of elephants is deemed morally unacceptable, however, ways must be found of increasing the income generated by non-consumptive means, and a proportion of this income must benefit the people who live alongside the elephants.

21.9   AESTHETIC VALUES

It is impossible to put a value on the beauty of an elephant, or on the pleasure many people derive from knowing that elephants still roam the wilds of a country they may never visit. Nevertheless, it is to a large extent the aesthetic value which lies behind the tourist trade, and wildlife artists who paint and sculpt elephants derive a living from our perception of beauty in the unlikely form of the elephant. This value is, however, one which is most frequently expressed by people who are well fed and who do not have their crops raided by hungry elephants, or their children terrorized on their way to school by angry elephants. To paraphrase a Zimbabwean farmer, 'When we have plenty to eat, elephants are beautiful; when we are hungry, elephants are meat.' Clearly, beauty is in the stomach as well as the eye of the beholder and so conservation will be most successful when it improves the lot of the local people who share their land with the wildlife.

21.10   CONSERVATION

The current crisis facing both elephant species represents the final stages of a process which began thousands of years ago. During the period of recorded history, the elephant's numbers and distribution have steadily diminished as a direct result of human activity. It is no exaggeration to say that elephants were once the dominant animal species in the ecology of Africa and Asia. The African species was found over virtually the whole continent, except for the waterless deserts (which were much smaller before human-induced desertification). Just how many elephants roamed Africa in previous ages is unknown, but an order of magnitude calculation gives us some indication. Africa is roughly 11.5 million square miles (30 million km^2) in total; deduct 3 million square miles for the Sahara and a bit more for other non-elephant habitats, and we are left with, roughly, 7 million square miles (18 million km^2). Elephant densities are known to vary according to habitat from less than one to more than

10 per square mile (see for example Table 4.2 in Eltringham, 1982; Stromayer and Ekobo, 1992), and so there may once have been between 7 million and 70 million African elephants.

During the twentieth century, the roles of elephant and human have been reversed. Much of Africa is now a cultivated landscape, criss-crossed by roads and fences and dotted with modern cities. There are still islands of elephants in this sea of humanity, but those which are not protected by law are rapidly dwindling in area. Inevitably, the increases in human population, agricultural land development and escalating rates of natural resource exploitation have changed the landscape over much of Africa. Wherever such changes have taken place, there has been a corresponding decrease in elephant populations. It is often said that one cannot share an acre with an elephant. This competition for space is heightened by the fact that, in terms of climate and vegetation type, elephants and humans have similar requirements. Agriculture and forestry in Africa require a good rainfall and an absence of extreme temperatures. A map showing areas with such conditions shows a clear correlation with the distribution of elephant populations, and also with human population density (Burrill and Douglas-Hamilton, 1987).

Elephants are still found in 37 African and 13 Asian countries, though only just in several cases. By virtue of their size, intelligence and long-evity, both species present special problems when it comes to protecting their habitat. Given the human population explosion it is likely that the only elephant habitat in the future will be within the boundaries of national parks, reserves and private conservancies.

The question of how big a protected area must be in order to provide all the needs of a population of elephants is a contentious issue (see, for example, Hanks, 1979). Radio tracking has recorded movements of up to 200 km (124 miles) in northern Botswana (Spinage, 1994), and an individual elephants' home range might be more than 3000 km^2 (1150 square miles) in marginal habitat (Leuthold and Sale, 1973). Because marginal habitat is the least desirable for agriculture, that is where parks and reserves tend to occur. Drawing a line around an area which appears, on limited observations, to support a population of elephants may not take into account such movements. Each elephant population has a cultural knowledge of its environment, which is passed on from generation to generation (Moss, 1988). This may include such information as where to find water in the event of a severe drought – an event which may only happen once in decades, and may involve migrating to another area. Thus, to protect elephants, a detailed knowledge of their movements over several decades is required before park boundaries are finally gazetted. This has seldom been the case, and so elephants restricted by man-made boundaries may consis-tently attempt to leave their park. When the park is hemmed in by agri-culture, they are then branded 'crop-raiders' and may be legally shot.

In the context of habitat preservation, the elephant is a key species. A park or reserve which is large enough for a viable population of elephants, will automatically ensure protection of all sympatric smaller species. The converse is not true (Shoshani, 1987). Small parks may protect other wildlife, but elephants cannot survive in them without extensive intervention by human management. Elephants are, however, very popular; conservationists list them amongst the 'flagship species', whose role is to attract funds and resources to their habitat in order that they, and the other species therein, can be protected.

21.11 CONCLUSIONS

Africa was once a continent of elephants, as indeed was much of Asia. Its habitats were shaped by elephantine appetites, its seedlings planted in countless piles of dung, and the resulting trees pruned by millions of trunks. Scattered throughout the elephant's domain were small tribes of upright apes. They carved out islands of cultivation around their villages, from which elephants were excluded. The footpaths which linked those villages were often elephant roads, and elephants were hunted by some tribes for meat, skins and tusks. Gradually, the cultivation crept along the roads, and fields around neighbouring villages began to join up. Over the millennia, there was a reversal of fortune between proboscideans and people. Today we have islands of elephants in a sea of cultivation, and one by one those islands are being swamped by the rising tide of the human population explosion. Understanding the ecological value of elephants brings about the realization that simply 'saving' the two species of elephants is not the goal of conservation, if being 'saved' is defined as the continued existence of at least one population of each species. We must instead adopt a more functional, ecosystems approach, which ensures that elephants survive in naturally maintained densities in every habitat which evolved around these two super-keystone species.

Thus, the only hope for the elephants' survival is if the swarming millions of humans can appreciate the true values of their elephant neighbours, and learn to accommodate their needs in return for the benefits they bring.

ACKNOWLEDGEMENT

I am grateful to Tusk Force for sponsoring my participation in The Exploitation of Mammals Symposium.

REFERENCES

Alexandre, D.-Y. (1978) Le rôle disséminateur des éléphants en Forêt de Tai, Côte-d'Ivoire. *Terre Vie*, **32**, 47–72.

Barbier, E.B., Burgess, J.C., Swanson, T.M. and Pearce, D.W. (1990) *Elephants, Economics and Ivory*, Earthscan Publications, London.

Baron, R. (1993) Apes and ethics. Letters, *New Scientist*, 26th June 1993, p.48.

Brown, G. and Henry, W. (1989) *The Economic Value of Elephants*. LEEC Discussion Paper 89–12, London Environmental Economics Centre.

Burrill, A. and Douglas-Hamilton, I. (1987) *African Elephant Database Project*, GRID Case Study Series No.2, WWF/EWAA/GEMS/UNEP, 91 pp., 12 maps.

Cavalieri, P. and Singer P. (eds) (1993) *The Great Ape Project, Equality beyond Humanity*, Fourth Estate, London.

Chadwick, D.H. (1992) *The Fate of the Elephant*, Viking, p. 177.

Child, G. and White, J. (1988) The marketing of elephants and field-dressed elephant products in Zimbabwe. *Pachyderm*, **10**, 6–11.

Currey, D. and Moore, H. (1994) *Living Proof; African Elephants – the Success of the CITES Appendix I Ban*, Environmental Investigation Agency, London, UK and Washington DC, USA.

Diamond, J. (1991) *The Rise and Fall of the Third Chimpanzee*, Radius, London.

Douglas-Hamilton, I. and Douglas-Hamilton, O. (1975) *Among the Elephants*, Collins, London.

Eltringham, S.K. (1982) *Elephants*, Blandford Mammal Series, Blandford Press, Poole, Dorset.

Galdikas, B.M.F. (1995) *Reflections of Eden*, Victor Gollancz, London.

Hanks, J. (1979) *A Struggle for Survival: the Elephant Problem*, Country Life Books, London.

ITRG [Ivory Trade Review Group] (1989) *The Ivory Trade and the Future of the African Elephant*. A Report for the 7th CITES Conference of the Parties, coordinated by Stephen Cobb, International Development Centre, Oxford, UK.

Laws, R.M. (1966) Age criteria for the African elephant, *Loxodonta africana*. *East African Wildlife Journal*, **4**, 1–37.

Leader-Williams, N. (1993) The cost of conserving elephants. *Pachyderm*, **17**, 30–34.

Leuthold, W. and Sale, J.B. (1973) Movements and patterns of habitat utilization of elephants in Tsavo national park, Kenya. *East African Wildlife Journal*, **11**, 369–384.

Lloyd, R. (1992) *The Role of Elephants in the Ecology of African Rainforests and the Effects of their Diminishing Numbers*. Report on a pilot study for the African Ele-Fund and PCFN, unpublished.

Moss, C. (1988) *Elephant Memories*, Elm Tree Books, London.

Parker, S.T., Mitchell, R.W. and Boccia, M.L. (1994) *Self-awareness in Animals and Humans – Developmental Perspectives*, Cambridge University Press, Cambridge, UK.

Pilgram, T. and Western, D. (1984) Managing elephant populations for ivory production. *Pachyderm*, **4**, 9–11.

Redmond, I. (1989) *Trade in Gorillas and other Primates in the People's Republic of the Congo*. Unpublished report for the International Primate Protection League, 42 pp. plus 3 appendices.

Redmond, I. (1991) With elephants underground, in *Elephants – Saving the Gentle Giants*, (ed. R. Orenstein), Bloomsbury Press, UK (published as *Elephants – the Deciding Decade* in the USA and Canada by Key Porter Books, Toronto), pp. 114–129.

Redmond, I. (1992) Erosion by elephants, in *Elephants – Majestic Creatures of the Wild*, (ed. J. Shoshani), RD Press/Weldon Owen, Australia; Simon & Schuster, UK, pp. 128–130.

Redmond, I. (1995) The ethics of eating ape. *BBC Wildlife*, **13**(10), 72–74.

Redmond, I. and Shoshani, J. (1987) Mount Elgon's elephants are in peril. *Elephant*, **2**(3), 46–66.

Sabater Pi, J. (1980) Exploitation of gorillas *Gorilla gorilla gorilla*, Savage & Wyman 1847 in Rio Muni, Republic of Equatorial Guinea, West Africa. *Biological Conservation*, **19**, 131–140.

Santiapillai, C. and Jackson, P. (1990) *The Asian Elephant: An action plan for its conservation*. IUCN, Gland, Switzerland.

Shoshani, J. (1987) Elephant research in a historical perspective. *Proceedings of the 7th Annual Elephant Workshop, held in Calgary Zoo September 25–27 1986*, pp. 63–76.

Shoshani, J. (ed.) (1992) *Elephants – Majestic Creatures of the Wild*, RD Press/Weldon Owen, Australia; Simon & Schuster, UK, pp. 128–130.

Smart, N. (1989) *The World's Religions*, Cambridge University Press, Cambridge, UK.

Spinage, C. (1994) *Elephants*, T & AD Poyser, London, p. 207.

Stromayer, K.A.K. and Ekobo, A. (1992) The distribution and number of forest dwelling elephants in extreme southeastern Cameroon. *Pachyderm*, **15**, 9–14.

Styles, C. (1995) The elephant and the worm. *BBC Wildlife*, **13**(3), 22–24.

Thomsen, J.B. (1988) Recent US imports of certain product from the African elephant. *Pachyderm*, **10**, 1–5.

Tudge, C. (1991) *Last Animals at the Zoo*, Hutchinson Radius, London, pp. 9–10.

Western, D. (1989) The ecological value of elephants: a keystone role in African ecosystems, in *The Ivory Trade and the Future of the African Elephant*. Report for the 7th CITES Conference of the Parties, coordinated by Stephen Cobb, International Development Centre, Oxford, UK.

22

Human disturbance of cetaceans

Peter G.H. Evans

SYNOPSIS

As our coastal seas become used increasingly for commercial, industrial and recreational purposes, whales and dolphins face threats not only directly from physical damage by vessels but also indirectly from the sound disturbance that vessels may cause.

Concerns about possible effects of disturbance were first directed at seismic activities during oil and gas exploration off Alaska in the 1970s and their potential impact upon bowhead and gray whale behaviour. At the same time, the rapid rise in whale watching in North America caused concern particularly for gray whales in the Gulf of California and humpbacks along the north-eastern seaboard of the United States. Results of various studies were sometimes conflicting and often ambiguous; it became clear that vessel type, vessel behaviour and the nature of the disturbance each have an important influence on whether a cetacean species will react negatively, as does the particular ecology and behaviour of that species at the time. Those differential effects are examined here in some detail.

Two small cetacean species, the bottlenose dolphin and harbour porpoise, are amongst those which most frequently come into contact with coastal human activities. Results of recent studies of the reactions of the two species to a variety of craft are also presented, with particular attention to aspects of the ecology of those species which may influence their vulnerability. From all these studies, a set of recommendations is made with the aim of minimizing human disturbance. Whale watching and dolphin watching are being promoted as an alternative to the lethal utilization of cetaceans; as this activity becomes more popular, procedures for regulating those activities and monitoring negative effects will need to be addressed.

22.1 INTRODUCTION

For about 10 million years, representatives of modern orders of Cetacea, both Mysticeti (baleen whales) and Odontoceti (toothed whales, dolphins and porpoises), have evolved in a relatively quiet world, accompanied only by the sounds of other animals and the occasional earthquake or some other natural perturbation. Only in the last hundred years have our oceans and rivers been disturbed in any substantial way with additional sounds created by mankind. In particular, human use of the marine environment for commercial, industrial and recreational purposes has led to a considerable increase in the level of background sound (for reviews, see Haines, 1974; Ross, 1976; Urick, 1983, 1986).

Animals such as cetaceans which live entirely within an aquatic environment rely heavily on sound both to acquire information about their environment and for communication (Evans, 1987; Martin, 1990). Additional sounds may therefore cause disruptions to the lives of cetaceans, distracting, annoying or even frightening them, as well as providing the potential for causing behavioural and physiological upset. To date, we have scarcely any idea of the long-term effects of such sounds upon cetaceans at the species, population or even individual level. This chapter reviews current knowledge with emphasis upon those areas of likely conflict, and attempts to identify some lines of inquiry required if we are to have a better understanding of the problem.

22.2 NATURAL AND MAN-MADE SOUNDS IN THE OCEAN AND THEIR PROPAGATION

When sound is propagated underwater from a source, it will inevitably be diminished or distorted by spreading, reflection from the sea surface or the seabed, scattering and absorption, and refraction. Initially, sound radiates spherically, but in arctic regions the stability of the surface layer acts as a channel which confines the vertical extent of the expanding sound wave, resulting in cylindrical spreading and the sound carrying over greater distances. Cylindrical spreading also occurs in deep water from 700 to 1500 m depth, often referred to as the SOFAR (Sound Fixing and Ranging) channel, or deep sound channel. Here, sound travels more slowly than in any other part of the water column, thus increasing the range over which it may be heard. Higher frequencies (1 kHz or above) are absorbed more readily than low frequencies (100 Hz or less), and so do not carry such great distances (Urick, 1982, 1983). It is, therefore, only with low frequency sound that the deep sound channel has any significant effect. In real life, propagation losses rarely exactly mirror these theoretical relationships, being influenced by pressure, salinity and temperature gradients (Greene *et al.*, 1985; Gilders, 1988).

22.2.1 Natural sounds

Cetaceans will be primarily influenced by human sounds and those made by conspecifics, when their intensities exceed the ambient noise level at the corresponding frequencies. Natural ambient noise is defined here as those sounds produced in water whose origin is either physical or biological but which does not emanate from human activities.

There are many sources of natural sounds (for reviews, see Wenz,

1962; Tavolga, 1967). These include wind, wave and ice action, precipitation, natural seismic events and a diversity of aquatic animal life. The frequencies of sound they produce vary according to the source between 10 Hz and more than 100 kHz.

Microseismic activity is a source of low-frequency sound (< 100 Hz), as may be turbulence caused by deep ocean currents. Surface waves and wind on the surface cause noise in the near region from 500 Hz to 30 kHz. Sound source levels at wind force 1 (sea state 0) is 45 dB (in deep water) at 1 kHz; whereas at wind force 7 (sea state 6) this increases to 70 dB (Wenz, 1962; Ross, 1976; Urick, 1982, 1983). In coastal waters, tidal streams cause noise on the seabed, particularly when it is composed of rough and unstable material, and such sounds may occur at frequencies as high as 300 kHz. In polar seas, whereas noise levels under ice may be less than 30 dB re 1 µPa at 1 m (at 100 Hz), levels may exceed 130 dB (at 200 Hz) at the ice edge and along ice pressure ridges. Finally, marine life in the sea is also responsible for quite a lot of noise. Snapping shrimps, for example, are well known for the amount of noise they can create – a more or less continuous crackling or rasping sound over the frequency band 0.5–25 kHz. Cetaceans themselves may generate sounds of considerable intensity. Source levels for the low frequency pulses (12–25 Hz) of fin and blue whales, for example, have been estimated at between 170 and 200 dB re 1 µPa at 1 m (declining to *c.* 140 dB at 100 m) (Richardson *et al.*, 1991).

22.2.2 Man-made sounds

Since the industrial era, humans have developed a number of highly intense sources of sound (Ross, 1976; Urick, 1983, 1986). Indeed, Ross (1976) estimated that between 1950 and 1975 ambient noise had risen by 10 dB in areas where shipping noise dominates, and he predicted it would rise a further 5 dB by the end of the twentieth century as shipping traffic increased. The more powerful the engine that a vessel possesses, the greater the amount of sound (at least at low frequencies) it will produce. Supertankers, in particular, produce sound intensities of between 187 dB (at 50 Hz) and 232 dB (at 2 Hz) re 1 µPa, at very low frequencies (particularly < 10 Hz) (Cybulski, 1977; Leggat *et al.*, 1981).

Acoustic oceanographers also use intense sounds mainly in the low-frequency range (< 1 kHz) to study both the physical properties of the ocean (Spindel and Worcester, 1990; Worcester *et al.*, 1993) and marine organisms such as zooplankton in the deep scattering layers (Mauchline, 1980). Recently, there has been much debate over the potential impact upon cetaceans and other marine life that could arise from a project referred to as the Acoustic Thermometry of Ocean Climate (ATOC) project which proposes to measure the speed of sound in the ocean repeatedly over time in order to determine whether the oceans, which are the main heat sink, are warming (Mulroy, 1991; National Research Council, 1994). For that project, high intensity (*c*. 190 dB re 1 µPa) low-frequency sound (mainly 60–90 Hz) would be generated at depths of around 900 m over long-distance undersea paths such as the SOFAR channel.

Besides propeller and engine sound generated by vessels during commercial, military and recreational activities, surface vessels and submarines employ active sonar which uses sonic or ultrasonic waves to locate submerged objects, at the same time introducing brief, high-intensity pulses into the marine environment that sometimes may be transmitted over great distances. Source levels of sound are *c*. 200–250 dB re 1 µPa at frequencies up to 200 kHz. High resolution sidescan sonar (generally below 14 kHz) is also used in geophysical seismic surveys, particularly during oil and gas exploration, along with lower resolution explosive techniques (airguns, sleeve exploders, etc.) mainly at frequencies below 500 Hz (Richardson *et al.*, 1991; Evans *et al.*, 1993).

Most of the sounds generated from these maritime activities (with the exception of sonar) are at frequencies lower than 1 kHz. However, when a surface vessel travels at high speed, the propeller may cavitate and produce much higher frequency sound (between 2 and 20 kHz) (Evans *et al.*, 1992). Measurements of various small craft (up to 15 m length, 240 hp engine) indicated source levels ranging from 100–125 dB re µPa at 2 kHz and 60–105 dB re µPa at 20 kHz. Cavitation is also more likely to occur when the propeller is damaged.

22.3   CETACEAN SOUND PRODUCTION AND HEARING

Our knowledge of the hearing capabilities of cetaceans and the mechanisms they use for receiving and interpreting sounds remains very limited. Underwater hearing abilities have been studied experimentally in only a few odontocete species and in no mysticetes. Subjects have been studied in captivity and this imposes constraints upon the species and size of cetacean involved. Where experimental data do not exist, some inference of the sound frequencies which are important to cetaceans can be made from the characteristics of the sounds they produce, and from the structure of their hearing organs.

The auditory sensitivities of those porpoises, dolphins and smaller toothed whales that have been examined are greatest at very high frequencies – between 10 and 150 kHz, with a hearing threshold of about 40 dB at those frequencies, increasing to around 100 dB at 1 kHz and 120 dB at 100 Hz, at least for those species for which data are available (Richardson *et al.*, 1991). Although there is no quantitative information on the auditory sensitivities of mysticetes, tentative audiograms for the gray whale and bowhead whale are presented by Moore *et al.* (1984) and Dahlheim and Ljungblad (1990). They suggest that greatest hearing sensitivities occur between 100 Hz and 5 kHz, on the assumption that whales will hear approximately over the same frequency range as the sounds they produce. Using this argument, we would expect fin whales to be very sensitive to frequencies of 20 Hz and blue whales to 10–20 Hz.

The sounds produced by odontocetes (Popper, 1980) may conveniently be divided into:

- pure tone whistles, generally in the frequency range 500 Hz–20 kHz, used mainly for communication;
- pulsed sounds or clicks varying from 500 Hz to 150 kHz, used mainly for echolocation.

Source levels for both types of sound are estimated usually to be between 150 and 200 dB, although pulsed sounds for non-echolocatory purposes may be produced at source levels of 115 dB, mainly in the frequency range below 20 kHz. Most of these measurements were made in captivity and it should be noted that animals can modify their sound production (particularly its intensity) in confined situations, and indeed also do so in open water.

The sounds produced by mysticetes may be classified into four types (see Thompson *et al.*, 1979):

- low-frequency moans, typically with frequencies of 12–500 Hz and of 0.4–36 seconds duration;
- grunt-like thumps and knocks with most sound energy concentrated between 40 and 200 Hz;
- chirps, cries and whistles at frequencies between 1 and 10 kHz;
- clicks or pulses at frequencies up to 20–30 kHz and lasting from 0.5 to 5 milliseconds.

Sound source levels range between 150 and 200 dB, at frequencies of 500 Hz or less.

To summarize, most odontocetes can hear sounds over a wide range of frequencies from 75 Hz to 150 kHz, with greatest sensitivity around 20 kHz (although low frequency hearing of odontocetes has not been fully investigated), whereas the hearing of mysticetes probably ranges from frequencies of 10 Hz to 10 kHz, with greatest sensitivity usually below 1 kHz (this is based on sound production levels since no audiograms exist). Major differences in hearing between baleen and toothed whales is further supported by anatomical differences between the hearing organs of these two groups (Ketten, 1991).

22.4 POSSIBLE EFFECTS UPON CETACEANS OF ANTHROPOGENIC SOUNDS

By combining information on average ambient noise levels with the putative hearing ranges of both odontocetes and mysticetes, and a knowledge of source levels of anthropogenic noise and likely propagation conditions, it is possible to make some predictions of where sound interference is most likely to occur (Richardson *et al.*, 1991; Figure 22.1). Of course, all these values depend upon a variety of factors: the range and intensity of sound sources in the area, local properties of sound propagation, novelty of the sound, and the species of cetacean concerned. However, it does serve to highlight that dolphins are more likely to be affected by surface vessels with cavitating propellers (this can include large vessels) and their sonar. Other sources of man-made sound in the marine environment are most likely to impinge upon baleen whales. It has been suggested that toothed whales and dolphins may not even be able to detect continuous sounds with received levels around 115 dB re 1 µPa (National Research Council, 1994). Although low frequency hearing of odontocetes has been poorly studied, the fact that belugas have been shown to strongly avoid icebreakers in the Canadian arctic 40 km away, albeit probably in rather special conditions of sound propagation (Finley and Davis, 1984), emphasizes that they can also be affected over long distances by vessel transport.

In humans, sounds become excessively loud at 126–146 dB (i.e. 100–120 dB above threshold level) at 1 kHz (Kryter, 1985). For an odontocete such as the bottlenose dolphin, this would correspond to a sound level of approximately 150–170 dB at 10–100 kHz, the frequency range of greatest sensitivity. For a mysticete, the corresponding value could be around 140–170 dB in the frequency range 100 Hz to 1 kHz. However, it is unlikely that human criteria are applicable when considering the question of cetacean discomfort. Cetaceans themselves produce sounds above these source levels. Mysticete whales (e.g. bowhead, blue and fin whales) have been recorded making sounds of between 130 and 200 dB. Similarly, odontocetes (e.g. beluga and killer whale) can have source levels of between 180 and 220 dB. This means that, at least at some frequencies, neighbouring cetaceans could receive sound levels from conspecifics as high as 160 dB even at distances of 100–150 m.

There are two ways in which cetaceans might cope with excessive loudness from whatever source:

- They might use a mechanism referred to as the middle ear reflex, which in harp seals, for example, has been shown to shut down effective hearing in response to loud impulsive sounds.
- They may undergo frequency shifts both in sound production and

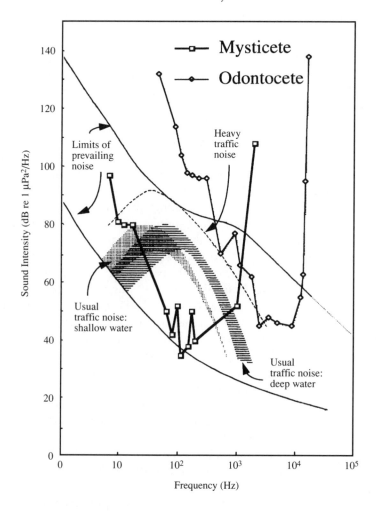

Figure 22.1 Background shipping noise in relation to generalized odontocete and hypothetical mysticete audiograms. Adapted from Wenz (1962); Gilders (1988); NRC (1994).

perhaps in hearing such that they are able to communicate outside the frequency of maximum sound energy which they are experiencing.

Although evidence for either of these occurring in cetaceans is generally lacking, Au *et al.* (1985) have observed frequency shifts in beluga sound production from 40–60 kHz in San Diego Bay to 100–120 kHz in Kaneohe Bay, Hawaii, where they also increased source levels by 18 dB, in this case apparently in response to the noisier environment created by the snapping shrimp.

Following offshore oil and gas industrial activities (seismic surveys, drilling, oil production platforms, dredging and support traffic) in North America during the late 1970s and early 1980s, various studies were conducted to determine reactions of baleen whales to the sounds produced. These results (mainly on bowheads and gray whales) suggested that whales tend to move away when received sound levels are around 115–120 dB re 1 µPa for continuous sounds (as produced by vessels, offshore drilling, etc.) and about 160–170 dB re 1 µPa for intermittent noise impulses (as produced by seismic air guns, etc.) (Malme *et al.*, 1983, 1984, 1988; Richardson *et al.*, 1985, 1986, 1990; Ljungblad *et al.*, 1988; Richardson and Malme, 1993).

Experimental playbacks of low frequency noise (150–170 dB re 1 µPa source levels at 800–900 Hz) have also been carried out recently in the presence of an odontocete species, the bottlenose dolphin, which showed little response either to transmitted impulses or continuous sounds (Tyack *et al.*, 1993).

The paucity of quantitative data makes it very hard to determine not only the distances at which a particular sound intensity may be propagated but also whether sound levels such as those given above actually have any biological impact.

22.5   EFFECTS OF ECOTOURISM DEVELOPMENT UPON CETACEANS

In the last 25 years, the amount of sound generated by both commercial and recreational vessels must have increased enormously, given the increase in quantity of traffic in many areas. As noted earlier, this will not only have increased ambient noise levels at frequencies below 1 kHz, but also, particularly in coastal areas where speedcraft operate, it will have introduced significantly higher levels of high frequency sound. This could potentially affect the daily lives of inshore toothed whales and dolphins such as the bottlenose dolphin and harbour porpoise.

During the 1970s, concern was expressed that gray whales were being displaced from their breeding lagoons in Baja California, Mexico, by the large numbers of speedboats and low-flying aircraft visiting the area for whale-watching purposes (Reeves, 1977). Although later studies found no correlation between changes in whale distribution in Baja California and that of human activities (Jones and Swartz, 1984), nor any difference in swimming speeds or respiration patterns of migrating whales in the presence of boats in southern California (Sumich, 1983), a longer-term impact upon biological parameters such as reproductive success could not be excluded.

In Hawaii, Kaufman and Wood (1981) found that the presence of humpbacks varied inversely with the amount of daily boat traffic and with days on which military bombing practice took place. In Alaska, clear and graded changes in the behaviour of humpbacks in response to vessel traffic were observed even at distances of over 3 km (Baker and Herman, 1989). These included longer dives, shorter periods at the surface, movements away from the paths of vessels and temporary displacement of individuals from preferred feeding areas (Baker *et al.*, 1982, 1983). With a single approaching vessel, aerial behaviours and surface-feeding patterns did not change, but with several vessels in the area, the number of breachings was reduced.

In the Gulf of Maine, fin whales are often approached by whale-watching vessels, and a recent shore-based study investigated whether this affected their respiration rates and time spent at the surface (Stone *et al.*, 1992). Although no differences were found in the overall percentage of time spent at the surface, fin whales made significantly shorter dives in the presence of boats, and the number of blows made on surfacing was fewer, resulting in shorter surfacing sequences.

Most studies of reactions of cetaceans to vessels have concentrated upon baleen whales. However, in recent years, increasing attention is being paid to the possible impact that ecotourism and other recreational activities may be having upon odontocetes (toothed whales and dolphins). Because of the higher sound frequencies at which species of this group operate, one might expect rather different responses if sound alone was the stimulus eliciting a reaction. In Kaikoura, New Zealand, sperm whales are visited frequently by whale-watching vessels. A study by Gordon *et al.* (1992) in this area found that resident sperm whales had shorter surface times and made fewer blows in the close presence of motorized whale-watching vessels compared with their own relatively silent sailing vessel. On disturbance, the whales were also more inclined to dive without throwing their tail flukes up into the air and to change their acoustic behaviour on diving. In western Norway (off Andenes), Eberhardt (1993) found that lone sperm whales (but not groups of two to five animals), resting at the surface, were affected by boats approaching to within 50 m, showing avoidance behaviour and changes in the intervals between blows.

Of smaller odontocetes, one of the species most exposed to whale-watching activities is the killer whale, particularly in the vicinity of Vancouver Island (Canada) and Puget Sound, Washington State. Shore-based studies by Kruse (1991), Duffus and Dearden (1992) and Otis (in Phillips and Baird, 1993) have generally found little if any effect of the presence of boats on killer whale behaviour. The only possible influence occurred on occasions when several boats came within 400 m of the whales, who then tended to increase their swimming speed by around 50% above that of their undisturbed speeds (Kruse, 1991).

A study of reactions by bottlenose dolphins to the presence of speedboats in Cardigan Bay, west Wales, also showed shorter periods at the surface, longer dives and movements away from vessel paths (Evans *et al.*, 1992). These occurred over a range of 150–300 m at sea states 3–4. A noise transmission simulating a motor boat rapidly approaching at a distance of 150 m produced the most marked negative response, presumably the result of a startle effect, but this may reflect the rather artificial mode of sound transmission. Sound characteristics of various craft were also measured and indicated that dolphins should first hear a jet ski 450 m away, an inflatable at 1 km, and larger motorized vessels (up to 240 hp engine) between 1.1 and 3.1 km distance (above a background sea state of 3).

It is quite possible that, over a period of time, cetaceans become habituated to the presence of particular vessels. The study by Gordon *et al.* (1992) of the effects of whale-watching activities on the behaviour of sperm whales off Kaikoura, New Zealand, for example, found that transient individuals were less tolerant (spending shorter periods at the surface and having shorter blow intervals) than residents. In Johnstone Strait (British Columbia) and Puget Sound (Washington State), the resident killer whale pods appear to have habituated to the presence of boats and either ignore or even approach vessels (Phillips and Baird, 1993). In the Gulf of Maine (north-east United States), Beach and Weinrich (1989) found that humpback whales were less inclined to respond negatively to whale-watching boats than in south-east Alaska, where they experience much lower levels of boat traffic. On the other hand, bowhead whales in the Beaufort Sea, where they experience about three times as much shipping as in Baffin Bay, exhibit shorter dives and surfacings on feeding grounds, with less tail fluking, and fewer sexual interactions (Richardson *et al.*, 1995). Although these differences in behaviour may reflect adaptive changes by individuals, it is difficult to determine the long-term effects on survival and reproductive success given all the other environmental variables prevailing.

At Cape Cod (north-east United States), whale watching of humpbacks, fin, minke and right whales has seen an enormous growth since the late 1970s. A study of around 12 000 logbook entries from research boats over the 25-year period 1957–1982 carried out by Watkins (1986) indicated systematic changes from a situation where whales generally ignored or avoided vessels in the years prior to the development of whale watching, to positive reactions with whales actually approaching vessels in the later years. Following repeated contact with boats, most species have changed their reactions from avoidance behaviour and reduced vocalization to showing general disinterest. Resting whales were more difficult to approach without being disturbed, whilst actively feeding or socializing animals tended to ignore the presence of boats. However, responses did vary between species. Humpbacks tended to respond positively to the increasing presence of whale-watching vessels, sometimes changing their behaviour to make an approach or engage in predictable surface behaviours. Fin whales changed from strongly negative reactions to ships to ignoring them, continuing to feed unless approached to within 30 m, although usually they would fall silent. Minke whales were more inquisitive in the early years, changing later to little or no reaction with increased exposure to vessels, whilst right whales showed little change in behaviour over the years, except for becoming more silent in the more coastal areas. Some of the more positive changes in behaviour may of course reflect the more careful behaviour of whale-watching operators once guidelines were introduced. However, the frequent reports of cetaceans of various species (e.g. bottlenose dolphins and gray, humpback and minke whales) behaving in a 'friendly' manner may reflect an increasing acceptance by them of the close presence of humans and their vessels.

Bottlenose dolphins and other small odontocetes not infrequently bow-ride vessels including jet skis, and have been observed even at the bow of seismic vessels. Minke whales may ride either the bow or the stern of boats. These regular occurrences suggest that the sounds produced by them are certainly not always aversive. A study of reactions by harbour porpoises to a variety of craft in the Shetland Islands showed different responses depending upon various factors (Evans *et al.*, 1994). Porpoises were more likely to respond negatively to speedboats and a large ferry, both of which they experienced only infrequently, compared with sailing boats and a small daily ferry. They also showed less of a negative response in groups than when singly or as adult–calf pairs. Finally, whereas porpoises tended to move away from vessels early in the summer, as the season progressed their reaction changed so that by early autumn the majority of individuals actually approached vessels (Figure 22.2), possibly reflecting the lower vulnerability of their growing young or their greater curiosity as they became more actively social, or perhaps to some degree of habituation.

Besides whale-watching activities, the other form of human disturbance that should be considered is that of researchers themselves, either by close approach to one or more cetaceans for purposes of individual photo-identification, or for taking biopsy samples by darting. Where studies have examined the effect of biopsy darting of humpbacks, little or no observable reaction was observed in 35–60% of cases (35% on feeding grounds; 60% on breeding grounds), the remainder showing little more than a temporary moderate response in the form of hard tail flicks and trumpet blows (Weinrich *et al.*, 1991; Clapham and Mattila, 1993). Generally, a marked behavioural response was observed on 5% or less of occasions, and the researchers concluded that darting results in only a brief disturbance of the animal's behaviour.

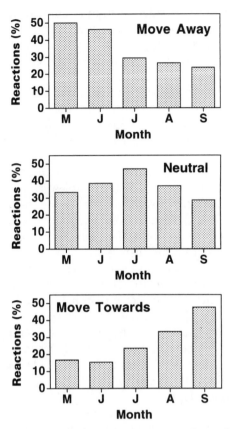

Figure 22.2 Seasonal changes in the reactions to vessels made by harbour porpoises in South-East Shetland (based on 89 interactions) (from Evans *et al.*, 1994).

22.6 CONCLUSIONS AND FUTURE NEEDS

The possible effects of underwater noise upon cetaceans have been reviewed in a number of publications (see, for example, Myrberg, 1978; Gilders, 1988; Richardson *et al.*, 1991; Reeves, 1992; National Research Council, 1994; Richardson *et al.*, 1995). They emphasize the paucity of information from which to draw any firm conclusions. Mysticetes and odontocetes appear to have rather different hearing sensitivities, the former being most sensitive at low frequencies below 5 kHz and the latter above 10 kHz. This makes the former group more vulnerable to sounds produced by shipping, oil and gas operations, military detonations and oceanographic experiments; and the latter group more vulnerable to speedcraft of various types. However, ships, seismic surveys and the like may also generate some sound in the high frequency range (including sidescan sonar up to 200 kHz), the effects of which have been scarcely examined, whilst lower frequency sounds emitted at 500 Hz to 1 kHz may interfere with communication in odontocetes.

The mixed responses that have been observed from odontocetes in the presence of vessels of various types may simply reflect whether they perceive the craft as a physical threat rather than that the sounds emitted are harming them in some way. With most sound energy generated at frequencies below their main hearing sensitivities, the choice of a toothed whale or dolphin to bow-ride, ignore or flee from a vessel could well be dictated more by that vessel's behaviour and whether the animal has vulnerable young.

Several studies have documented behavioural responses by cetaceans to underwater sounds (for three typical examples see Figure 22.3). Usually these have taken the form of some avoidance action, any general impact being represented as a deflection from a migration path or displacement from a feeding or breeding area. The long-term consequences of such actions remain speculative. Just as humans prefer to avoid loud noises but will tolerate them at least for short periods without any detrimental effect, so might cetaceans habituate to sounds and learn to live with them. Equally, a lack of reaction does not prove there is no long-term effect. At present we have little idea whether the presence of loud, low-frequency sounds in the oceans is leading to hearing damage and gradual deafness for individuals, populations or even species. It is conceivable that whales are slowly growing deaf as their auditory thresholds become degraded to cope with this additional noise, given the short time in evolutionary terms that they have had to adapt.

Clearly an empirical approach with more field observations is badly needed. More experimental data are required with measures of both received and source levels of sound, their frequency spectra and the behaviour of animals before, during and after sound transmission. There should also be sufficient replicates and controls to allow for adequate assessment of potential confounding factors. Much more information is also needed about the auditory sensitivities of different cetacean species (particularly mysticetes) and different individuals within species. Since underwater sounds may also affect other marine organisms that are prey to cetaceans, the results of studies of these should also be taken into account. Finally, attempts should be made to monitor locally the possible long-term effects of human disturbance by determining not only changes in behaviour such as movement, respiration and dive patterns, but also changes in annual usage of particular areas for feeding or breeding, reproductive rates and, where possible, survivorship.

If the 'precautionary principle' is to be adopted, as is becoming increasingly widely proposed (see, for example, Declaration from the 1992 Rio Earth Summit Conference), then we should not continue potentially detrimental activities when we lack sufficient data to evaluate their

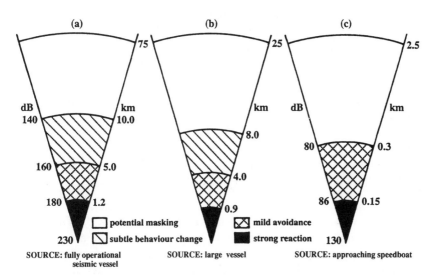

Figure 22.3 Zones of influence. (a) Gray whale (adapted from Gilders, 1988, using data from Malme *et al.*, 1983; Greene *et al.*, 1985). (b) Humpback whale (adapted from Gildlers, 1988, using data from Baker *et al.*, 1982, 1983). (c) Bottlenose dolphin (from Evans *et al.*, 1994)

impact. Damage that may already be occurring could be reduced if there were restrictions on the intensities of sounds (particularly when continuous) emitted by, for example, supertankers, oceanographic experiments, and seismic surveys. US regulations (Marine Mammal Protection Act, 1972) have used the 120 dB criterion as the threshold above which acoustic effects on marine mammals might occur. However, as a National Research Council (1994) report pointed out, and as indicated here, many maritime activities result in sound emissions well above this intensity and the majority of these have not been limited. The problem is that, with so little hard data, unequivocal evidence does not exist that such sound intensities have any long-term negative effects, or necessarily any negative effect at all in certain situations. Can we, indeed, afford the precautionary principle when this might stop not only all merchant shipping but also recreational activities, including whale watching – options which are unlikely to find favour with many people?

Some compromise may be necessary that is acceptable to industry and commerce but which adopts a more guarded approach to the use of high intensity sounds. As a minimum, this should involve initial surveys of the area, with any subsequent activity confined to locations or periods when potential conflict with marine mammals is minimized. Particular concern should be paid to received sound source levels of 160 dB or above (this value may be modified in the light of further information).

Any negative impact from disturbance associated with the increase in recreational activities (including whale watching and dolphin watching) could be mitigated by more widespread adoption of codes of conduct by water users. Simple rules that could readily be adopted include:

- Avoid suddenly altering vessel speed or direction.
- Avoid steering directly at cetaceans, and particularly avoid pursuing them.
- Limit the number of vessels in close proximity.
- Limit the length of time of the encounter.
- Be sensitive to the cetacean's response to the encounter; where possible, allow it to initiate the interaction.

The distances which a vessel should keep from a cetacean and the numbers of vessels that should remain within a certain radius are measures with little scientific evidence to support them. They are generally 'common sense' measures, and values vary in different regions. For example, boats watching humpback whales are recommended to come no closer than 400 m in Alaska compared with 30 m in New England. Most regulations around the world impose a distance of between 100 and 300 m from a vessel (cf. 100–500 m in the case of right whales). It should also be borne in mind that most people are unlikely to estimate distances at all accurately, so the extent to which such values are likely to be observed with any precision would seem small. Nevertheless, they do give a rough guideline.

It is likely that when cetaceans avoid vessels, they do so because either there is a threat of physical damage or some interference with an important activity such as feeding. A baffle around the vessel's propeller would provide some protection from the former. Maintenance of the propeller so that it is not chipped would also markedly reduce the amount of high frequency sound generated by cavitation.

Sensitivity by boat owners to behavioural reactions from local cetaceans would also go a long way to reducing potential negative effects. Educational material, in combination perhaps with a self-policed flexible permit system overseen by some independent watchdog, may be the way forward here. At Kaikoura in New Zealand, the permit scheme introduced by the local Maoris might be a good example to follow.

ACKNOWLEDGEMENTS

I thank Jonathan Gordon for comments upon a draft of this paper, Jim Heimlich-Boran for help with preparation of the figures, and the following persons for their help with my own fieldwork reported here: Mick Baines, Bill Bailey, Richard Bartley, Peter Canwell, Quentin Carson, Paul Fisher, Philip Hodge, Emily Lewis, Rachael Limer, David Mason and Ian

Rees. Grants towards this fieldwork have kindly been provided by World Wide Fund for Nature, Scottish Natural Heritage, Shetland Amenity Trust and the Whale & Dolphin Conservation Society. This review was made possible by funding to the author from the UK Department of the Environment.

REFERENCES

Au, W.W.L., Carder, D.A., Penner, R.H. and Scronce, B.L. (1985) Measurements of Beluga echolocation signals in two different ambient noise environments. *Abstracts of Fifth Biennial Conference on Biology of Marine Mammals*, p. 3.

Baker, C.S. and Herman, C.S. (1989) *Behavioral Responses of Summering Humpback Whales to Vessel Traffic: experimental and opportunistic observations*. Technical Report NPS-NR-TRS89–01. National Park Service, Alaska Regional Office, Anchorage, Alaska.

Baker, C.S., Herman, L.M., Bays, B.G. and Stifel, W.F. (1982) *The impact of vessel traffic on the behavior of humpback whales in southeast Alaska*. Contract No. 81-ABC-00114. Report to the National Marine Fisheries Service, Seattle, WA. 39 pp.

Baker, C.S., Herman, L.M., Bays, B.G. and Bauer, G.B. (1983) *The impact of vessel traffic on the behavior of humpback whales in southeast Alaska – 1981 season*. Contract No. 81-ABC-00114. Report to the National Marine Fisheries Service, Seattle, WA. 39 pp.

Beach, D.W. and Weinrich, M.T. (1989) Watching the whales. *Oceanus*, **32**, 84–88.

Clapham, P.J. and Mattila, D. (1993) Reactions of humpback whales to skin biopsy sampling on a West Indies breeding ground. *Marine Mammal Science*, **9**, 382–391.

Cybulski, J. (1977) Probable origin of measured supertanker radiated noise spectra, in *Oceans '77 Conference Record*, Inst. Electrical and Electronic Engineering, New York, NY, pp. 15C-1 to 15C-8.

Dahlheim, M.E. and Ljungblad, D.K. (1990) Preliminary hearing study on gray whales (*Eschrichtius robustus*) in the field, in *Sensory Abilities of Cetaceans. Laboratory and Field Evidence*, (eds J.A. Thomas and R.A. Kastelein), Plenum Press, New York, pp. 335–346.

Duffus, D.A. and Dearden, P. (1992) Whale, science and protected area management in British Columbia, Canada. *The George Wright Forum*, **9**, 79–87.

Eberhardt, F. (1993). Surface behaviour of sperm whales *Physeter macrocephalus* under the impact of whale watching vessels off Andenes, Norway. *Zoologica Institutionen*, **17**, 1–24.

Evans, P.G.H. (1987) *The Natural History of Whales and Dolphins*, Christopher Helm, London.

Evans, P.G.H., Canwell, P.J. and Lewis, E.J. (1992) An experimental study of the effects of pleasure craft noise upon bottle-nosed dolphins in Cardigan Bay, West Wales, in *European Research on Cetaceans – 6*, (ed. P.G.H. Evans), European Cetacean Society, Cambridge, England, pp. 43–46.

Evans, P.G.H., Fisher, P. and Lewis, E.J. (1993) *Potential impact of seismic activities upon cetaceans in the Irish Sea*. Report to Marathon Oil UK, Sea Watch Foundation, Oxford.

Evans, P.G.H., Carson, Q., Fisher, P. *et al.* (1994) A study of the reactions of harbour porpoises to various boats in the coastal waters of S.E. Shetland, in

European Research on Cetaceans – 8, (ed. P.G.H. Evans), European Cetacean Society, Cambridge, England, pp. 60–64.

Finley, K.J. and Davis, R.A. (1984) Reactions of beluga whales and narwhals to ship traffic and ice-breaking along ice edges in the Eastern Canadian High Arctic: 1982–1984. Overview. Unpublished report by LGL Limited, Toronto, for Canada Department of Indian Affairs and Northern Development, Ottawa.

Gilders, M. (1988) The influence of anthropocentric acoustic disturbance of cetaceans. BSc Honours Thesis, University of Oxford.

Gordon, J., Leaper, R., Harley, F.G., and Chappell, O. (1992) *Effects of Whale-watching Vessels on the Surface and Underwater Acoustic Behaviour of Sperm Whales off Kaikoura, New Zealand*. Science & Research Series No. 32, NZ Dep. Conserv., Wellington, New Zealand.

Greene, C.R., Engelhardt, F.R. and Paterson, R.J. (1985) *Proceedings of the Workshop on Effects of Explosive Use in the Marine Environment, January 29–31 1985, Halifax*. Canada Oil and Gas Lands Administration, Environmental Protection Branch, Technical Report No. 5.

Haines, G. (1974) *Sounds Underwater*, David & Charles, Newton Abbot, Devon.

Jones, M.L. and Swartz, S.L. (1984) Demography and phenology of Gray Whales and evaluation of whale-watching activities in Laguna San Ignacio, Baja California Sur, Mexico, in *The Gray Whale*, (eds M.L. Jones, S.L. Swartz and S. Leatherwood), Academic Press, Orlando, FL, pp. 309–374.

Kaufman, G. and Wood, K. (1981) Effects of boat traffic and military activity on Hawaiian Humpback Whales. *Abstracts of Fourth Biennial Conference on Biology of Marine Mammals, San Francisco, CA*, p. 67.

Ketten, D.R. (1991) The marine mammal ear: specializations for aquatic audition and echolocation, in *Evolutionary Biology of Hearing*, (eds D.B. Webster, R.R. Fay and A.N. Popper), Springer Verlag, Berlin, pp. 717–750.

Kruse, S. (1991) The interactions between killer whales and boats in Johnstone Strait, British Columbia, in *Dolphin Societies*, (eds K. Pryor and K. Norris), Univ. of California Press, Berkeley, pp. 149–159.

Kryter, K.D. (1985) *The Effects of Noise on Man*, Academic Press, Orlando, FL.

Leggat, L.J., Merklinger, H.M. and Kennedy, J.L. (1981) LNG carrier underwater noise study for Baffin Bay, in *The Question of Sound from Ice-breaker Operations*, (ed. N.M. Peterson), Proceedings of a Workshop Arctic Pilot Project, Petro-Canada, Calgary, Alberta, pp. 115–155.

Ljungblad, D.K., Würsig, B., Swartz, S.L. and Keene, J.M. (1988) Observations on the behavioral responses of bowhead whales (*Balaena mysticetus*) to active geophysical vessels in the Alaskan Beaufort Sea. *Arctic*, **41**(3), 183–194.

Malme, C.I., Miles, P.R., Clark, C.W., Tyack, P. and Bird, J.E. (1983) *Investigations of the potential effects of underwater noise from petroleum industry activities on migrating gray whale behavior*. BBN Rep. No. 5366, Bolt Beranek & Newman Inc., Cambridge, MA.

Malme, C.I., Miles, P.R., Clark, C.W., Tyack, P. and Bird, J.E. (1984) *Investigations of the potential effects of underwater noise from petroleum industry activities on migrating gray whale behavior. Phase II: January 1984 migration*. BBN Rep. No. 5586, Bolt Beranek & Newman Inc., Cambridge, MA.

Malme, C.I., Würsig, B., Bird, J.E., and Tyack, P. (1988) Observations of feeding gray whale responses to controlled industrial noise exposure, in *Port and Ocean Engineering Under Arctic Conditions*, Vol. II, (eds W.M. Sackinger, M.O. Jeffries, J.L. Imm and S.D. Treacy), Geophysical Inst., Univ. of Alaska, Fairbanks, AK, pp. 55–73.

Mauchline, J. (1980) The biology of Mysids and Euphausiids. *Adv. Mar. Biol.*, **18**, 1–681.

Martin, A.R. (ed.) (1990) *Whales, Dolphins and Porpoises*, Salamander Press, London.

Moore, S.E., Ljungblad, D.K. and Schmidt, D.R. (1984) Ambient, industrial and biological sounds recorded in the Northern Bering, Eastern Chukchi and Alaskan Beaufort Seas during the seasonal migrations of the bowhead whale, 1979–1982.

National Research Council (1994) *Low-Frequency Sound and Marine Mammals. Current Knowledge and Research Needs.* Committee on Low-Frequency Sound and Marine Mammals Ocean Studies Board. National Academy Press, Washington.

Mulroy, M.J. (1991) Munk's experiment. *Science*, **253**, 118–119.

Myrberg, A.A., Jr (1978) Ocean noise and the behavior of marine animals, in *Effects of Noise on Wildlife*, (eds J.L. Fletcher and R.G. Busnel), Academic Press, New York, pp. 169–208.

Phillips, N.E. and Baird, R.W. (1993) Are killer whales harassed by boats? *The Victoria Naturalist*, **50**, 10–11.

Popper, A.N. (1980) Sound emission and detection by delphinids, in *Cetacean Behavior: Mechanisms and Processes*, (ed. L.M. Herman), J. Wiley and Sons, New York, pp. 1–51.

Reeves, R.R. (1977) *The Problem of Gray Whale Harassment: at the breeding lagoons and during migration*, US NTIS, PB Report, PB-272 506, 1–6. Marine Mammal Commission 76–06.

Reeves, R.R. (1992) *Whale Responses to Anthropogenic Sounds: a literature review.* Sci. & Res. Ser. 47, NZ Dep. Conserv., Wellington, New Zealand.

Richardson, W.J. and Malme, C.I. (1993) Man-made noise and behavioural responses, in *The Bowhead Whale*, (eds J.J. Burns, J.J. Montague and C.J. Cowles), Spec. Publ. 2, Soc. Mar. Mamm., Lawrence, KS., pp. 631–700.

Richardson, W.J., Fraker, M.A., Würsig, B. and Wells, R.S. (1985) Behaviour of bowhead whales *Balaena mysticetus* summering in the Beaufort Sea: reactions to industrial activities. *Biol. Conserv.*, **32**, 195–230.

Richardson, W.J., Würsig, B. and Greene, C.R. Jr (1986) Reactions of bowhead whales, *Balaena mysticetus*, to seismic exploration in the Canadian Beaufort Sea. *J. Acoust. Soc. Am.*, **79**(4), 1117–1128.

Richardson, W.J., Würsig, B. and Greene, C.R. Jr (1990) Reactions of bowhead whales, *Balaena mysticetus*, to drilling and dredging noise in the Canadian Beaufort Sea. *Mar. Environ. Res.*, **29**(2), 135–160.

Richardson, W.J., Greene, C.R. Jr, Malme, C.I. and Thomson, D.H. (1991) Effects of noise on marine mammals. OCS Study MMS 90–0093. LGL Rep. TA834–1. Unpublished report from LGL Ecol. Res. Assoc. Inc., Bryan, TX for US Minerals Manage. Serv., Atlantic OCS Reg., Herndon, VA. NTIS PB91–168914.

Richardson, W.J., Finley, K.J., Miller, G.W., Davis, R.A. and Koski, W.R. (1995) Feeding, social and migration behavior of bowhead whales, *Balaena mysticetus*, in Baffin Bay vs. the Beaufort Sea – regions with different amounts of human activity. *Mar. Mammal Sci.*, **11**, 1–45.

Richardson, W.J., Greene Jr, C.R., Malme, C.I. and Thomson, D.H. (1995) *Marine Mammals and Noise*, Academic Press, San Diego, CA, USA.

Ross, D. (1976) *Mechanics of Underwater Noise*, Pergamon, New York.

Spindel, R.C. and Worcester, P.F. (1990) Ocean acoustic tomography. *Sci. Am.*, **263**(4), 94–99.

Stone, G.S., Katona, S.L., Mainwaring, A., Allen, J.M. and Corbett, H.D. (1992) Respiration and surfacing rates of fin whales (*Balaenoptera physalus*) observed from a lighthouse tower. *Rep. Int. Whal. Commn*, **42**, 739–745.

Sumich, J.L. (1983) Swimming velocities, breathing patterns and estimated costs of locomotion in migrating gray whales (*Eschrichtius robustus*). *Can. J. Zool.*, **42**, 739–745.

Tavolga, W.N. (1967) Noisy chorus of the sea. *Natural History*, **76**, 20–27.

Thompson, T.J., Winn, H.E. and Perkins, P.J. (1979) Mysticete sounds, in *Behavior of Marine Animals, Vol. 3, Cetaceans*, (eds H.E. Winn and B.J. Olla), Plenum Press, New York, pp. 403–431.

Tyack, P.L., Wells, R.S., Read, A., Howald, T. and Spradlin, T. (1993) Experimental playback of low frequency noise to bottlenose dolphins, *Tursiops truncatus. Abstracts of Tenth Biennial Conference on the Biology of Marine Mammals, Galveston, Texas*, p. 3.

Urick, R.J. (1982) *Sound Propagation in the Sea*, Peninsula Publishing, Los Altos, CA.

Urick, R.J. (1983) *Principles of Underwater Sound*, McGraw-Hill, NY.

Urick, R.J. (1986) *Ambient Noise in the Sea*, Peninsula Publishing, Los Altos, CA.

Watkins, W.A. (1986) Whale reactions to human activities in Cape Cod waters. *Mar. Mammal Science*, **2**, 251–262.

Weinrich, M.T., Lambertsen, R.H., Baker, C.S., Schilling, M.R. and Belt, C.R. (1991) Behavioural responses of humpback whales (*Megaptera novaeangliae*) in the Southern Gulf of Maine to biopsy sampling. *Rep. Int. Whal. Commn* (special issue 12), 91–97.

Wenz, G.M. (1962) Acoustic ambient noise in the ocean: Spectra and sources. *J. Acoust. Soc. Am.*, **34**(12), 1936–1956.

Worcester, P.F., Lynch, J.F., Morawitz, W.M.L. *et al.* (1993) Evolution of the large-scale temperature field in the Greenland Sea during 1988–1989 from tomographic measurements. *Geophys. Res. Lett.*, **20**(20), 2211–2214.

Index

Page numbers appearing in **bold** refer to figures and page numbers appearing in *italic* refer to tables.